Managing Bee Health

A Practical Guide for Beekeepers

Managing Bee Health

A Practical Guide for Beekeepers

JOHN CARR

B.V.Sc., Ph.D., D.P.M., Dipl E.C.P.H.M., M.R.C.V.S

First published 2016, revised reprint 2021

Published by
5M Books Ltd,
Lings, Great Easton,
Essex CM6 2HH, UK,
Tel: +44 (0)330 1333 580
www.5mbooks.com

A Catalogue record for this book is available from the British Library

ISBN 978-1-910455-03-6

Book layout by
Keystroke, Station Road, Codsall, Wolverhampton

Printed by Short Run Press, Exeter

Photos by John Carr unless otherwise credited

Contents

Acknowledgements

Writing a book is never achieved without considerable help from friends and colleagues. I would like to acknowledge the help of Professor C. L. Scudamore, Royal Veterinary College London and Iain MacMillan, Glasgow University, Dr Ke-Jein Chen, the staff at the Animal Technology Research Institute Taiwan in particular Professor Shih Ping Chen and Dr Chuan Hsing Chang for the preparation of the stained bee sections.

To the Nakasongola Apiculture Training Centre, Gulu Road, Nakasongola, Uganda for their very generous help with their bees; they could not do enough to help me with this and other projects. If any enthusiast is passing please call in and visit the institute.

To Jakob Hoffer, Starlite Colony, Manitoba for lending me a hand during the making of this book.

To Howells Veterinary Services and the North Park Veterinary Group.

To Sunjin Korea and the Philippines for allowing me to work with their hives.

To Dan Martin, DPI Victoria, Australia who showed me some of Australia's extensive biosecurity preparations to protect their bee (farmed and native) populations.

To Alice, our queen for allowing me to take numerous photographs of her and her subjects.

To Cath Fraser MRCVS who provided critical comment when required.

To all the staff at 5M who have helped put this book together.

Thanks to Linda for putting up with me over the years.

Image credits

I am extremely grateful for all the friends who allowed me to photograph their bees. In particular I would like to acknowledge the following contributors:

Author
Anderson, Denis, Bees Downunder
Arroio, Agnaldo
Barker, Colin (CC BY-SA 3.0)
Bksimonb (CC BY-SA 3.0)
Carreon, Anna
Chrumps (CC BY 3.0)
Core, A., Runckel, C., Ivers, J., Quock, C., Siapno, T., et al. (2012) (CC BY 2.5)
Don, Dmitri (CC BY-SA 3.0)
Friedman, Jerry (CC BY-SA 3.0)
Gagnon, Eric G. (Tobrook) (CC BY-SA 2.0)
Galipedia (CC BY-SA 3.0)
Goodwin, R.
Hardyplants (Public Domain)
Hectonichud (CC BY-SA 3.0)
Hoyland, Sean (Public Domain)
Isipeoria (CC BY-SA 3.0)
JoJan (CC BY-SA 3.0)
Kalmia Latifolia (CC BY-SA 3.0)
Karwath, André (CC BY-SA 2.5)
KENPEI (CC BY-SA 3.0)
Koley, Surajit (Public Domain)
Lindsey, James (CC BY-SA 3.0)

Author
Martin, D.
Moonlight0551 (CC BY 2.0)
PiccoloNamek (CC BY-SA 3.0)
Rasbak (CC BY-SA 3.0)
Shebs, Stan (CC BY-SA 3.0)
Siegmund, Walter (CC BY-SA 3.0)
The Animal and Plant Health Agency (APHA) UK Crown Copyright
Walker, Ken, Museum of Victoria, Pest and Diseases Image Library
Wraith, Fell (Public Domain)
Zelenko, Eugene (CC BY-SA 3.0)

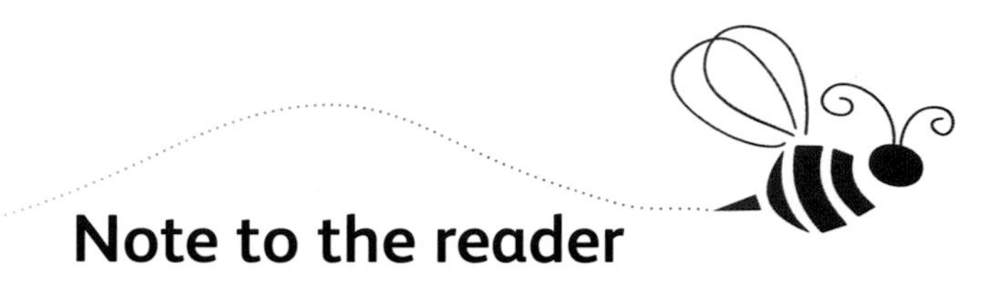

Note to the reader

The methods of treatment and control of conditions discussed in this book are guidelines only. Any recommendations given and so used are the responsibility of the beekeeper, and the advice of his or her veterinarian or bee inspector should be sought in case of doubt. No responsibility is accepted by the author or publishers for any application of the advice given in the book because each area and region is different and responses cannot be predicted. Some trade names and their chemical compounds are used throughout. No endorsement is intended, nor is any criticism implied of similar products not named. Note, some compounds may be illegal in your area and it is imperative that you discuss any used with your veterinarian and bee inspector.

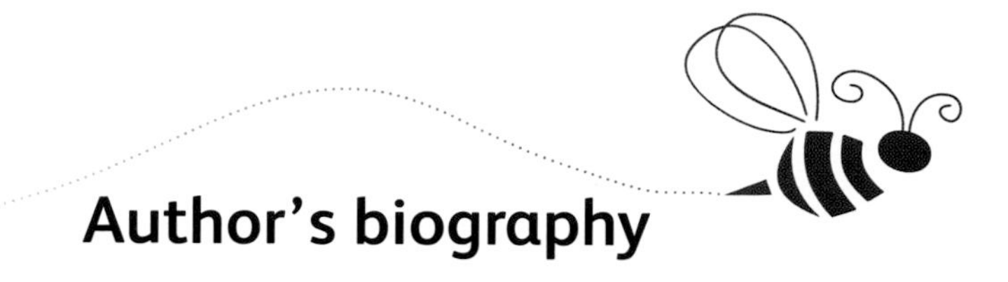

Author's biography

Dr John Carr started with his interest in bees at primary school at Leconfield, East Riding of Yorkshire, UK when the children were encouraged to collect and process honey from the school's hive. He is a veterinarian who qualified from Liverpool University in 1982 and, after being a Leverhulme resident, obtained his PhD from Liverpool University in 1990. He has been fortunate to be able to examine bees throughout the world together with his international interest in population medicine. He has taught bee medicine at universities, at the Royal Veterinary College, London, UK and Murdoch, Perth, Australia. He is a member of the British Bee Veterinary Association.

This book is part of his interest in understanding Colony Collapse Disorder and bee anatomy and pathology. It is hoped this book will stimulate a new generation of veterinarians to contribute to the health and welfare of bees and other invertebrates under their care.

John Carr
BVSc, PhD, DPM, Dipl ECPHM, MRCVS
May 2015

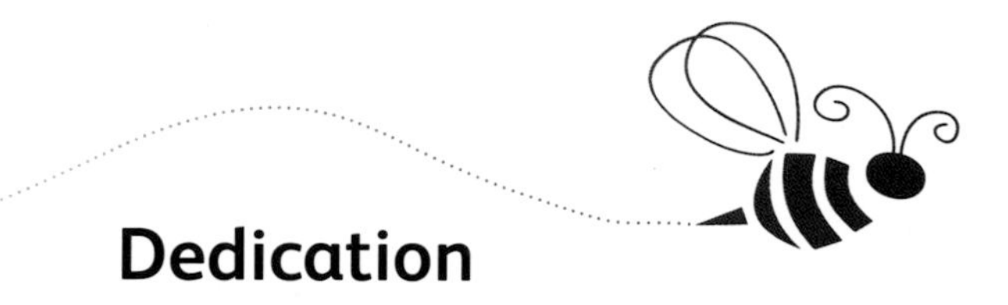

Dedication

This book is dedicated to

Geoff Berry – a naturalist and beekeeper

Chris Goodwin – Biology teacher and Head Master of Beverley Grammar School who encouraged a lifelong interest in biology and taught me not to believe everything I read.

1

Introduction and anatomy of bees

Apart from providing mankind with honey and a whole host of products, the crops that we rely on are pollinated by bees to provide us food. In parts of the world, bee pupae may also be harvested as a food source.

Figure 1.1 Honey bees working almond blossom

The annual almond crop in the USA alone requires 1.3 to 1.5 million hives – that's 200,000 million bees!

Honey bees have been introduced around the world, most notably in North America and Australia, and they have become vital to our food source.

They also aid in providing us the beauty of our flowering gardens.

Native Americans knew that the Europeans were coming when they started seeing *Apis mellifera* on flowers. They called our majestic bee 'white man's flies'.

Bees are farmed around the world and their products are used on a worldwide basis. With the changing world of economics and globalisation there has been a similar change in honey production around the world over time. As countries develop so will their honey production.

Table 1.1 Honey production as a percentage of total global production

	1400s	1700s	1970	2000	2010
Africa	35	30	13	11	11
Americas	0	5	40	27	21
Asia	35	35	22	36	45
Europe	30	30	25	23	22
Oceania	0	0	3	2	2

Source http://faostat3.fao.org

In 2010 the world produced 1,547,000 tonnes of honey. This was a 23% increase over 2000. This is about 200g per person per year!

Evolution of bees

How many species of insects are there?

There are nearly 1 million described species of insects. It is believed that there could actually be 5 to 30 million individual species of insects. They are highly successful animals, and are vital for most ecological systems and the sustainability of current life on this planet.

Evolution of the insect

The earliest definitive insects can be found in the Devonian period – around 480 million years ago and by 400 million years ago insects were already well developed. Insects are in the Arthropod phyla but the exact relationship between the various Arthropod groups, the Crustacea, Spiders and Insects is not clear.

All insects appear to have evolved on land and then, where applicable, returned to the water. In all insects the respiratory tract is a closed tracheal system, which is designed for a terrestrial existence. Insects and their nymphs, which live in the water have gill systems (tracheal gills), which originate from the terrestrial tracheal system they have readapted to an aquatic existence. This is in comparison with Crustacea who have gill systems, which are not tracheal, and have evolved from an aquatic ancestor.

One issue of early insect evolution is the evolution of the wing. The earliest insects discovered already had formed wings. In fact some extinct insects actually had 3 pairs of wings (Palaeodictyoptera).

Amber (plant sap) has captured many insects over the millennia and has preserved them in exquisite detail allowing an insight into the ancient world of insects and dinosaurs and so allowing the investigation into the evolution of insects and their relationship with other animals and plants.

Figure 1.2 A reconstruction of a three winged early insect

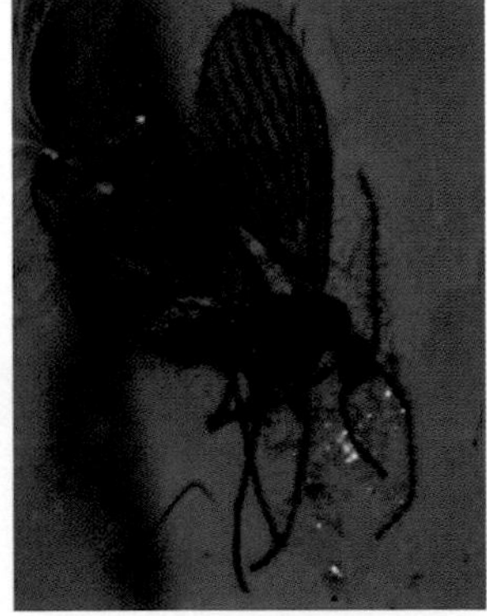

Figure 1.3 A midge in amber

The wings may have evolved from tracheal gills. In the Mayfly the tracheal gills are very similar in structure to wings. If wings evolved in insects on the land, their early use may have been for gliding from the tops of plants back onto the ground. Once wings were formed, the insects that were able to fold their wings would have an advantage in beginning to be able to exploit smaller niches. Today the neopterous (wing folding insects) are the dominant form of insect. Many orders of insects have a very long heritage – Mayflies (Ephemeroptera), Cockroaches (Blattodea), Dragonflies (Odonata), and Grasshoppers (Orthoptera) are all evident in the Carboniferous period. Other modern orders have evolved with the more recent development of the flowering plants – Butterflies (Lepidoptera), flies (Diptera) and beetles (Coleoptera).

Major orders of insects and the approximate global numbers

Coleoptera	38%
Lepidoptera	16%
Hymenoptera	13% – the order to which *Apis*, *Bombus* and Meliponini belong.
Diptera	12%
Others	21%

The major developments in the evolution of bees within the insect world could be characterised by:

- The development of the ovipositor first seen in the Orthoptera (Grasshoppers) in the female to assist laying of eggs in cavities.
- The presence of hooks (hamuli) on the hindwing, which connect with a fold on the forewing allowing both wings to work as one. This defines the order Hymenoptera.
- The development of the petiole (waist) making the abdomen very mobile from the thorax.
- Development of poison glands at the ovipositor base.
- The development of a new oviduct allowing the ovipositor to become a sting with the poison glands.
- Move from carnivorous feeding habits to a vegetarian lifestyle based on flowers.
- The tongue elongates to allow access to nectar in deep flowers.
- Hairyness to collect pollen.
- Pollen brush and baskets to transport the pollen.

The evolution of the flower must be considered alongside the evolution of the bee, as the two go hand in hand.

There is one Family that will be considered in more detail – Apidae. This family includes the Bumblebee (*Bombus*) and the Honey bee (*Apis mellifera*). In addition, the stingless bee (Meliponini) is also discussed.

Family Apidae

There are many species of insects in the Apidae family. In the UK alone there are 250 species.

Species of Apis

There are 8 species of *Apis*:

Subgenus *Micro Apis*: dwarf honey bees
Apis andreniformis
Apis florea

Subgenus *Apis*
Apis mellifera
Apis nigrocincta
Apis cerana
Apis koschevnikovi
Apis nuluensis

Subgenus *Mega Apis*: giant honey bees
Apis dorsata
Apis dorsata laboriosa is sometimes described separately.

Many insects can look alike. Each country has its own types of bees. In the UK, for example, the honey bee (*Apis mellifera*) is most likely to be confused with members of the Andrena group. There are 6 species of Andrena in the UK (*Andrena cineraria, haemorrhoa, florea, fulva, flavips* and *marginata). A. mellifera* however can be distinguished by the appearance of the wing where she has a long marginal cell and 3 submarginal cells on the forewing. Also, *A. mellifera* has a long tongue. This will be explored later in the chapter.

The honey bee's DNA code has been examined and is 236,000,000 base pairs long.

Distribution of the major *Apis* species on the planet

Apis mellifera has been deliberately introduced by man into the Americas in the seventeenth century and then Australasia in the 19th Century. It now has a world-wide distribution and plays a vital part in the human occupation of Earth.

Anatomy of the bee

As discussed, there are many thousands of bees. To introduce their anatomy the honey bee (*Apis mellifera)* is chosen as an example.

The western honey bee *Apis mellifera* is defined by its 236 million base pairs DNA (Mbps) organised into 16 chromosomes. As a comparison *Bombus terrestis* has 274 Mbps to its DNA makeup. The bee utilises haplodiploid genetics – the females have two sets of chromosomes (diploid), whereas the males have only one set of chromosomes (haploid). The males are from unfertilised eggs.

The castes of bees

Honey bees have three castes – the male is a called a drone. The females are separated into a single queen, (which is the only fully fertile female) and a number of sterile (or subfertile) female workers. Each of these castes can be readily recognised from their appearance.

Figure 1.4 Queen

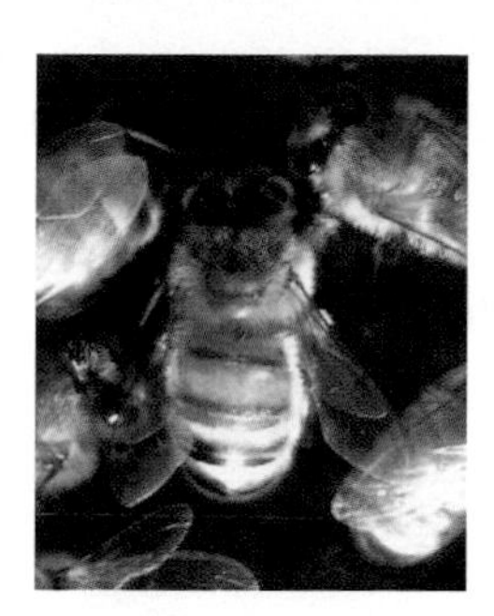

Figure 1.5 Worker

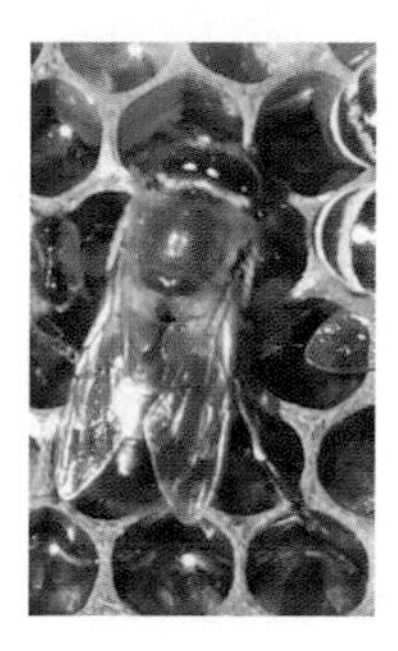

Figure 1.6 Drone

Note the queen is large with a long tapered abdomen and the wings are shorter than the body. There will be only one queen in the hive.

In *Apis mellifera,* the drone can be recognised by being bigger than the worker bees in the colony. When you look at a drone you will easily see his very large eyes. The eyes cover most of the head. The drone's wings are as long as their bodies. They have a blunt tipped stocky abdomen with prominent hairs. They have no sting. Note the great size difference between drones and workers, seen in *A. mellifera* does not occur in the Giant honey bee *Apis dorsata.*

The workers will be most of the visible bees. There may be a few different colour stripes within the group.

The queen mates with 20 or so males during her mating flight. This means that there will be around 20 different half-sister worker genetic groups within the hive. It is probable that the different half-sisters even recognise each other and behave differently when meeting full sisters and half-sisters. The drones are all full brothers as they are from unfertilised eggs so carry only the queen's genetics.

At peak production, in the summer time the hive can contain 1 queen, 50,000 adult workers and 500 drones. There will be no drones over the winter period. There could be around 20,000 capped brood, 9,000 uncapped brood and 6,000 eggs.

Figure 1.7 Bees surrounding the queen including workers and drones

Visualisation of the honey bee anatomy

The anatomy of the honey bee can be easily visualised by most enthusiasts using a simple dissecting microscope and a good light source. For more detail, stained (through your local animal health laboratory) and unstained sections can be made and examined down a light microscope. Enthusiasts can spend a lot of money on a light microscope but many good models can be obtained second hand. The images are then captured with a camera. Honey bees can be dissected after mounting in bee's wax to hold the body, with small scissors and a steady hand. See Chapter 12 for further details.

General organ layout of the adult *Apis mellifera*

Orientation around the bee

When describing the anatomy of the honey bee various specific terms are used to orient the location of the organ.

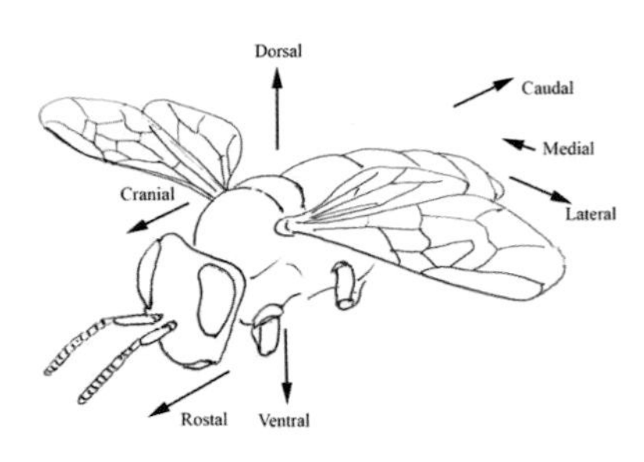

Figure 1.8 Orientation around the adult bee. Medial is towards the centre of the body, lateral is away from the centre of the body

Other terms:

Proximal – towards the body
Distal – away from the body
Axial – inside the leg (medial)
Abaxial – outside the leg (lateral)
Dorsal – front of the leg
Palmar – back of the leg

External anatomy

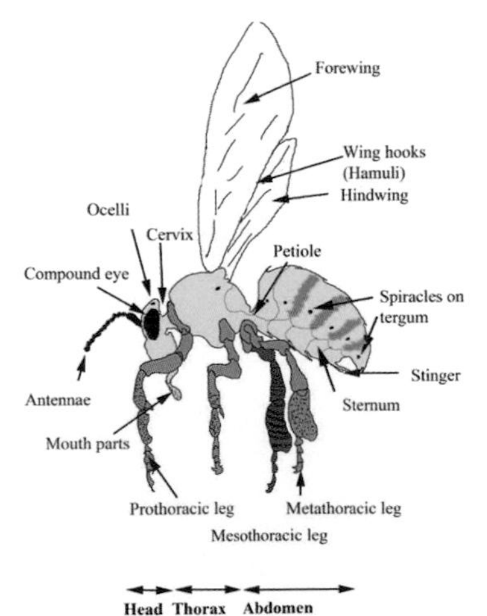

The bee larva has a head and 13 (three thoracic and 10 abdominal) visible segments and as the bee matures these segments become modified into the three major body parts: the head, thorax and abdomen.

The head has the eyes, antennae and mouthparts and internally the brain and various glands. The head is attached to the thorax by the cervix (neck).

The thorax carries the first three segments and to provide extra space for the wing muscles also the first abdominal segment. The thorax has three legs and two wings. The thorax is attached to the abdomen via the petiole (waist).

The abdomen carries the rest of the segments. Only five abdominal segments are visible (as stripes) on the abdomen (II to VII). Segments VIII and IX are internal and are used to construct the sting apparatus. Segment X carries the anus and is situated under visible tergum VII.

Figure 1.9 The basic surface anatomy of the honey bee – *Apis mellifera*

General anatomy of the honey bee *Apis mellifera*

The relative small size of the honey bee allows for the whole bee to be examined down the microscope. There are three major orientations to consider when examining the anatomy of the bee: the sagittal plane, the transverse plane and the frontal plane.

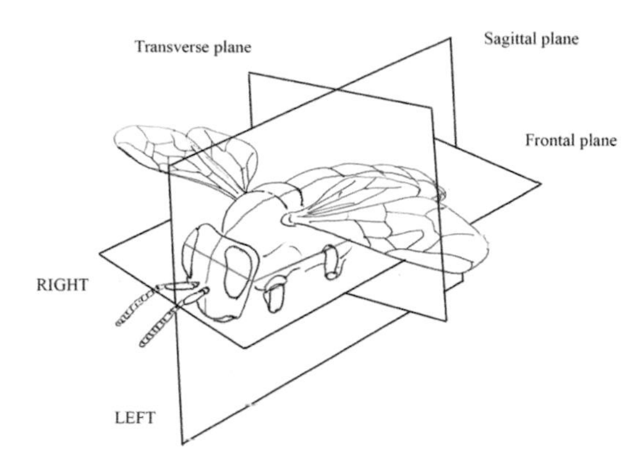

Figure 1.10 The major dissection planes of the bee

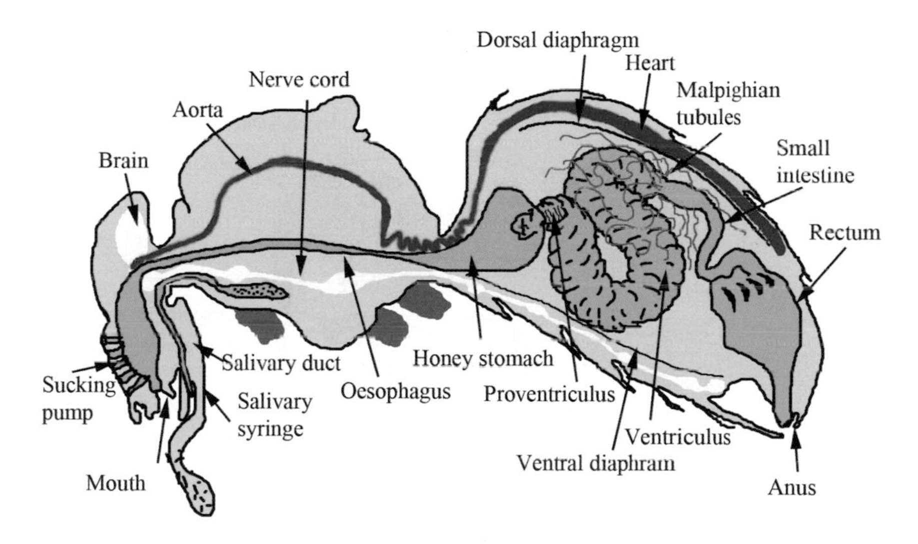

Figure 1.11 Drawing of the major organs of the honey bee sagittal view

When stained with two stains, Haematoxalin and Eosin (referred to as H&E) and cut into thin slices (3-5μm) the honey bee's internal anatomy can be illustrated.

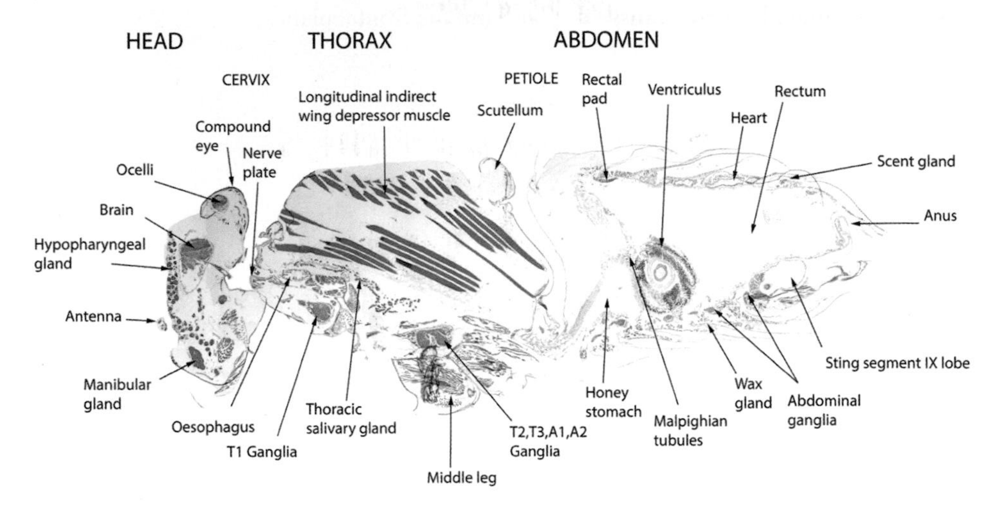

Figure 1.12 A midline sagittal section of a worker honey bee

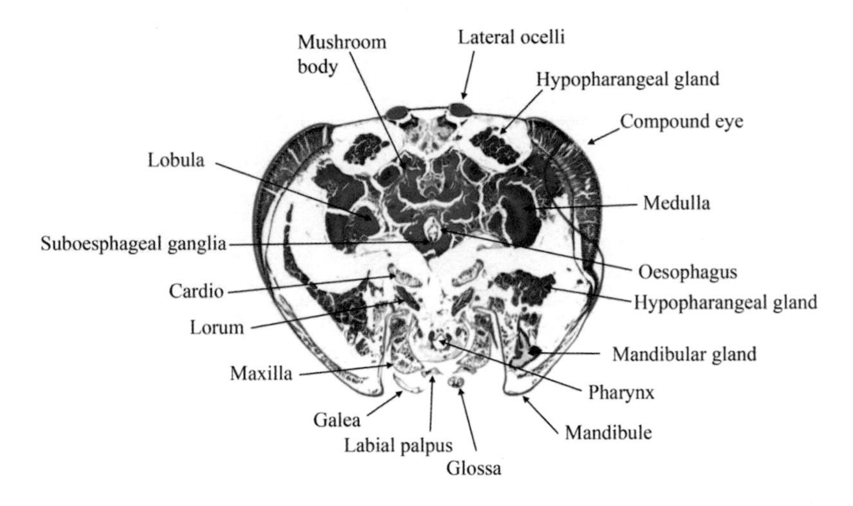

Figure 1.13 A transverse section of the head of a worker

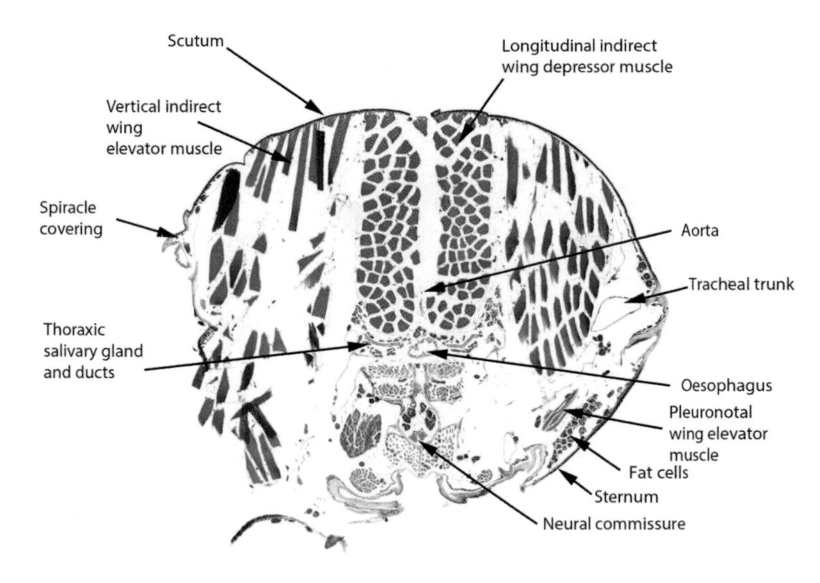

Figure 1.14 A transverse section of the thorax of a worker bee

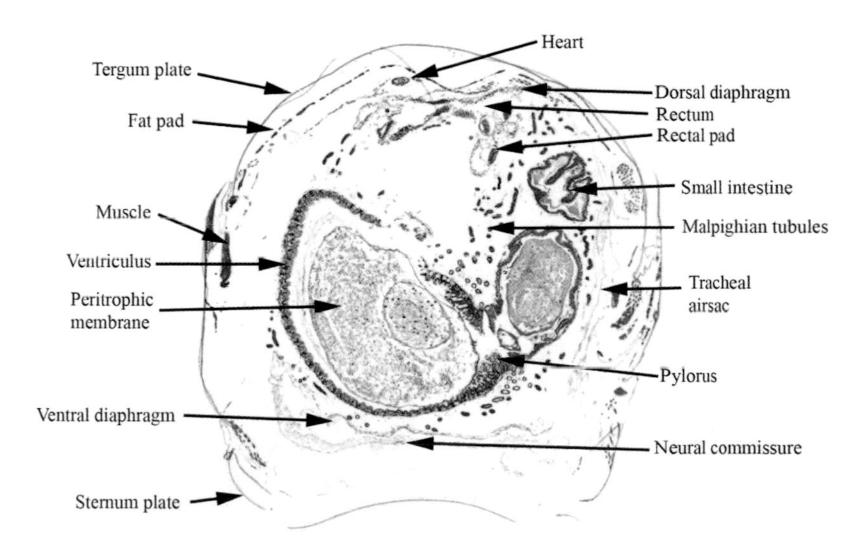

Figure 1.15 A transverse section of the mid-abdomen of a worker bee

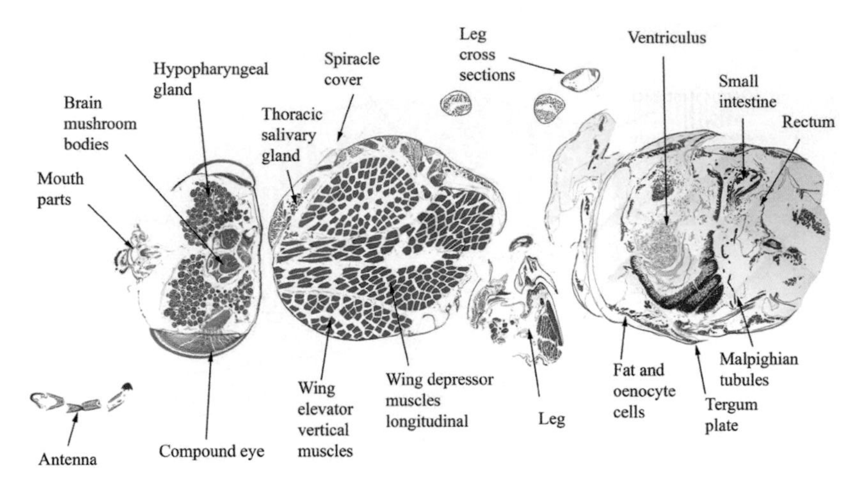

Figure 1.16 A frontal section of a worker honey bee

The general anatomy of the honey bee through its life cycle

Egg

The queen lays eggs in the cell. She measures the size of the cell with her back legs. If the cell is a large cell (7mm in diameter) the queen does not fertilise the egg and the egg develops into a drone. If the cell is normal (5mm in diameter) she will fertilise the egg to produce a female egg – this could become a worker or a queen. This depends on the workers. The sperm enter the egg through the micropyle.

The honey bee egg is 1.5mm long. The egg is laid with the caudal area attached to the base of the cell. The egg has a slight curve – the concave surface will become the ventral surface, the convex surface will become the dorsal surface of the future bee.

During peak laying the queen can lay 2000 eggs per day.

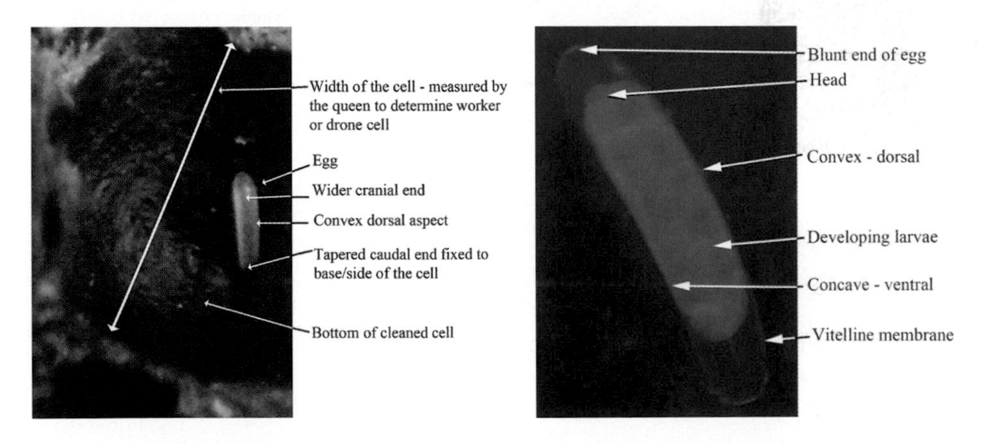

Figure 1.17 The egg laid in the cell

Figure 1.18 The anatomy of the developing egg

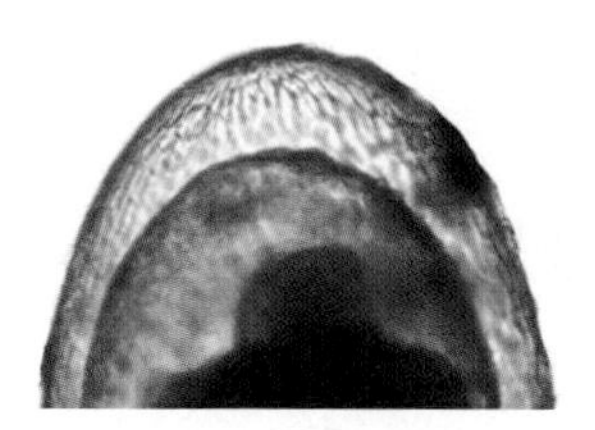

Figure 1.19 The micropyle through which the single sperm enters

The larva hatches from the egg after 3 days.

Larva to pupa

An instar is the period between moults. The insect has a cuticle epidermis which can only grow when wet. Once the cuticle is dry it cannot expand. To grow, the insect has to shed its cuticle. In the honey bee there are 5 instars allowing the larva to grow. In the worker and queen female each of these instars take one day – thus at the end of the fifth instar the larva is 8 days from being laid. The drone period of growth is slightly longer. Each instar is characterised by a slightly different anatomy.

Larva instar 1

The larva hatches from the egg on day 3. The larva is laid on its side and is provided with royal jelly from the hypopharyngeal gland of the workers. This is interesting as the larva only has the uppermost surface spiracles open to the air and yet the whole body is oxygenated.

Dissection of the larva reveals it is predominately intestinal tract. The mouth leads into the stomodaeum (foregut), which opens into the ventriculus (midgut). At this stage the ventriculus does not have a connection with the proctodaeum (hindgut). This larval stage is therefore unable to defecate.

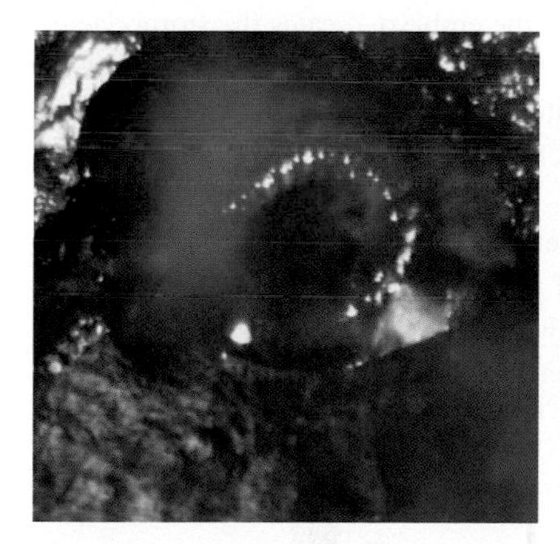

Figure 1.20 The L1 larva is 1.5 mm long. The first stage larva is quickly fed with a cloudy liquid – royal jelly

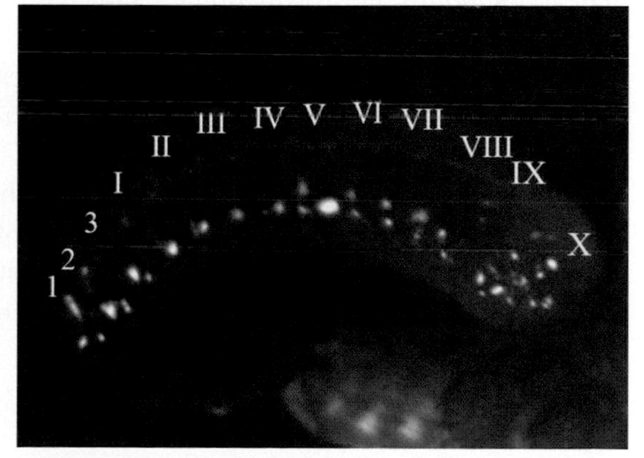

Figure 1.21 The L1 larva is transparent and the segments 1-X are clearly visible together with the tracheal network

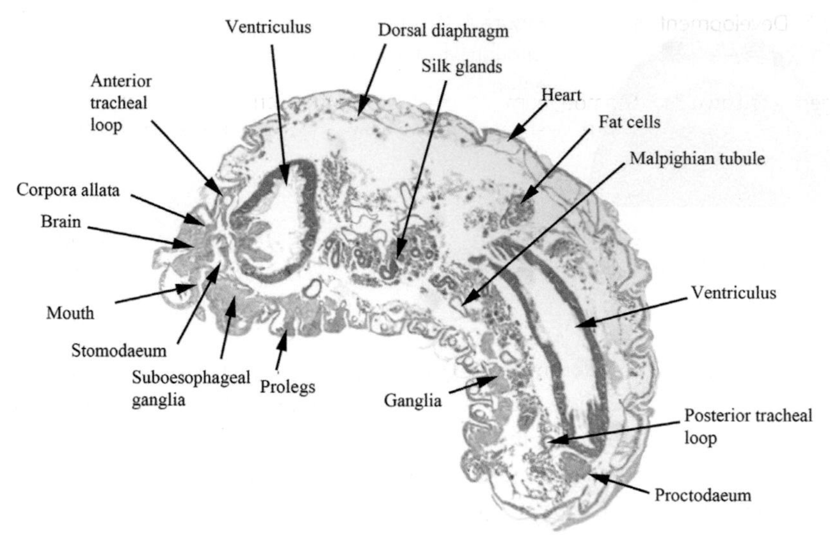

Figure 1.22 Midline sagittal section of L1 larva

The larva has a well defined stomodaeum and ventriculus. Note the procodaeum is evident but there is no passage between the ventriculus and procodaeum. There are four large Malpighian tubules. The silk glands, which will develop into the salivary glands of the adults are clearly visible. The brain and ganglia are clearly recognised. The fat cells are quite small. During the development the larva is fed 1400 times a day.

Development of the intestine in the insect

Unlike the mammal (vertebrates) the intestinal tract of insects is made from three components. The mouth, an invagination of epithelium of the embryo, creates the stomodaeum (foregut). At the same time, an anal invagination of epithelium creates the proctodaeum (hindgut). The tissues of the inside (mesoderm) create the ventriculus (midgut). These three components meet but do not necessarily (in the honey bee) have open connections at all times.

The importance of this development is that the external cuticle covers the surface of the stomodaeum and the proctodaeum portions of intestinal tract. The ventriculus is not protected by cuticle and while this can assist digestion it also allows for the ventriculus to easily become damaged. The ventriculus is protected by a secretion called the peritrophic membrane.

Table 1.2 Development of the worker honey bee intestinal tract

Stage			
Emergence to Larva 5	Stomodaeum connects to ventriculus	No connection between ventriculus or Malpighian tubule and proctodaeum	Eat No defecation or urination
L5 to propupa after capping	Stomodaeum connects to ventriculus	Ventriculus and Malpighian tubules connect to proctodaeum.	Defecation and urination
Propupa to 19 day stage pupa	No connection from stomodaeum to ventriculus	No connection between ventriculus and proctodaeum	Cannot eat No defecation No urination
20 day stage eye stage pupa and adult	Stomodaeum to ventriculus connection	Ventriculus connects to proctodaeum. Malpighian tubules connect	Eat Defecation Urination

Larva instar 2

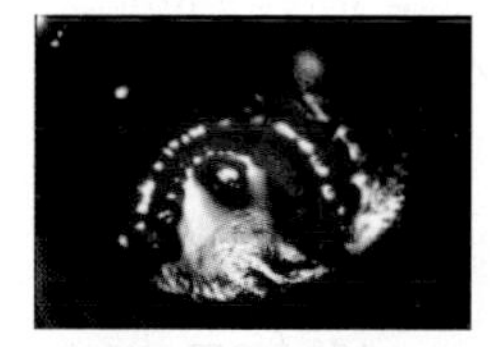

Figure 1.23 The L2 larva fills nearly half of the bottom of the cell at 4 mm long

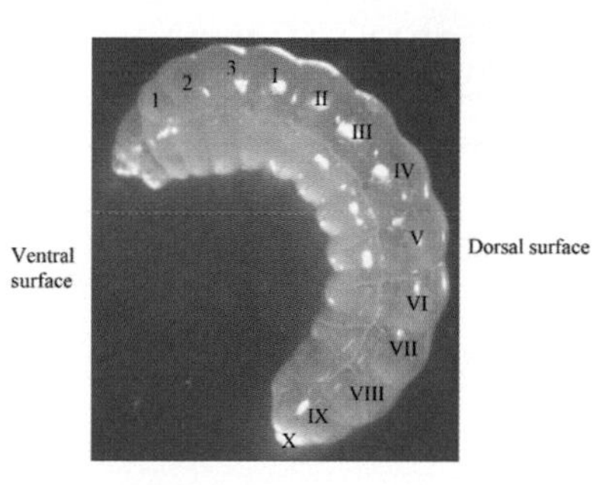

Figure 1.24 The L2 larva becomes more translucent but the segments are clearly visible with the tracheal truck. Head structures become more apparent particularly in queen larva

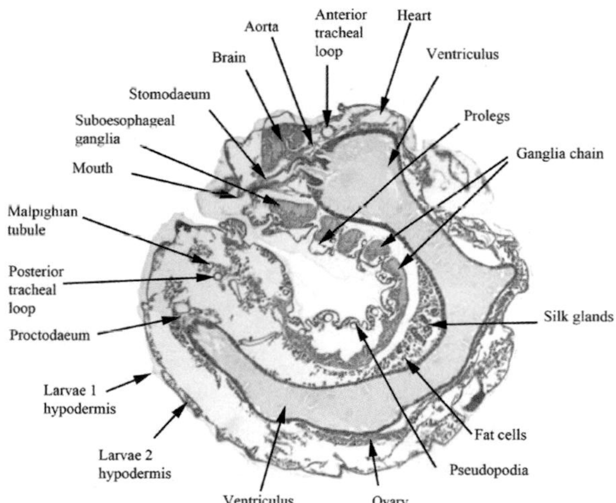

Figure 1.25 A midline sagittal section of an L2 larva

The ventriculus fills most of the larva. The brain and ganglia become clearly visible. Note the anterior and posterior tracheal loop. These structures can be used as landmarks for orientation.

Larva instar 3

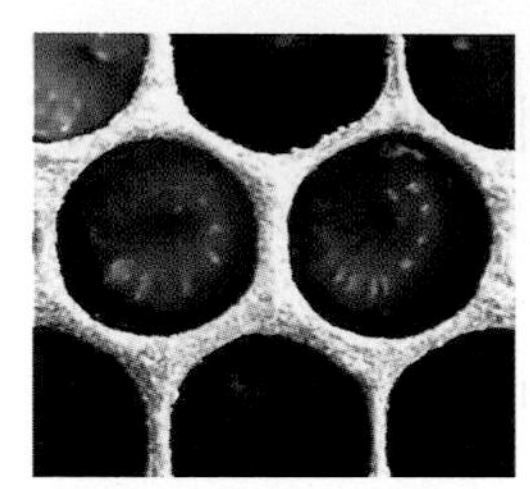

Figure 1.26 The L3 larva fills the bottom of the cell

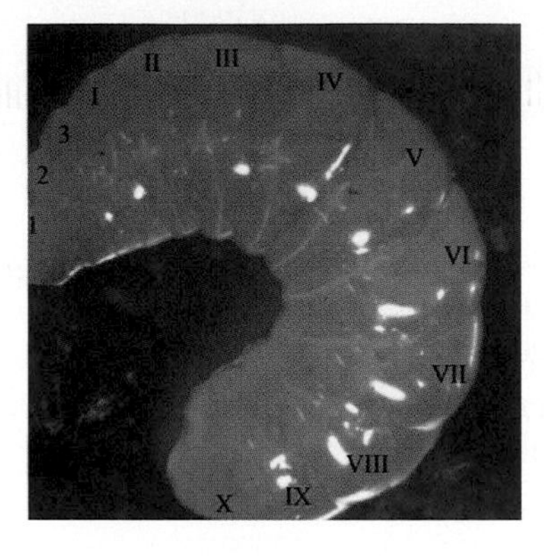

Figure 1.27 The L3 larva becomes white with the segments still clearly visible. The head end is more pointed

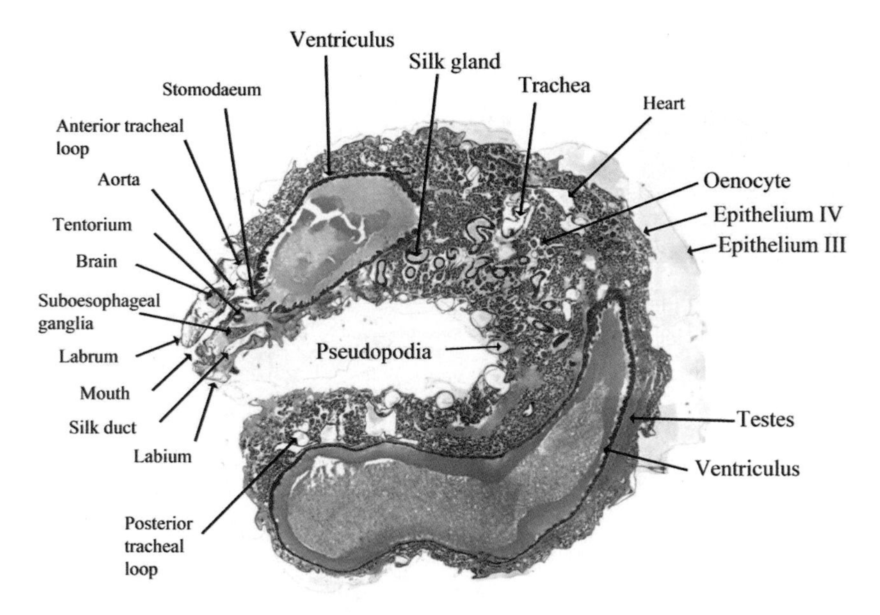

Figure 1.28 The midline sagittal L3 larva

The ventriculus fills the larva and majority of the rest of the body being fat cells. The silk glands and the Malpighian tubules become more apparent. Note the Malpighian tubules do not penetrate into the proctodaeum so urination is not possible.

Larva instar 4

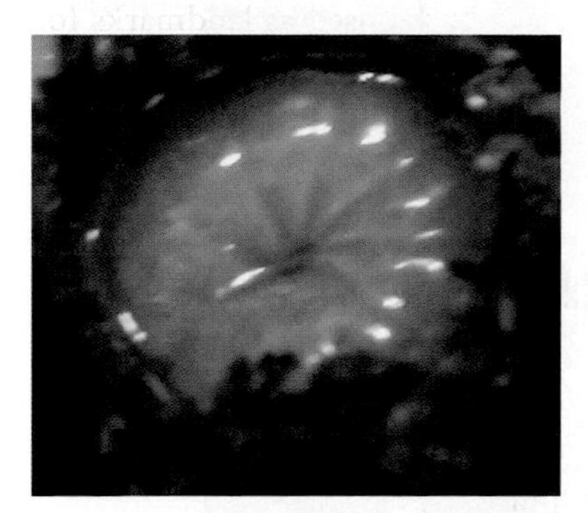

Figure 1.29 The L4 larva fills the entire bottom of the cell and over halfway up. The larva is in a tight circle with the head and anal region in contact with each other

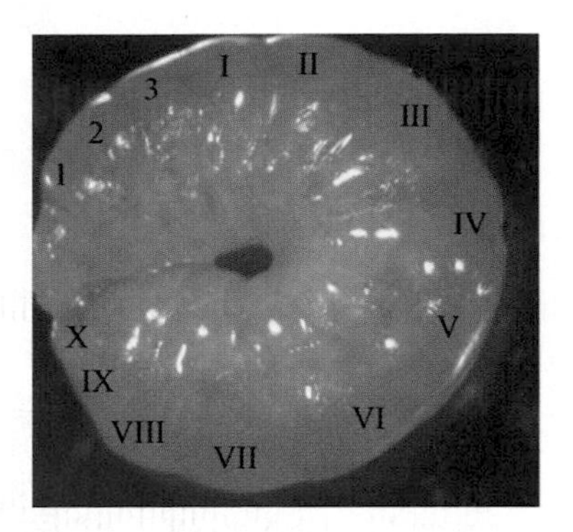

Figure 1.30 The larva is white in a tight circle. The segments are more difficult to make out. The ventriculus may be seen as a yellow line on the dorsal surface

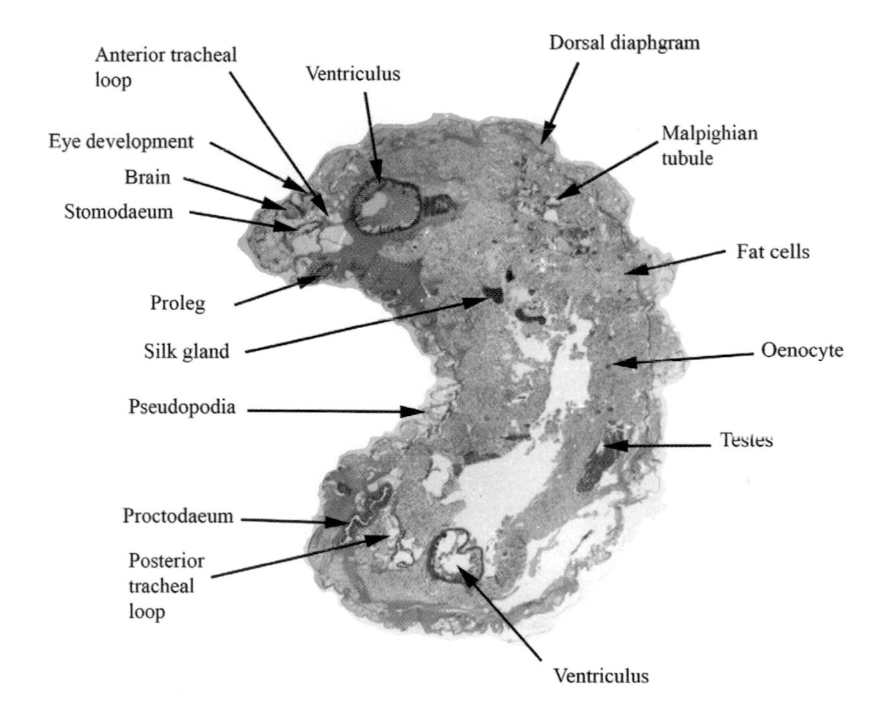

Figure 1.31 The sagittal section of an L4 larva male

The ventriculus continues to be the most prominent organ full of digesting food materials. The silk glands are very prominent. The oenocytes which appear to control a lot of the metabolism are clearly visible in conjunction with the fat cells.

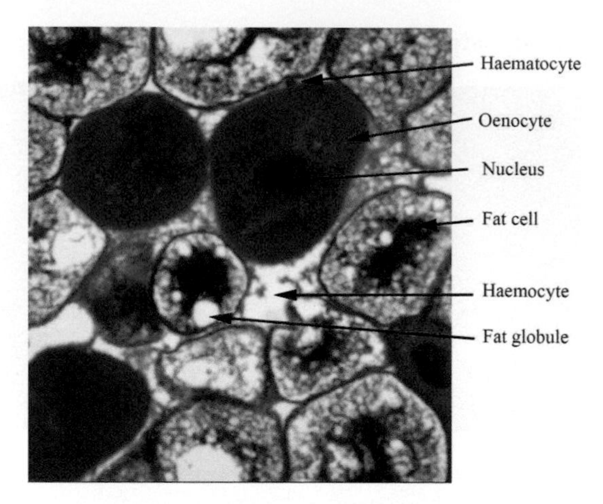

Figure 1.32 The oenocyte and fat cells from an L4 larva

Oenocytes and fat cells. The oenocytes originate from the epithelium (skin). They are generally large single cells, which are very prominent. The function of the oenocyte appears to coordinate the metabolism of the fat cells. In the adult worker they are generally found with fat cells and the wax mirror.

Larva instar 5

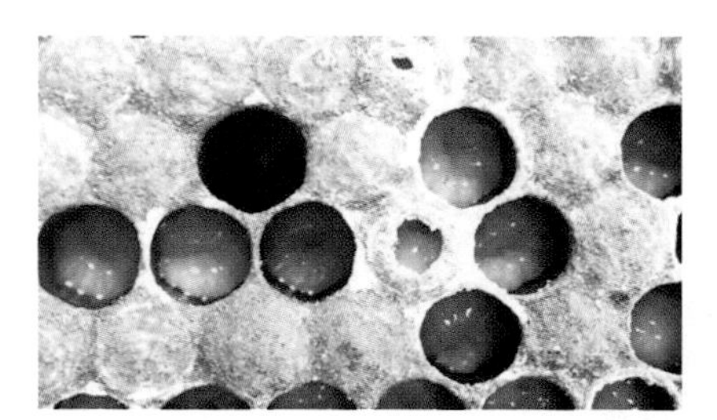

Figure 1.33 The centre cell with an L5 is being capped note surrounded by L4 larvae and one empty cell

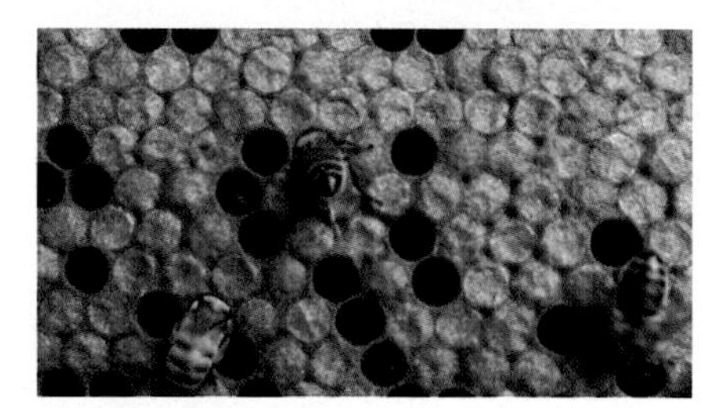

Figure 1.34 Capping complete

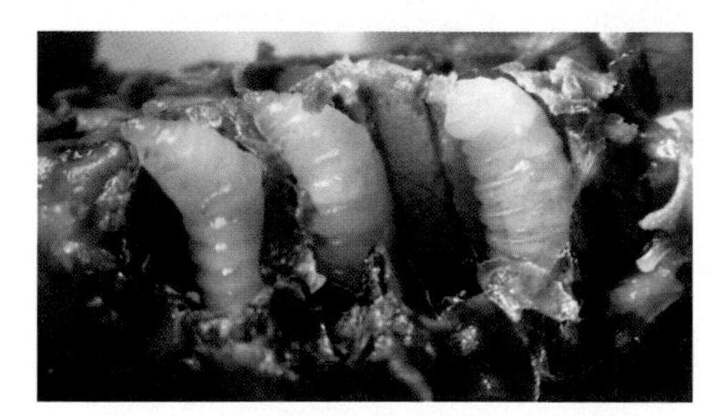

Figure 1.35 L5 with the wall of the cell removed showing them standing in the capped cell

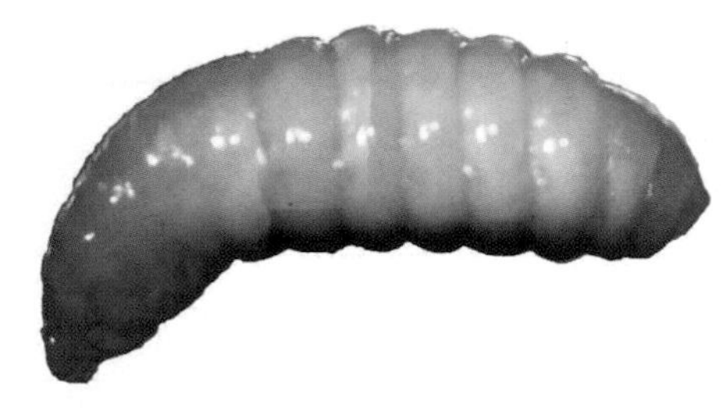

Figure 1.36 The lateral view of the L5 larva

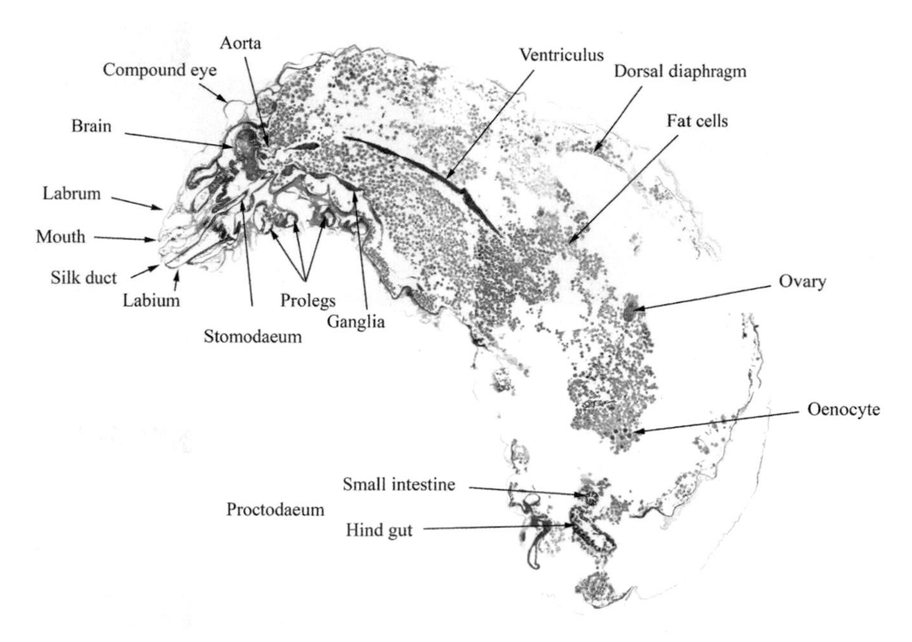

Figure 1.37 A midline sagittal section of an L5 larva after defecation

Initially the queen L5 faces towards the bottom of the cell where foods remains. The larva then turns around. The L5 larvae stands in the cell and the worker honey bees cap the cell. The L5 larva ventriculus opens into the proctodaeum and the four Malpighian tubules enter the small intestine. The larva then defecates and urinates. Once the ventriculus and Malpighian tubules are empty, the passage from the ventriculus and proctodaeum closes again. The passage from the stomodaeum to the ventriculus also closes. The larva does not eat and the cell is capped.

Prepupa

The honey bee is a little unusual having an intermediate stage between larva and pupa where the development from larva to pupa occurs within the cuticle of the L5 larva. This cannot be seen from inside the L5 larva outer cuticle. If the cuticle is dissected away the prepupa will be revealed. The term ecdysis is used to describe the transition between the L5 larva and the prepupa.

Pupa

The pupal stages can be broken down into the various eye colourations that are externally visible. After the eye colouration is complete the body turns brown and progressively black. The wings only expand on the final day.

White eye, day 13 worker, day 25 drone (from day of being laid)

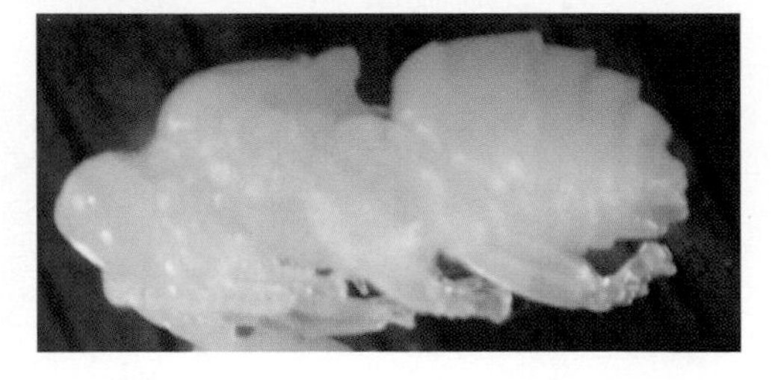

Figure 1.38 Capped pupae

Figure 1.39 Lateral view of a white eye pupa

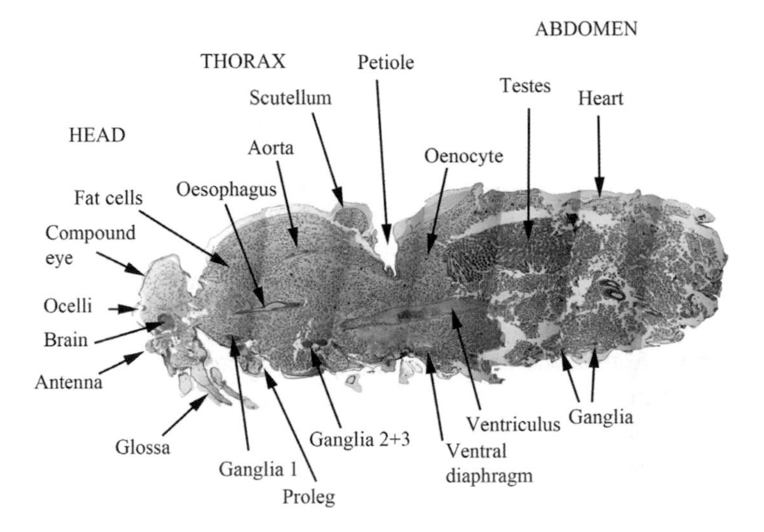

Figure 1.40 Midline sagittal section of a white eye drone pupa day 15

The cervix (neck) and petiole (waist) starts to develop and delineate the three main areas of the adult bee body. Note that the ventriculus is still within the thoracic 'area'. In dissections of the drone the testes are very large. The future outline of the surface organs become clear with the compound eyes, antennae and legs developing.

Pink eye, day 15 worker, day 16 drone (from day of being laid)

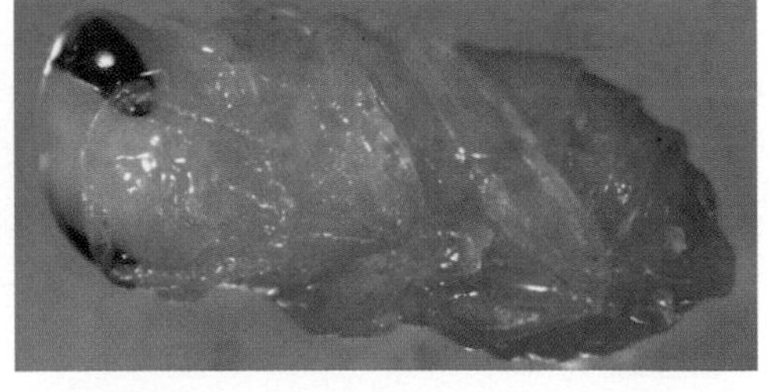

Figure 1.41 Capped pupae

Figure 1.42 Ventral surface of a pink eye pupae

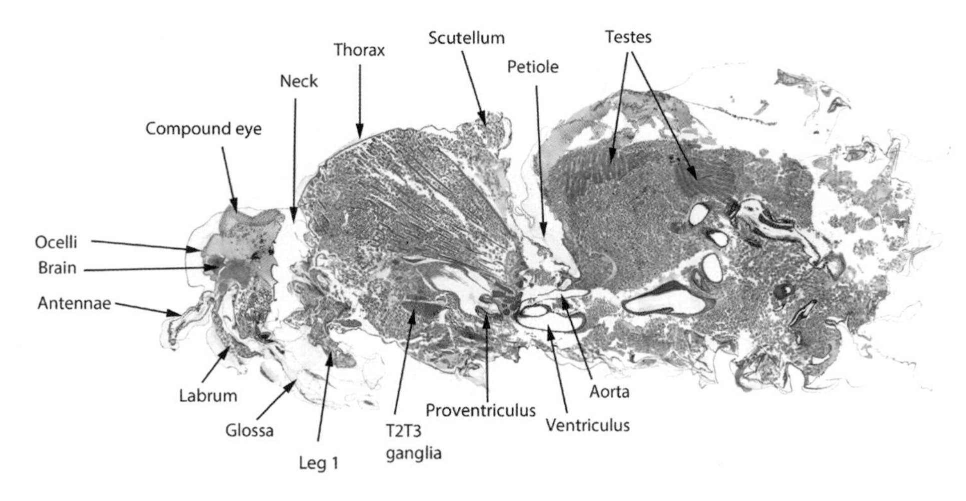

Figure 1.43 Midline sagittal section of a pink eye drone pupa day 16

Colour starts to develop in the eyes, particularly the ocelli. The outline of the future adult bee is clearly recognised. The body is encased in the pupal silk casing. This can be clearly visualised looking at the limbs, wings and antenna down the microscope.

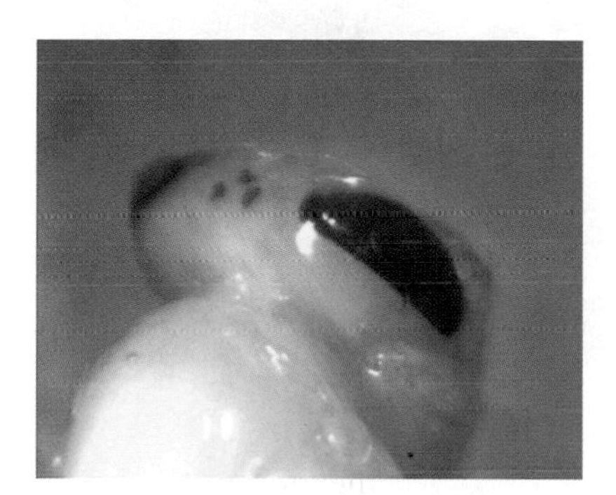

Figure 1.44 Gross head of a pink eye pupa showing the colour in the compound and three ocelli eyes

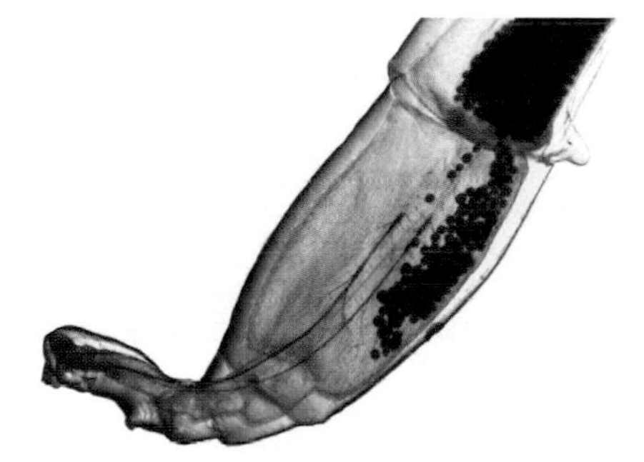

Figure 1.45 The developing hind leg in a pink eye pupa encased in the pupa silk cocoon. The back circles are fat bodies

Purple eye, day 17 worker, day 20 drone (from day of being laid)

Figure 1.46 Capped pupae

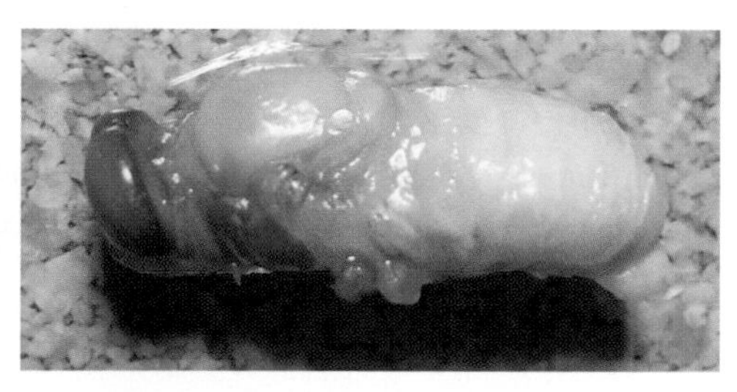

Figure 1.47 Purple eye drone pupa

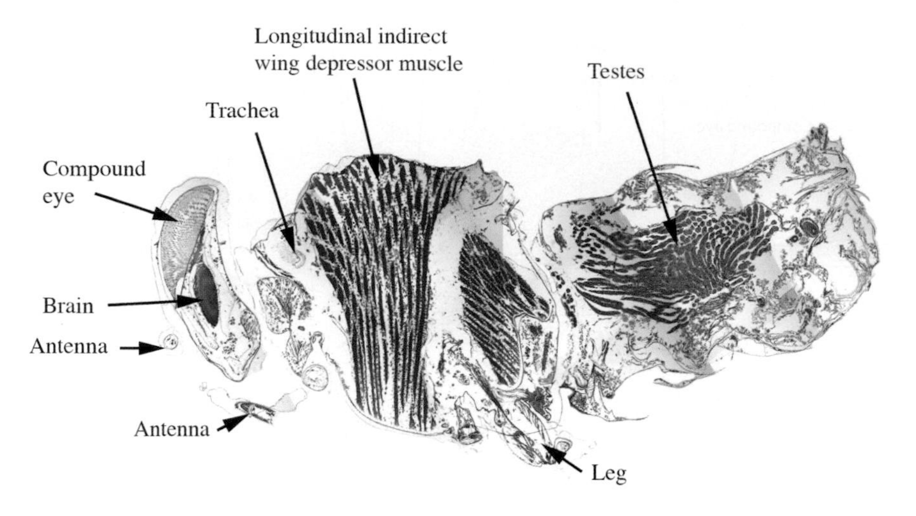

Figure 1.48 Sagittal section of a purple eye drone pupa day 19

Emergence

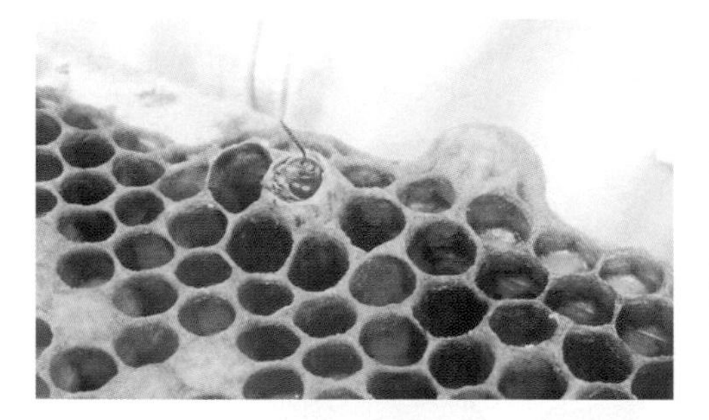

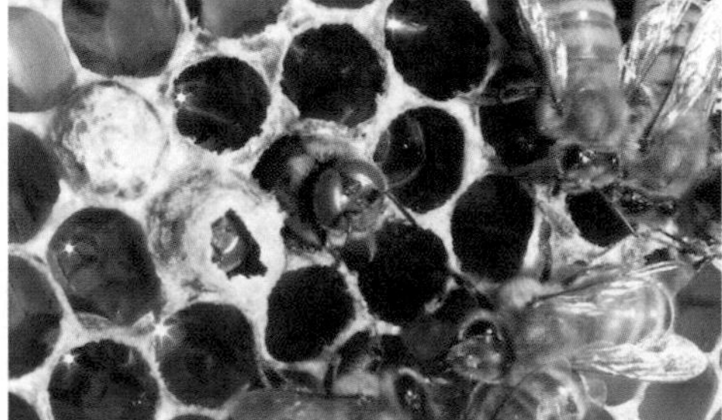

Figure 1.49 Emerging drone honey bees

Caste	Lay to emergence
Queen	16 days post laid
Worker	20–21 days post laid
Drone	24 days post laid

Key points to the anatomy changes

When you examine a pupa it is possible to age the pupa within about 24 hours looking at the change in colour to various parts of the anatomy. The eyes and ocelli in particular are useful guides.

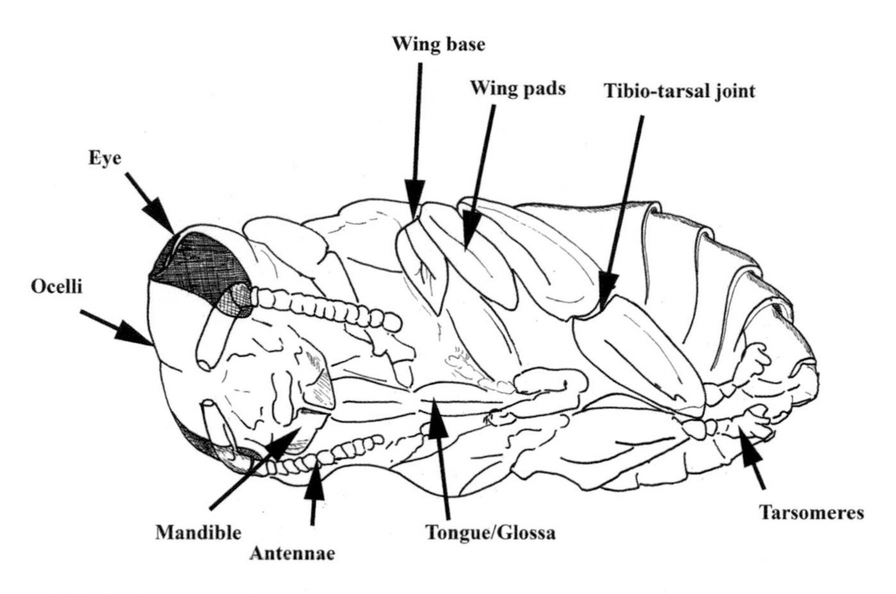

Figure 1.50 Ageing a pupa key anatomical change colour points

Table 1.3 Summary of the development of a queen, drone and worker *Apis mellifera*

Day	Queen	Drone	Worker
1	Egg	Egg	Egg
2	Egg	Egg	Egg
3	Egg	Egg	Egg
4	L1	L1	L1
5	L2	L2	L2
6	L3	L3	L3
7	L4	L4	L4
8	L5 Face down		L5 face up
9	Cell is capped Prepupa face down	L5 Face up	Cell is capped
10	Prepupa face up L5 moult		Prepupa
11	Light pink eyes	Cell is capped	L5 moult
12	Light pink ocelli Medium pink eye	Prepupa	White eye
13	Pink purple eyes Light yellow head, thorax and mandibles		
14	Purple pink eye and ocelli Dark brown mandibles Light yellow abdomen, legs antennae Light brown head, thorax, leg joints, tarsomeres	L5 moult	Slightly marked light pink eyes and ocelli

Table 1.3 *continued*

Day	Queen	Drone	Worker
15	Black eye, ocelli, flagella Dark brown leg joints, tarsomeres Medium grey head, thorax	White eye	Pink ocelli Light pink eye
16 QE	Wings expand Emergence	Light pink purple lower eye and ocelli	Slightly marked light brown head thorax Light brown tibio-tarsal joints, sutures, wing bases
17		Pink eye ocelli Darker lower eye	Purple eye and ocelli Light yellow abdomen and legs Light brown head and thorax Medium brown tibio-tarsal joints, sutures and wing bases
18		Dark purple lower eye	Black eyes Brown purple ocelli Light grey – wing pads Dark brown head and thorax Yellow abdomen, tongue and legs
19		Wing base light yellow	Medium grey wing pads Dark grey head and thorax Black ocelli, eye and flagella
20		Dark purple eye and ocelli Light brown tibio-tarsal joints, tarsomeres and mandible Slight yellow to abdomen and thorax	
21 WE		Medium brown leg joints, wing bases and sutures Light brown head, thorax Light yellow abdomen, tongue	Wings expand Emergence
22		Light grey wing pads Light brown tongue Medium brown head and thorax Dark yellow abdomen and legs Dark brown leg joints, wing bases, tarsomeres and mandibles	

Day	Queen	Drone	Worker
23		Dark grey wing pads Dark yellow abdomen and tongue Dark brown head and thorax Black tip of abdomen	
24 DE		Wings expand Emergence	

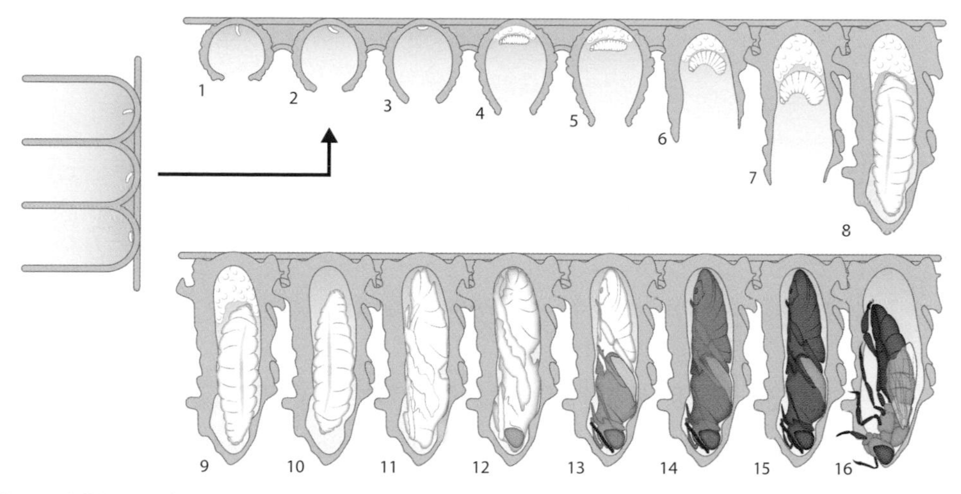

Figure 1.51 Development of the queen honey bee (illustration: Elaine Leggett)

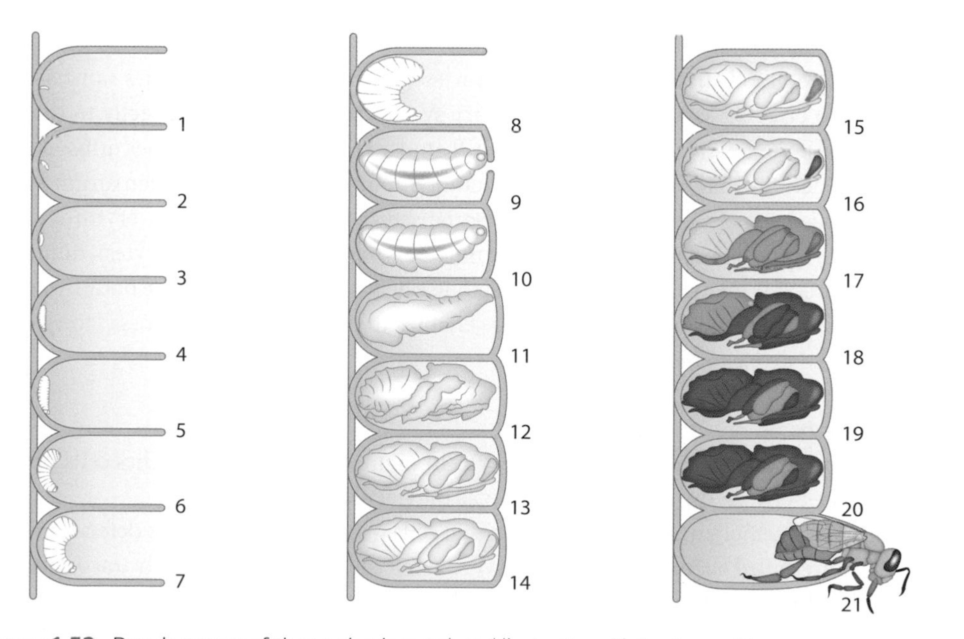

Figure 1.52 Development of the worker honey bee (illustration: Elaine Leggett)

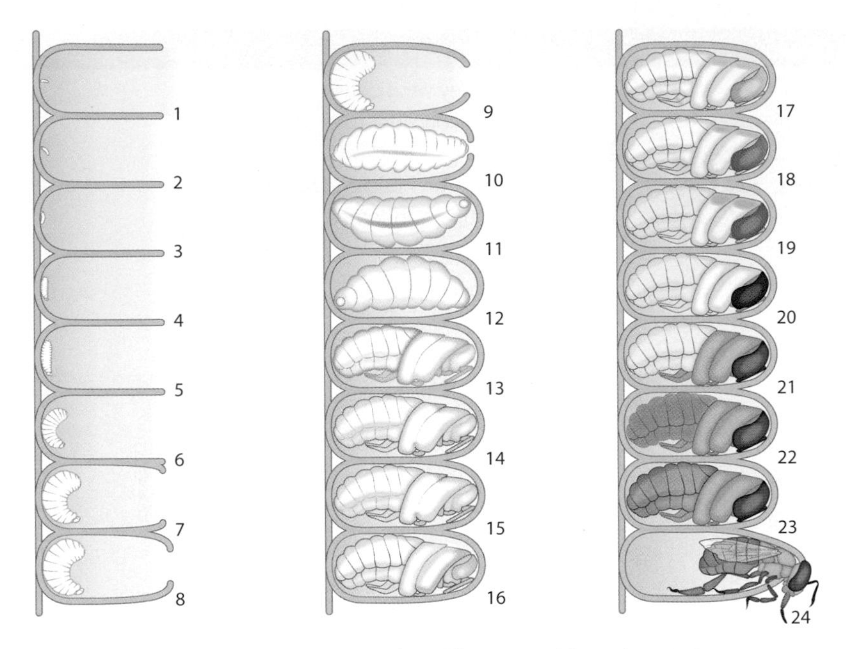

Figure 1.53 Development of the drone honey bee (illustration: Elaine Leggett)

The anatomy of the adult (imago) worker honey bee

The adult is also referred to as an imago. There are subtle differences between workers, queen and drone. But the basic design is the same and the following is a review of the basic anatomy of the honey bee (*A. mellifera*).

Circulatory system

The honey bee blood (haemolymph) is colourless, as the insect's blood does not have red blood cells to circulate oxygen. The respiratory system provides oxygen directly to the cells not the blood system. The honey bee, however, is unusual as an insect, as it does utilise a small amount of haemoglobin, dissolved in the haemolymph, to carry some oxygen around in its blood. But this haemoglobin is not kept in specialised cells (red blood cells – erythrocytes as in vertebrates). The blood is used to circulate nutrients, defence cells (haemocytes), dilute and circulate toxins (for removal), circulate and stabilise heat and help to move the body by hydrolytic pressure. The use of hydrolytic pressure enables the functions of the glossa (tongue) or eversion of the drone penis.

Haemolymph

The blood is not held within a closed system as in mammals. The blood bathes cells directly and the space between the cells is referred to as the haemocoele.

Within the haemolymph there are specialised cells which act as part of the defence mechanism; these are called haemocytes. They are very small cells. They are relatively uncommon in the adult bee but can be readily recognised in the various larval and pupal stages.

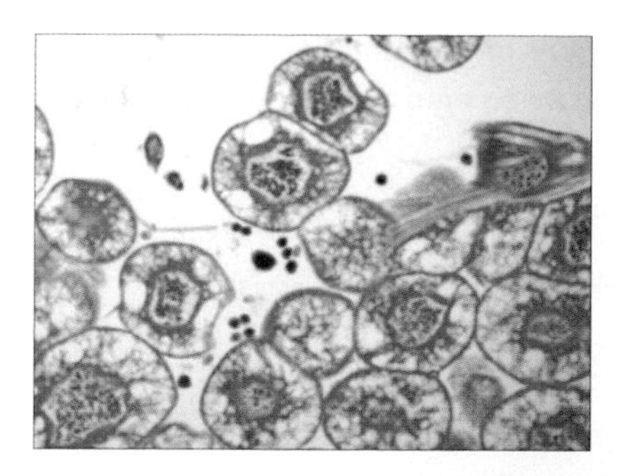

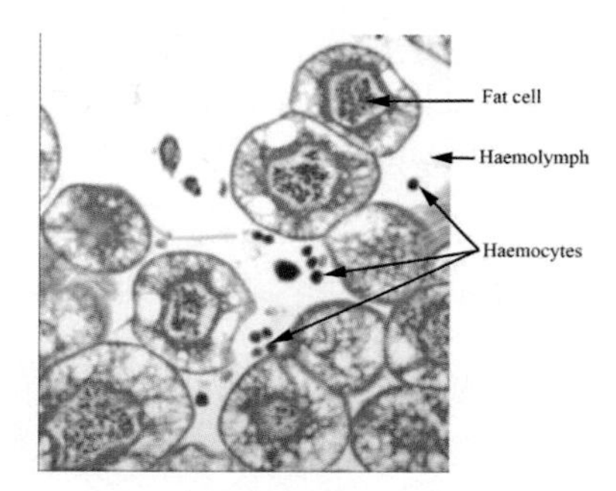

Figure 1.54 Various haemocytes in the haemolymph surrounding fat cells in an L3 queen larva

The honey bee circulates its blood (haemolymph) through a number of pump mechanisms including the heart, antenna pump, wing, legs and ventral diaphragm.

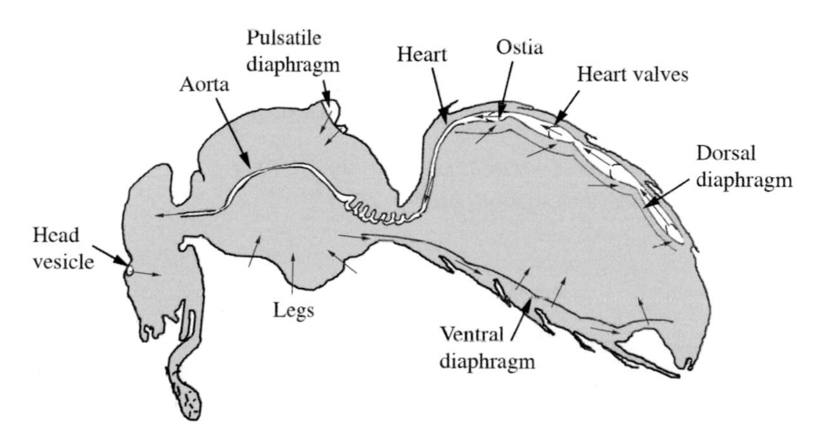

Figure 1. 55 The circulation of the haemolymph in the honey bee. The red arrows indicates the direction of the haemolymph flow

Heart

The heart creates a peristaltic wave starting at the caudal (rear) area, which moves cranially (forward towards the head). Blood is drawn in from the surrounding haemocoele through the dorsal diaphragm into the dorsal sinus and through the pores in the lateral walls of the heart. The blood backward flow is restricted by ostia (valves). The haemolymph travels then from the abdomen to the brain. At the caudal brain it leaves the aorta and moves around the head. There is a secondary pump in the base of the antennae. The haemolymph passes into the thorax. The circulation is assisted by secondary pumping mechanisms in the wings and circulation in the legs. The haemolymph then passes through the ventral diaphragm, which can also help to move the fluid, until the haemolymph is back into the abdominal haemocoele. There are 5 chambers to the honey bee heart.

The heart can be readily recognised in dorsal dissections of the adult honey bee. Once the dorsal tergum plates have been removed, the heart can be recognised as two lines running over the internal organs.

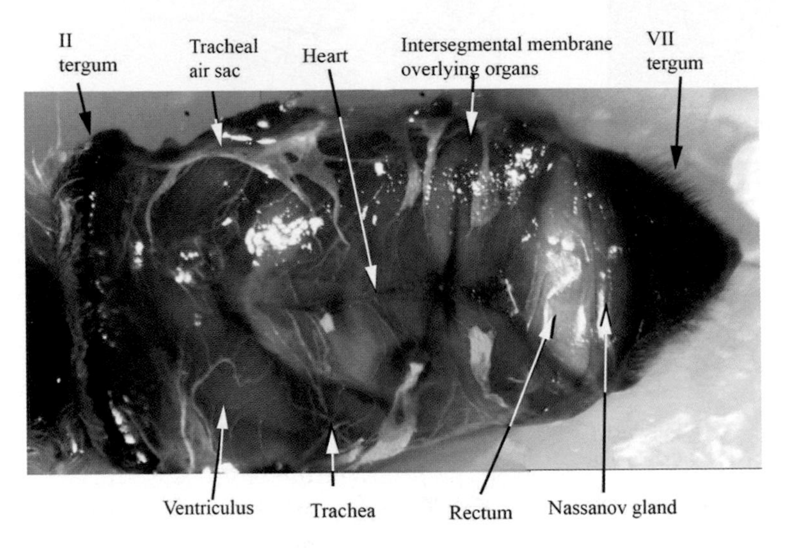

Figure 1. 56 Gross dissection of a worker honey with the III, IV, V and VI tergum plates removed revealing the heart and underlying organs *in situ*

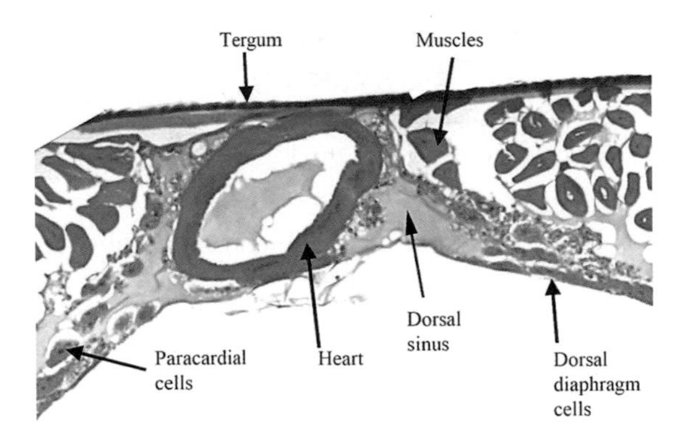

Figure 1.57 Transverse section of the heart in a drone

In sagittal histological sections the heart will only be visible in midline sections. However, in transverse sections the heart is very obvious as a muscular circle lying close to the dorsal tergum plates. This can be very useful to orientate the abdomen as the heart is situated at the dorsal point.

Dorsal diaphragm

The heart is suspended from the dorsal tergum plate by the dorsal diaphragm, which also acts as a filtering device removing particulate matter from the haemolymph.

Aorta

At the front of the abdomen the heart takes a sharp downward turn at the front of abdominal segment II to pass through the petiole. Proximal (forward) of this point, the heart is called the aorta. This has been predetermined in the larva by the position of the anterior tracheal loop. The aorta is thrown into a series of contortions in the caudal thorax. This acts as a counter current heat exchange system moving heat from the wing muscles. After passing through the petiole, the aorta again ascends towards the dorsal surface of the thorax passing between the two longitudinal indirect depressor flight muscles. In sagittal sections the aorta is extremely difficult to recognise, but in transverse sections can be readily identified in the middle of the thorax.

Ventral diaphragm

The ventral diaphragm extends from abdominal segment I (and therefore is in the thorax) through the petiole to the rear of the abdomen as a thin sheet dorsal to the neural commissures. This can be difficult to visualise, but is seen more clearly in transverse sections of the caudal abdomen. By muscular contraction this diaphragm can assist haemolymph circulation.

Cuticle

Most of the bee's skin is covered in a hard cuticle, which becomes an exoskeleton (outside skeleton), to which the muscles are attached. The cuticle structure varies over the bee's body depending on specific requirements. There are two forms of cuticle – the hard sclerotised exocuticle that forms plates and is non-flexible and other areas, which are non-sclerotised and are flexible. These non-sclerotised areas allow movement between fixed sclerotised parts. For example, where the head attaches to the thorax – the cervix (neck) or where the legs make contact with the thorax. This variable sclerotisation of the cuticle allows the insect to design a wide range of shapes and functions.

The adult bee's body still retains external evidence of the larva segmentation.

Table 1.4 Larval segmentation

Area in the adult	Segment
Head	
Thorax	Thorax 1,2,3 and Abdominal I
Abdomen External	II, III, IV, V, VI and VII
Abdomen Internal	VIII, IX (sting) and X (proctiger).

Figure 1.58 Midline sagittal section through the cervix (neck) of the worker honey bee

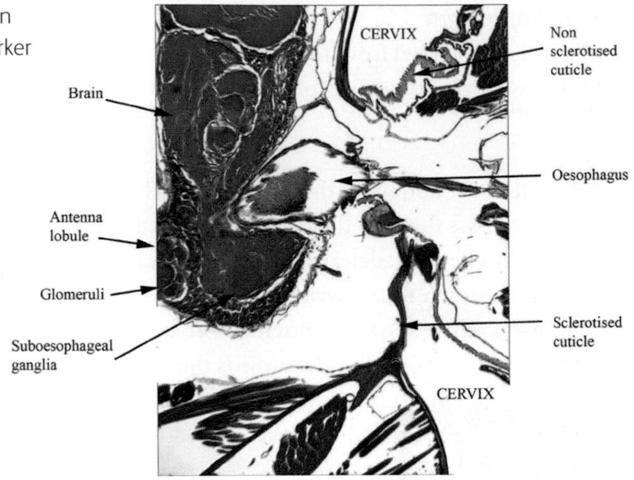

Movement between the abdomen cuticle plates

A flexible membrane called the secondary intersegmental membrane allows movement between the abdomen cuticle plates. Muscles attached to thickenings at the edge of each plate, called the antecosta, allow movement between the hard cuticle areas or plates. The movement between the abdominal plates is also important in the respiration of the honey bee.

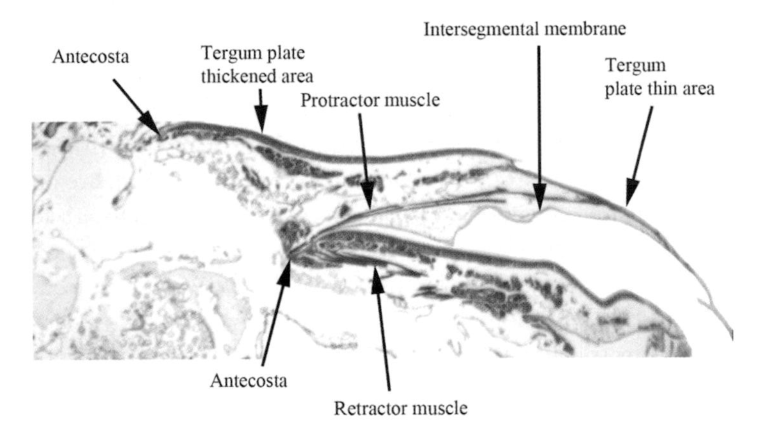

Figure 1.59 Two tergum plates and their longitudinal connections via the intersegmental membrane. The two tergum plates are moved by the actions of protractor and retractor muscles

The bee is covered in hairs. These are both straight and heavily branched. Many of these hairs are used by the bee to provide information on movement and airspeed. The branched hairs are used to capture pollen during visiting flowers. Pollen provides bees with their protein source. Around the ocelli hair also directs and controls the light direction entering the eye.

The cuticle also has specialised areas, which are designed for information gathering or for chemical/hormone or scent production or for the development of the sting apparatus.

Digestive system

Mouth

The digestive tract of the bee starts with the mouthparts.

There are three major components to the mouthparts and these are all modified limbs. The mouthparts allow the bee to both chew and suck-up liquids. The mandible can be used in defence and fighting.

Figure 1.60 Head of the queen bee illustrating the mouth parts

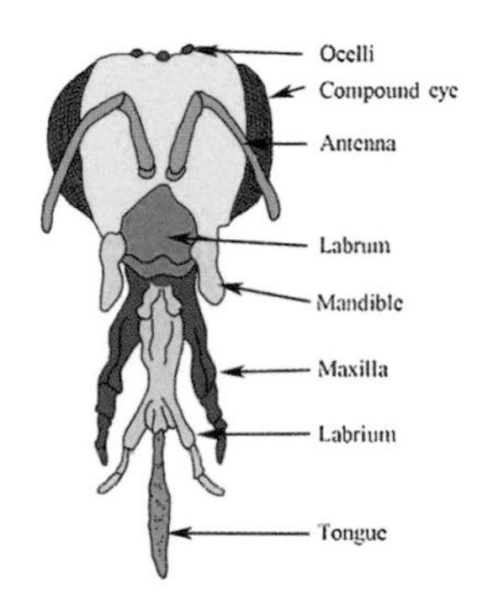

Figure 1.61 Drawing of the mouth parts of the worker honey bee

The mandible

The mandibles are the biting mouthparts and their shape is slightly different between the different castes. The worker bee has the largest mandibles. The mandible is a modified coxa. The other parts of the modified limb have been discarded. The mandibular gland's sections run down the mandible in a groove. Figure 1.60 illustrates the queen's mandible with the characteristic notch in the surface of the mandible. The photograph allows the reader to visualise the raw power within the mandibles as a chewing and cutting tool.

The proboscis

In the honey bee the two other adapted mouthparts are 'limbs'. The maxilla and labium combine to create the sucking mouthpart, the proboscis.

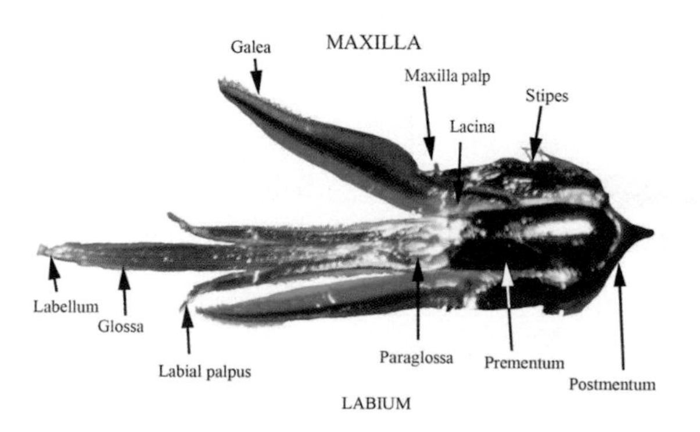

Figure 1.62 The dorsal proboscis. The proboscis is a combination of the maxilla and labium mouthparts

Maxilla

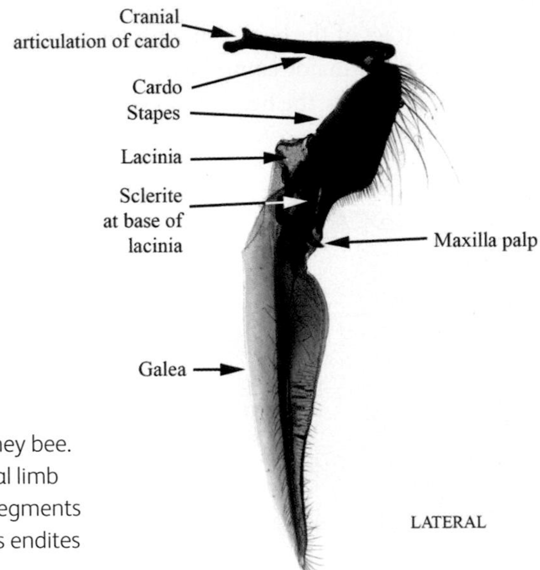

Figure 1.63 The maxilla of the worker honey bee. The cardo and stapes constitute the original limb coxa, the maxilla palp and the remaining segments of the leg. The galea and lacinia develop as endites from the medial coxa

The maxilla is a modified leg primarily the coxa. The remaining parts of the leg are simplified into the maxilla palp. Medially (inside) to the palp and at the end of the coxa there are two projections (coxal endites) resulting in the lacina (medial) and the galea (lateral). In the honey bee, the maxilla palp is very reduced with just two segments remaining. The galea is a long tapering blade like structure. The lacina in the honey bee is reduced to a membrane structure and plays a vital role in closure of the food canal of the proboscis.

Labium (underlip)

The labium is a fusion of two leg coxae. The previous lower leg segments are represented by the two labium palps. At the base of the coxa (as in the maxilla) there are two medial projections called the paraglossa and glossa.

In the honey bee the labium palps have retained the limb four distinct segments.

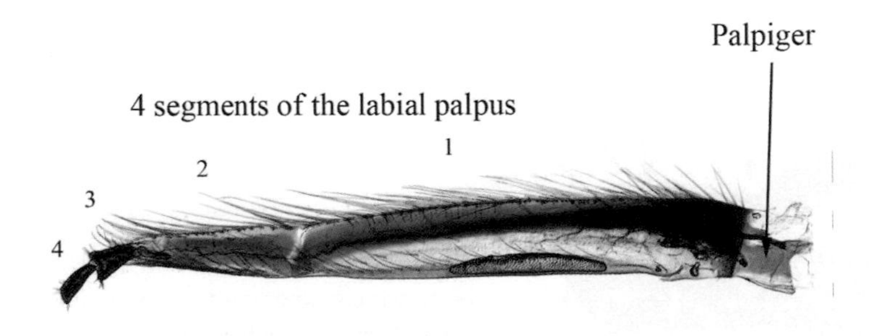

Figure 1.64 The labial palpus

The two paraglossae cover and support the proximal (top) of the glossa. The glossae are fused together to form the long hairy glossa (tongue). At the end of the glossa is the spoon shaped labellum. Using haemolymph, the bee is able to pressurise the glossa and cause it to extend.

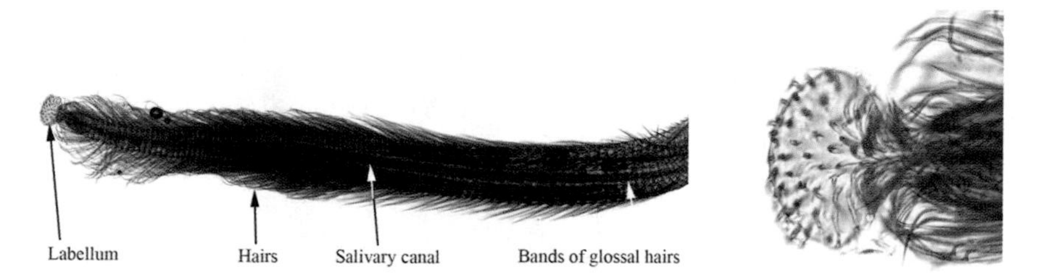

Figure 1.65 The glossa (tongue) of the honey bee with detail of the labellum

Figure 1.66 Detail of the labellum

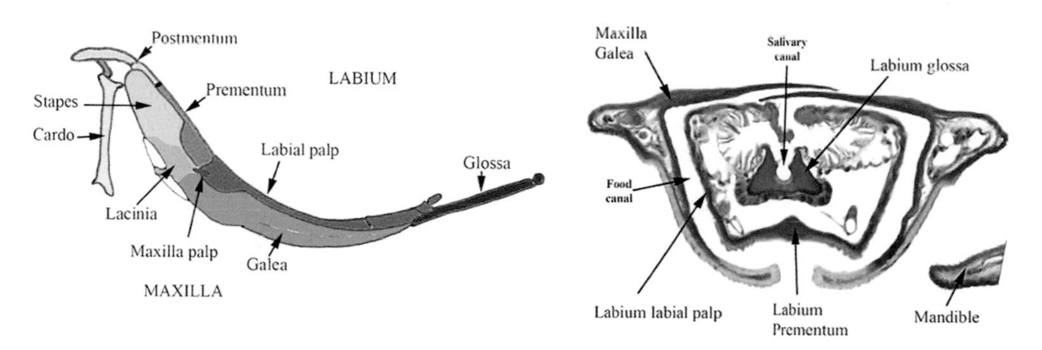

Figure 1.67 The proboscis mouthparts. Note the other mouthpart, the mandible, is for biting and chewing. The maxilla and labium lie parallel and embrace each other to form a close fitting tube

Figure 1.68 Transverse section of the posterior mouthparts demonstrating the embrace of the maxilla and labium to form a tube, by which liquids can be sucked into the mouth

The maxilla and labium mouthparts fit together to form a unit, the proboscis (tube or straw). The whole maxilla and labium swings around the pivot provided by the cardio, to move from the resting position, tucked under the head, to a forward vertical position in contact with the front of the mouth.

Once the food is moved into the mouth the material is softened and lubricated by the addition of saliva. This starts to break down the sugar materials within food.

Saliva glands

The adult bee has two saliva glands; one situated just behind the brain, the postcerebral salivary gland and the other in the thorax, the thoracic salivary gland. The salivary glands originate from the silk glands of the larva.

Pharynx and oesophagus

After entering the mouth the food passes into the pharynx and is swallowed into the oesophagus. This is a tube that runs the length of the thorax and passes through the narrowing of the petiole.

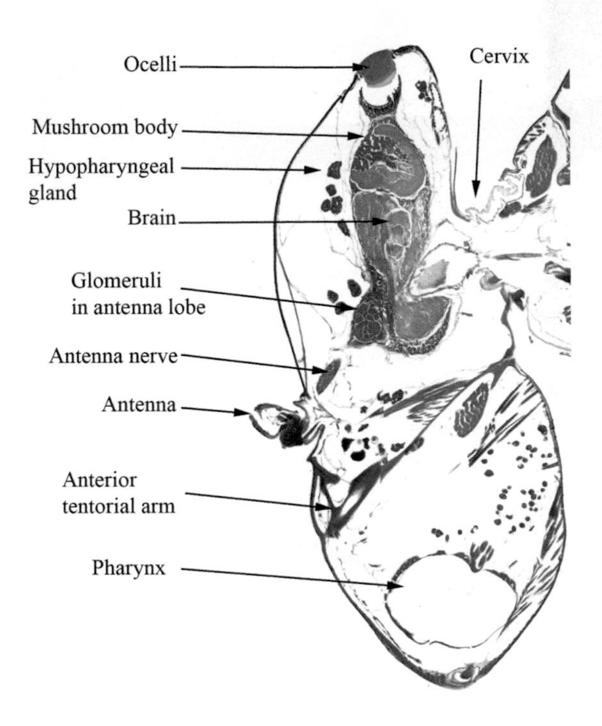

Figure 1.69 Transverse section of the head of a worker showing the pharynx

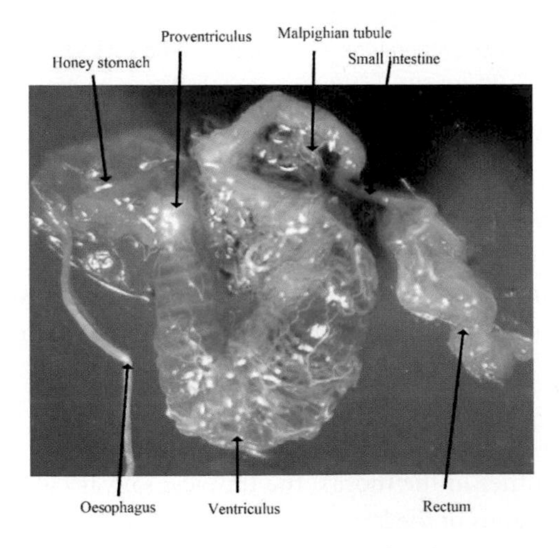

Figure 1.70 Gross dissection of the intestinal tract displayed to demonstrate the organs of the worker honey bee

The abdominal intestinal tract

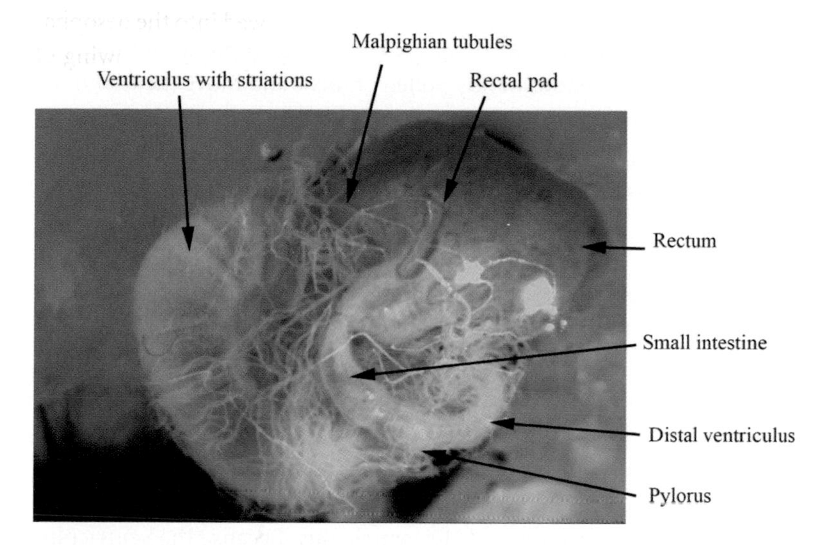

Figure 1.71 Gross dissection of the abdominal intestinal tract of the honey bee worker *in situ*

Abdominal oesophagus, honey stomach and proventriculus

After leaving the petiole (waist) the oesophagus becomes an extendable clear very thin walled bag. This is referred to as the honey stomach. Bees are able to store a large quantity of nectar in the honey stomach (30% of the bee's weight). The honey stomach is easily recognisable

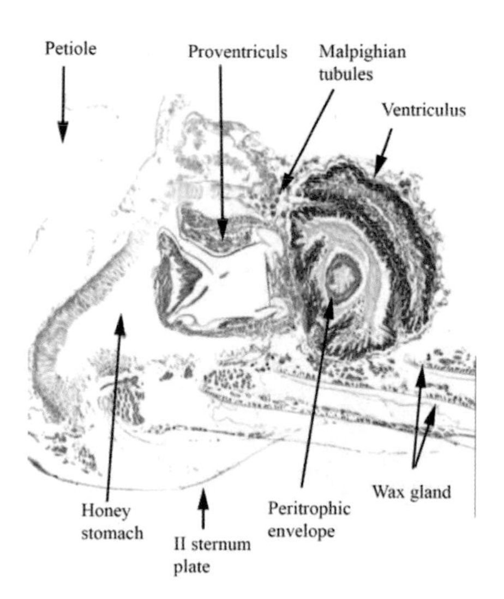

Figure 1.72 The oesophagus, honey stomach, proventriculus and ventriculus of a worker honey bee

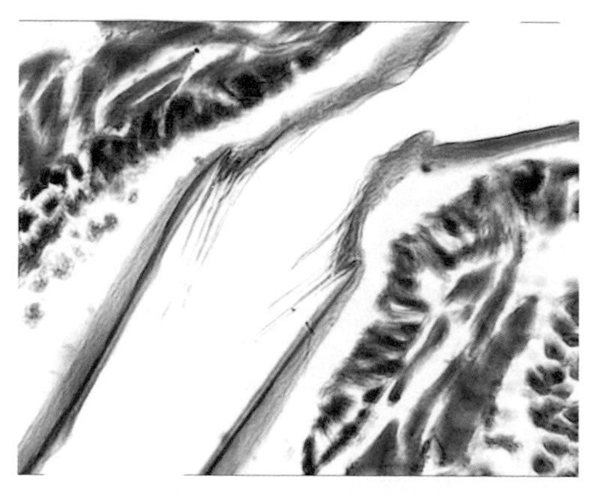

Figure 1.73 Entrance to the proventriculus with the backward facing hairs (arrowed) helping to move and strain pollen grains from the honey stomach to the ventriculus

in careful dissection of the bee's abdomen. Through the wall of the honey stomach a white thickening can be seen. This is the proventriculus. Careful examination will reveal the 4 lips of the proventriculus. The purpose of the proventriculus is to strain the contents of the honey stomach and remove large particles – predominantly pollen grains – and move them into the ventriculus, where digestion can continue. This straining is assisted by the presence of backward facing hairs on the surface of the proventriculus. Note the surface of the proventriculus, as part of the stomadaeum, is covered by cuticle and so the presence of hairs is not surprising.

Ventriculus

The ventriculus is a large slightly horseshoe shaped organ on the left hand side of the abdomen curving to the front and over to the right hand side of the abdomen. Its size can vary depending on how much food is present within the lumen. The ventriculus can be readily recognised by the clear striations on the surface. The ventriculus tends to be wider at the proximal (beginning) end. On histological examination the ventriculus is easily recognised by the series of concentric circles within the lumen. This is the peritrophic membrane, which acts as a molecular filter on the materials being digested within the ventriculus lumen. The peritrophic membrane also protects the surface of the ventriculus. Because the ventriculus originates from inside the bee, a cuticle does not cover the lining. In the drone and queen the ventriculus is often white or colourless.

Small intestine and pylorus

The ventriculus ends at the junction with the small intestine, commonly on the left hand side of the abdomen. At this point there is a thickening (pylorus) from where the adult Malpighian tubules originate in the L5 larva. The true small intestine then continues from this point. The small intestine is a thick walled thin tube easily seen leaving the distal (end) point of the ventriculus. On histology the small intestine is recognised by its narrow width and the characteristic 6 infolds of the serosa (inner) epithelium.

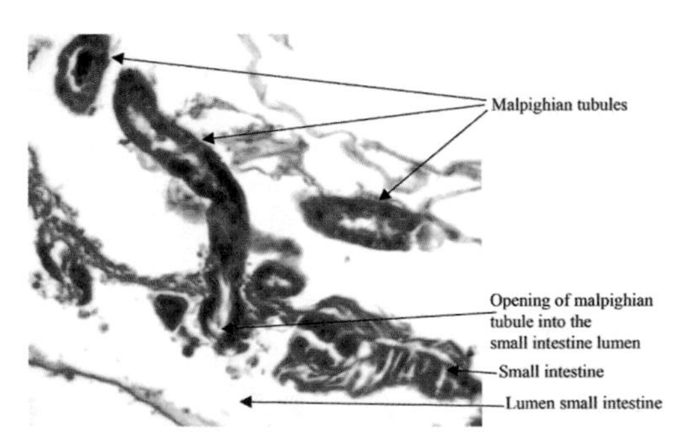

Figure 1.74 The pylorus with a Malpighian tubule entering the small intestine

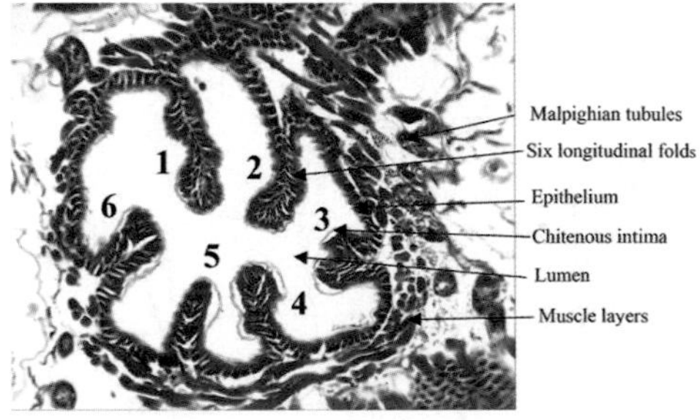

Figure 1.75 The small intestine with the characteristic 6 infolds

Rectum

In the honey bee (*Apis mellifera*), the small intestine is relatively short and enters a large often flaccid organ that easily ruptures in dissection; this is the rectum. The rectum is generally yellow/brown in colour and contains faecal material. The pH of the rectal contents is more acidic than the rest of the intestinal tract. The wall is thin and easily ruptured. On the surface of the rectum, six longitudinal lines will be seen – these are the rectal pads, which are important in electrolyte control. The rectum absorbs water from the digesta and acts as a storage organ. Overwinter bees store faeces in their rectum until their cleansing flights in the spring. Therefore, overwintered bees will often have an extremely full rectum, which appears to completely fill their abdomen. Part of this faecal content may be passed with the sting and poison gland when the bee stings its victim.

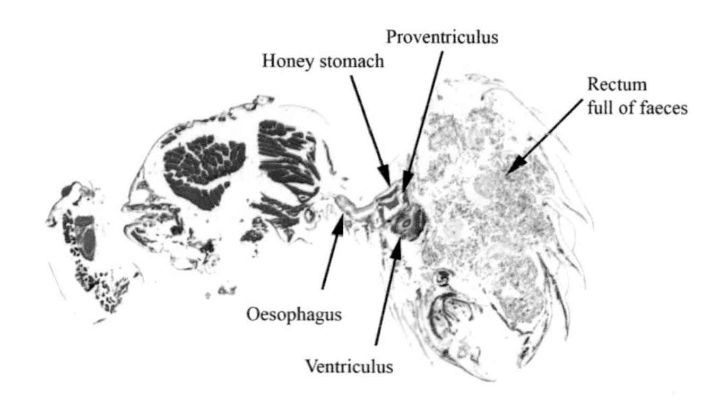

Figure 1.76 Midline sagittal section of the overwintered honey bee worker illustrating the full rectum

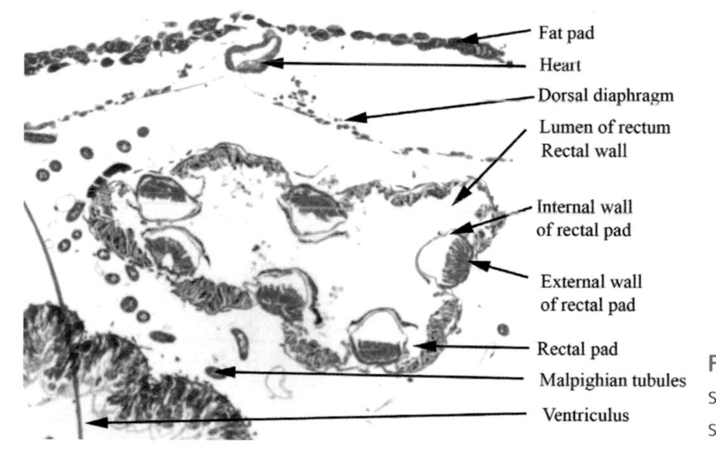

Figure 1.77 Transverse section of the rectum showing the six rectal pads

Anus

The anus exits segment X (the proctiger). This is situated immediately under tergum VII. Internally there is a sigmoid flexure to the intestinal tract, which allows for passive control of the rectal contents. The pressure from rectal contents push against the sigmoid flexure stops the passage of faeces out of the anus, until the bee actively raises tergum VII.

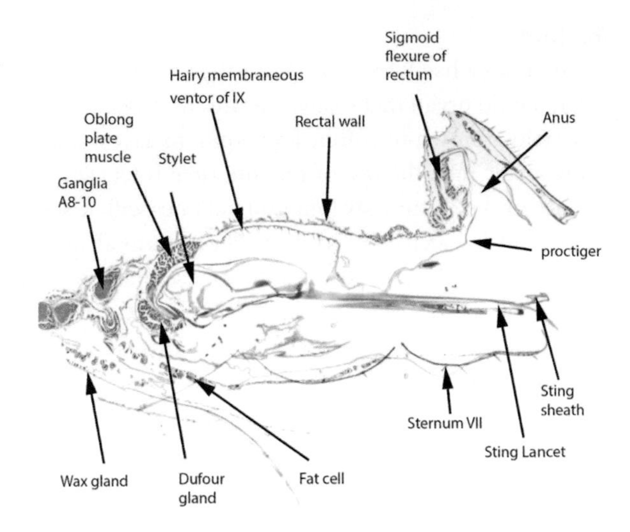

Figure 1.78 Midline sagittal section through the anus of the worker honey bee also showing the sting apparatus

Endocrine system

Endocrine glands produce chemicals, which control function and activity of other organs. The major endocrine organs of the bee control moulting and development. The corpora allata and corpora cardiac glands are major endocrine glands and are located behind the brain and are seen as a small circular ball of cells on suitable sagittal sections. The corpora allata are larger and more defined.

The corpora allata produces the juvenile hormone, which prevents and inhibits metamorphosis and is obviously important in the developing larva and pupa. In the adult it also has a role in developing guarding behaviour and aggression. The prothoracic gland, at the junction of the oesophagus and ventriculus produces ecdysone. This helps to control ecdysis.

The other major hormone is vitellogenin. This is stored in the fat bodies and triggers pupation. In the queen the hormone may also be a factor in long life.

Fat bodies

The adult honey bee stores fat mainly inside the abdomen to use as a reserve. The fat bodies are particularly well defined in the queen bee.

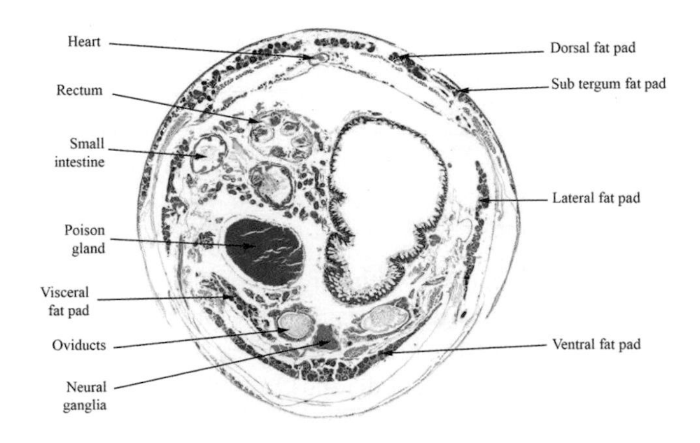

Figure 1.79 Transverse section of the abdomen to illustrate the fat pads of the queen honey bee

Glandular system

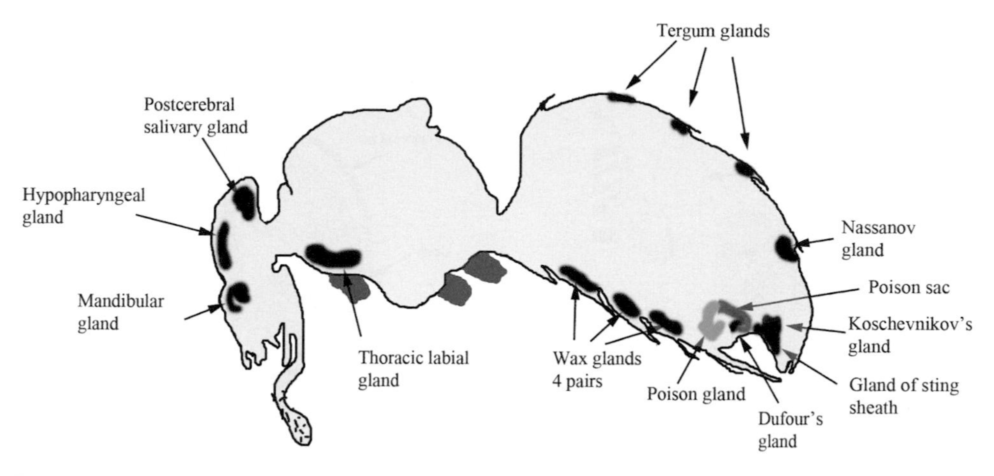

Figure 1.80 Position of the major glandular organs of the worker honey bee

Glands in the head

Mandibular gland

In young workers this gland produces the lipid-rich white substance mixed with the hypopharyngeal gland secretions resulting in royal jelly.

In an older worker this produces part of the alarm pheromone.

In the queen, this gland has a number of important functions – it produces the queen substance (queen mandibular substance) and is associated with:

- suppression of construction of emergency queen cells;
- inhibiting ovary development in the workers;
- attracting drones during the mating flight;
- attracting the attendant workers.

In the drone, the mandibular gland assists in the formation of drone gatherings – in drone congregation areas which appear (in *Apis mellifera*) in open fields in the air at about 10–20 metres altitude. The gland however is small in the drone.

Hypopharyngeal gland

This produces protein-rich sections (Royal jelly) when the worker is a nurse bee.

When the worker becomes a forager bee it produces amylase which helps break down starches into simpler sugars, di- and trisaccharides. The hypopharyngeal glands are absent in the queen and drone although remnants of the glandular system may occasionally be recognised.

Postcerebral salivary gland

Behind the brain and the caudal hypopharyngeal gland can be found the postcerebral salivary gland. This is more obvious in the drone as there is no hypopharyngeal gland.

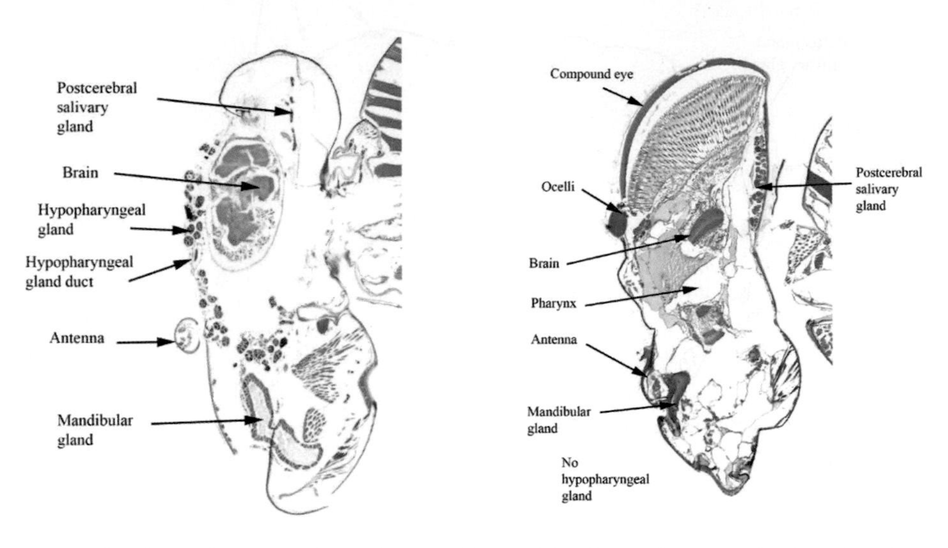

Figure 1.81 Sagittal section of a worker head

Figure 1.82 Sagittal section of a drone head

Glands in the thorax

With the development of the cervix (neck) in the pupa the larval silk glands are divided into two – those in the head and those in the thorax. In the head they change into the postcerebral salivary glands, in the thorax into the large thoracic salivary glands. They are readily seen on sagittal and transverse sections. Salivary glands produce invertase, which breaks down sucrose into fructose and glucose and lubricates the food prior to being swallowed.

Glands on the legs and feet

Pre-tarsus gland

As yet its function is not known.

Arnhart or footprint glands on each foot

These have a footprint odour which is used by the bees as part of their common hive smell and allows the bees to identify their hive.

Glands associated with the sting

Koschevnikov gland

This is a small clump of cells in the centre of each quadrate plate. This releases alarm pheromone, which is composed of several volatile compounds. The principle compound is isopentyl acetate. These chemicals encourage other bees to attack and sting the same part of the body of the offending animal.

In the queen this gland's products are responsible for the formation of the clusters of court bees that surround the queen.

Dufour's gland (alkaline gland)

The products of this gland line the entrance to the hive and may assist recognition of family or hive ownership.

Glands in the abdomen

Scent gland – VII Tergum – Nassanov gland

This gland is present in the worker but absent in the drone.

The gland produces a variety of chemicals, which the bee uses to assist identification of the entrance of the hive. The chemicals are excreted into the canal and released by depressing tergum plate VII.

Figure 1.83 A guard worker bee exposing the Nassanov (scent) gland

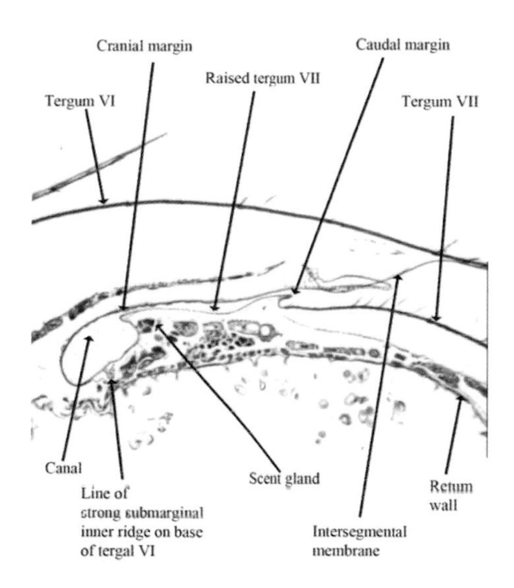

Figure 1.84 Histological detail of the Nassanov (scent) gland in a longitudinal section

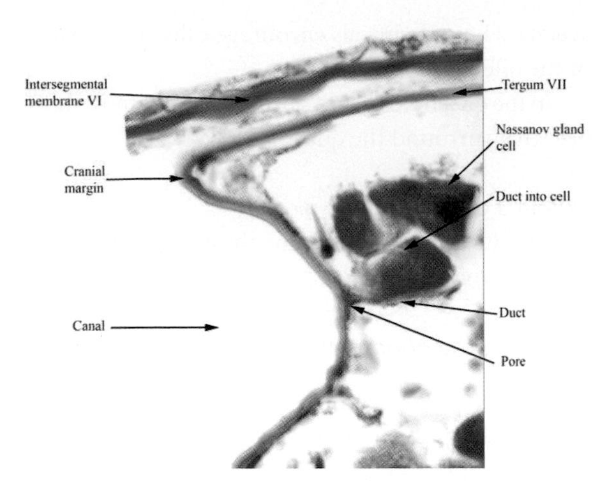

Figure 1.85 Sagittal section through the scent gland showing detail of the canal and duct system

Wax glands

In four of the sternum plates of the worker there are glands that produce wax. The wax glands are easily seen in sections as a line of columnar cells. They are particularly active and large in nurse bees actively producing wax for hive cleaning and development. The wax glands are very active and have a close association with fat cells and their accompanying oenocytes. These cells are absent in the queen and drone.

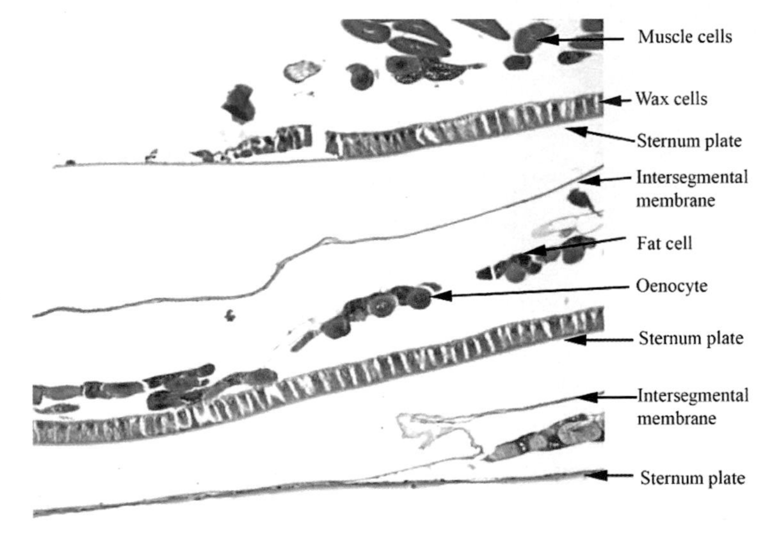

Figure 1.86 Wax cells in a sagittal section of a nurse worker bee. The height of the wax cells can be used to age the worker bee and determine its current status

Immune system

The honey bee has a variety of complex defence mechanisms. The haemocytes are free in the haemocoele to destroy foreign materials. The dorsal diaphragm acts as a filter for the haemolymph. The bee is also able to inactivate foreign materials by melanisation, which turns the foreign material black. This is commonly used for internal parasite defence.

The bee's defence mechanisms rely on three factors:

1 Physical barrier – the hard exoskeleton
2 Cellular immunity haemocyctes

- Prohaemocytes – stem cells;
- Granulocytes – which collect at the site of a foreign body and locally change the chemical makeup of the haemolymph;
- Plasmaocytes – able to phagocytose (consume) foreign materials.

3 Humoral immunity – production of chemicals such a melanin, quinones, hydrogen peroxide and other oxygen free radicals which all kill pathogens.

Once killed pathogen fragments are removed from the body via the Malpighian tubules and the dorsal diaphragm.

Muscular and locomotor system

In common with most insects, the adult bee has actually 12 legs. The first proximal three pairs are modified into the mouthparts. The last distal three pairs are clearly recognised as walking legs. However, each of these has multiple functions not just for walking and fighting.

The three walking legs originate from each of the three thoracic segments. While the first abdominal segment is part of the thorax (in the honey bee) it does not carry a leg. The three legs are referred to as the prothoracic, mesothoracic and metathoracic legs.

The mesothoracic and metathoracic segments also carry the wings.

The general anatomy of each leg is very similar in general design. Each leg is however, slightly modified to fulfil specific functions, which also differs between the castes. The legs provide the bee with a means of walking but also a complex tool box.

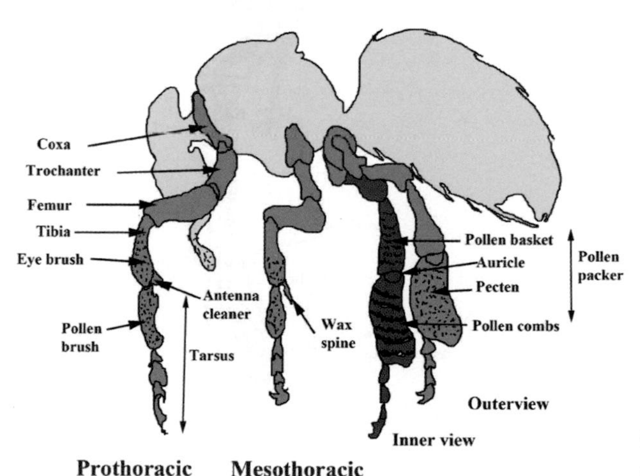

Figure 1.87 Drawing of the general anatomy of the honey bee legs

General anatomy of a leg

The leg is composed of five parts. The description starts proximal (towards the body) and works distally (away from the body) towards the 'foot'. Compare this description with the anatomy of the mandible, maxilla and labium mouthparts.

Coxa

The coxa articulates with the thorax and is a short square/rectangular shaped component. Each of the three different legs' coxa are set at slightly different angles to provide the bee with a slightly different plane of motion in each leg.

Trocanter

The trocanter is a short secondary segment. The segment can only move up and down.

Femur

The femur is a long third segment. The articulates slightly forwards and backwards against the trocanter.

Tibia

The tibia of the prothoracic and mesothoracic (1st and 2nd) legs is a slender segment about the same size as the femur. On the metathoracic (3rd) leg the tibia is greatly elongated especially in the worker.

Tarsus

The tarsus is divided into five subsegments called tarsomeres. The first segment is the largest called the basitarsus. This is then followed by three smaller segments.

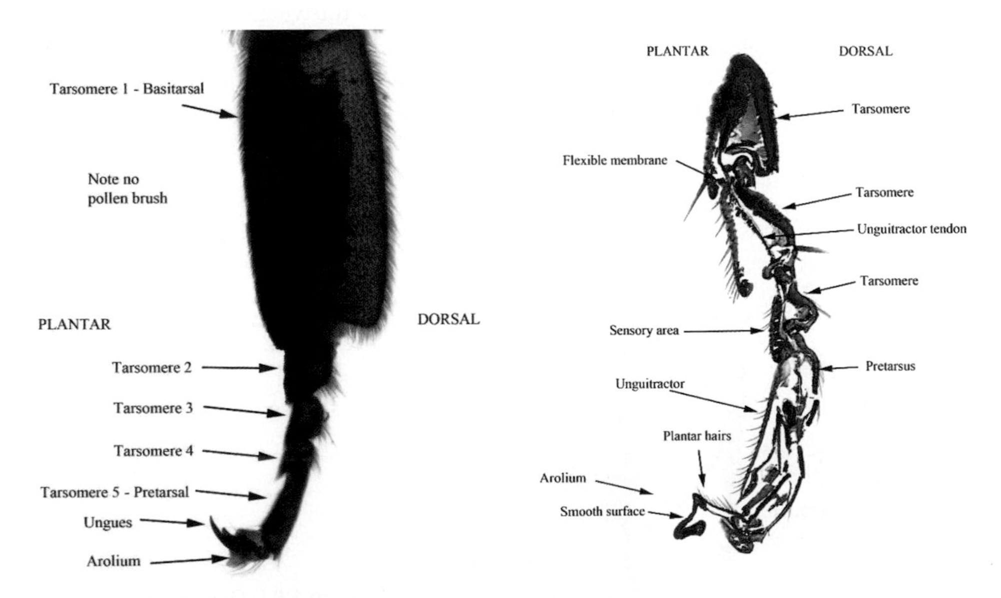

Figure 1.88 The abaxial aspect of the tarsus of the metathoracic leg of a queen bee

Figure 1.89 A sagittal section of tarsomere 2 to 5 of the honey bee

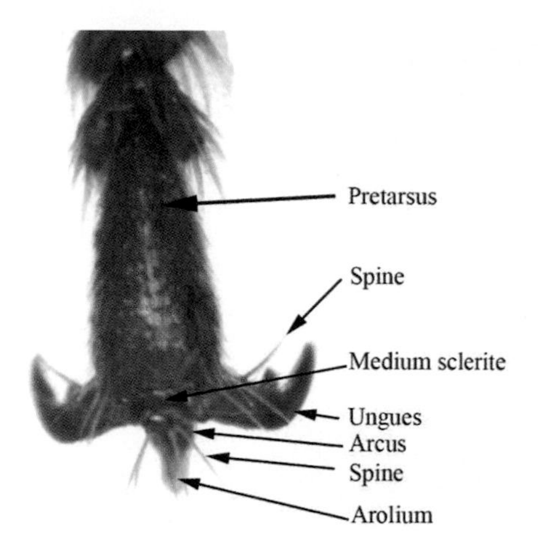

Figure 1.90 The dorsal aspect of the pretarsal (foot)

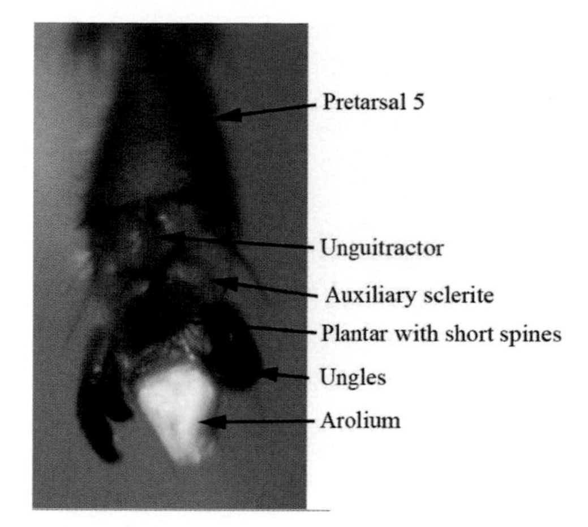

Figure 1.91 The plantar aspect of the pretarsal (foot)

The last tarsus segment is called the pretarsus (foot). The pretarsus is a complex sub-segment. The pretarsus has a long component and two claws which are articulated independently. In between the two claws there is a soft apical lobe: the arolium. This arolium acts as an adhesive organ allowing the bee to grasp onto surfaces otherwise too smooth to stand upon.

The claws on the last tarsomere allow bees to walk on rough surfaces (tree trunks) together with a soft pad (arolium) allowing them to walk on smooth surfaces (leaves).

The metathoracic (hind) legs in the worker are fringed with long, curved hairs and the space thus enclosed is called the pollen basket or corbicula.

Note when handling bees take care, especially with queen bees. Her legs can be easily damaged, and she needs to squat to lay her eggs and if her legs are damaged she will not be able to efficiently lay eggs. Nor will she be able to measure the size of the cell, vital to determine a worker or drone egg to be laid.

The specific tool set provided on each leg

Prothoracic leg (fore or first leg)

This leg is designed to clean the antenna, the major sense organ for the honey bee. Between the tibia and the basitarus is the antenna cleaner. The bee locks the antenna in the notch using the fibula and removes foreign material from the surface by running the antenna through the cleaner. The notch is provided with a brushing surface composed of some 70 stiff hairs.

The hairs of the tibia and basotarsus are also extensively used to clean the eyes and other body parts from pollen and other objects, which become attached to the bees.

Figure 1.92 The antenna cleaner

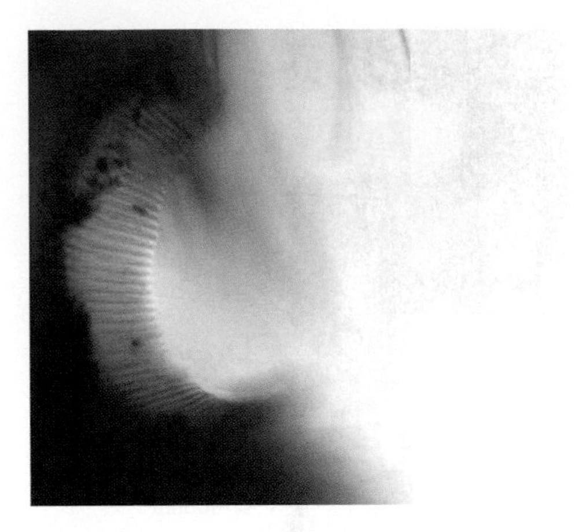

Figure 1.93 Detail of the antenna row of 70 stiff hairs that make the cleaner brush

Mesothoracic leg (middle or second leg)

The mesothoracic leg is provided with a wax spine. A handy spur of cuticle extends from the distal end of the tibia, which can be used for cleaning and defence. The spine is particularly large in the drone.

Figure 1.94 The wax spine in the worker

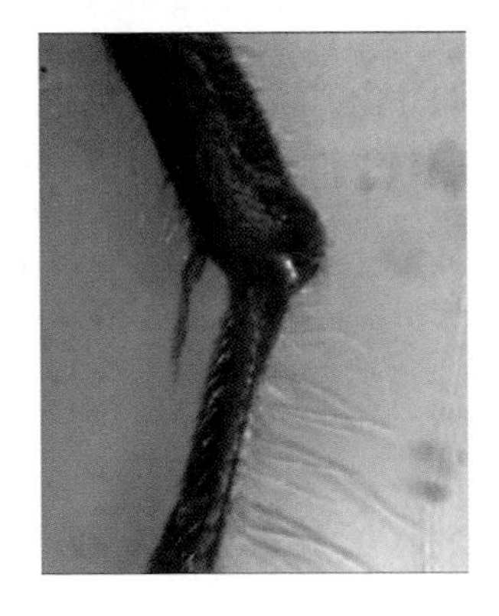

Figure 1.95 The wax spine in the drone

Metathoracic leg (hind or third leg)

This leg is the key to the honey bee's success. In this leg the basitarus is modified, in the worker, into the pollen press or corbicula.

The tibia abaxially (outside) has a smooth concave surface, fringed with long incurved hairs. This forms the corbicula (basket). By moving one metathoracic leg against the other, pollen, mixed with a small amount of nectar, is pressed into a ball.

The basitarsus is grossly enlarged in all three castes. In the worker the basitarsus is also modified to have nine transverse rows of long stiff spines at 45° to the surface. This area forms the pollen brush.

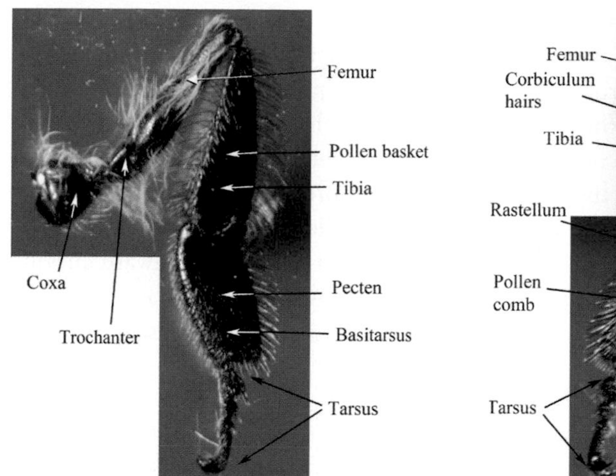

Figure 1. 96 Metathoracic (hind leg) abaxial (outside of leg) view

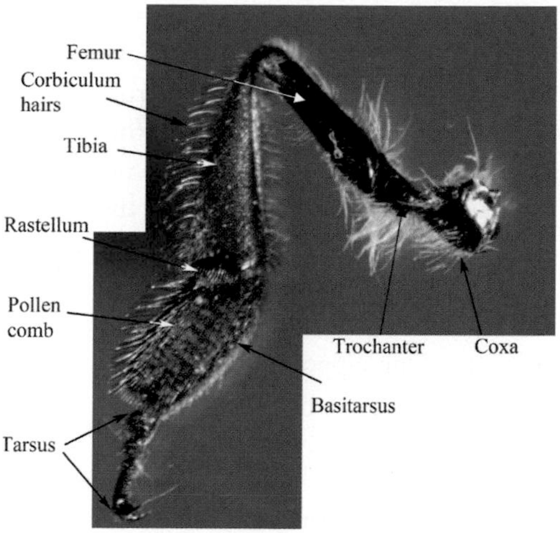

Figure 1.97 Metathoracic axial (inside of leg) view. There are nine rows of hairs in the pollen brush

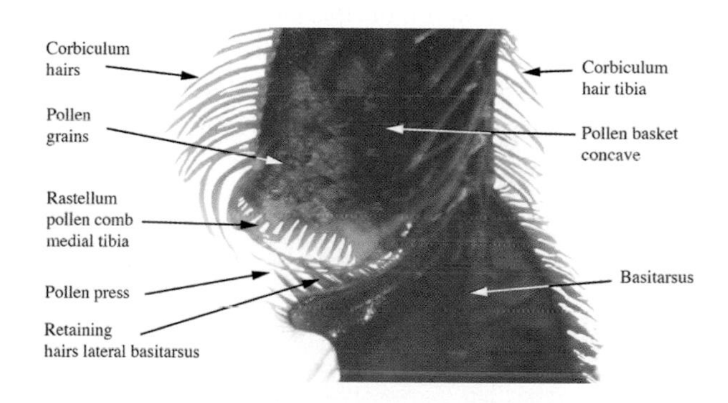

Figure 1.98 The anatomy of the pollen press in a worker

Figure 1.99 Pollen ball.

Note when full the double ball shape.

In the drone and queen, the basitarus is enlarged, but the pollen press is absent.

How the pollen ball is created and transported back to the hive

Pollen grains adhere to the numerous hairs covering the bee's body. With the prothoraxic (front) and mesothoracic (middle) legs the bee collects the pollen grains. The pollen brush on the medial surface of the basitarsus of the metathoracic (hind) legs remove the pollen grains from the other legs. The pollen grains are caught on the nine rows of bristles on the pollen brush.

The opposite metathoracic leg tibia moves down the basitarsus of the metathoracic leg. The stiff hairs of the rastellum (pollen comb) at the distal end of the tibia move (comb) between the hairs of the pollen brush removing the caught pollen grains. These are then directed up to the gap between the basitarsus and tibia where, combined with a little nectar, the pollen/nectar mass is compressed in the pollen press. Directing hairs (retaining hairs on the basitarsus) move the upward moving compressed pollen onto the lateral (outer) surface of the metathoracic tibia. As the sticky pollen emerges from the pollen press it is retained on the concave surface of the lateral tibia by the long inward curving corbiculum hairs. These hairs form part of the pollen ball. As the pollen basket fills one particular hair supports the pollen ball creating the characteristic double ball shape of a full pollen basket.

Wings

Bees fly at 20 km per hour, the wings beat 235 to 250 times per second. On the hind wing there is a series of hooks – hamuli – that attach to a fold on the forewing. This allows the bee to use the wings as one flight surface. The structure of the wing defines the honey bee as a member of the Hymenoptera order. There are often around 20 hamuli. However, the number of hamuli can be quite variable between bees and hives.

Examination of the wing anatomy allows differentiation of *Apis mellifera* from other bee species. The long marginal cell and three submarginal cells are used to identify the species. It is interesting that the wings do not actually beat, but that the flight muscles flex the thorax scutum and sternum plates. The wings are attached to these plates and therefore move up and down, resulting in flight.

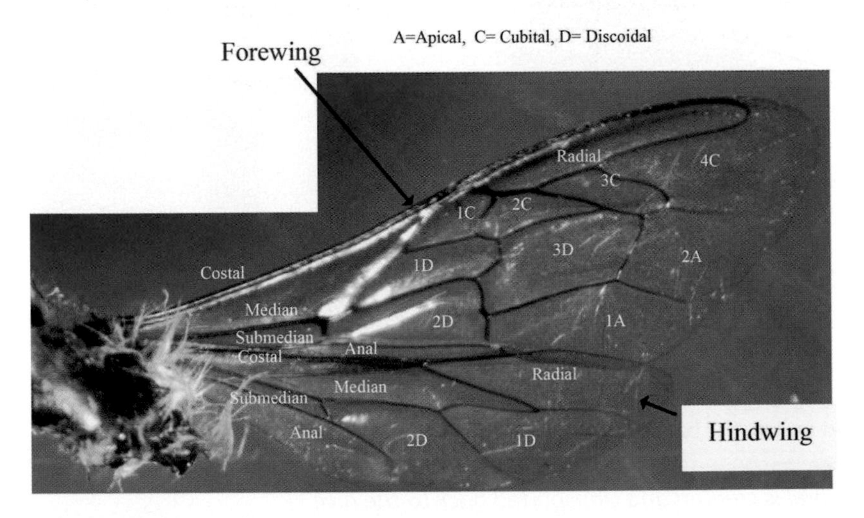

Figure 1.100 The wings of *A. mellifera*: note the marginal cell (radial) and three submarginal cells (4C, 3C and 2C) that can be used to identify *A. mellifera* from other bee species

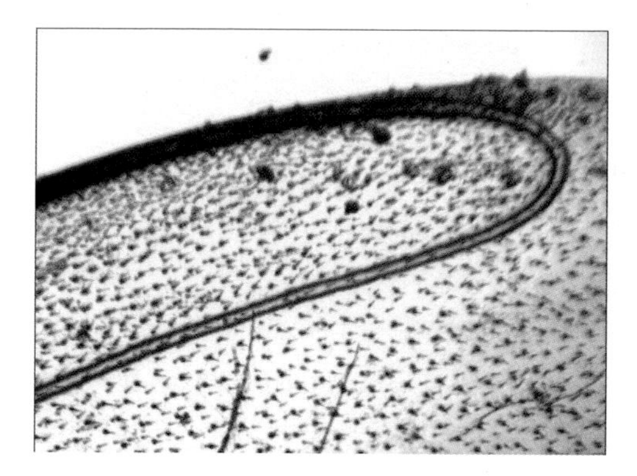

Figure 1.101 Detail of the leading edge of the forewing showing the vein and hairs (microtrichia) on the wing surface

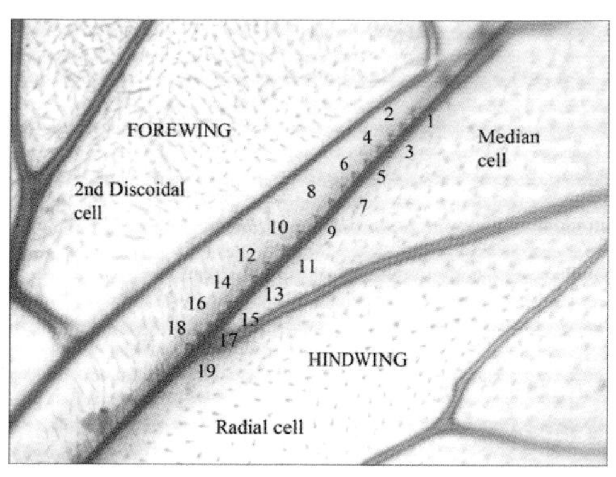

Figure 1.102 The 19 hamuli (in this specimen) on the hind wing attached to the fold on the forewing

Figure 1.103 Detail of the hamuli which is the characteristic anatomy of the Hymenoptera order

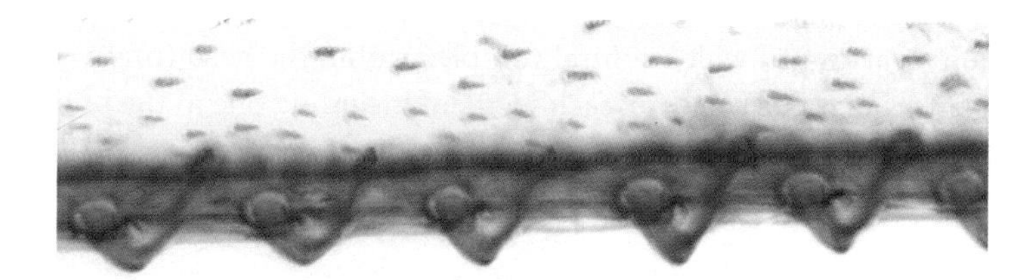

Cubital index

This is used as a means of differentiating between different races of *Apis mellifera* and to differentiate between the different *Apis* species.

Measure the distance between point A, B and then C in the wing, as illustrated in Figure 1.104. The cubital index is (bc/ab)*100. The example has a cubital index of 2.8.

Table 1.5 Cubital index table

Species	Cubital index
Apis florea	2.82
Apis mellifera	1.53–3.60
Apis cerana	3.98
Apis dorsata	7.25

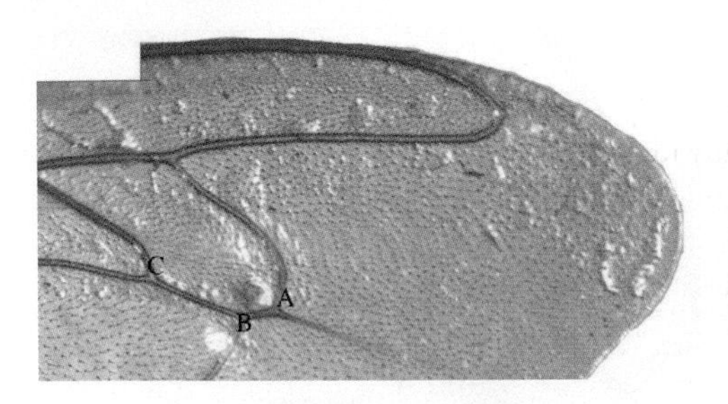

Figure 1.104 The three vein branch cubital points A, B and C

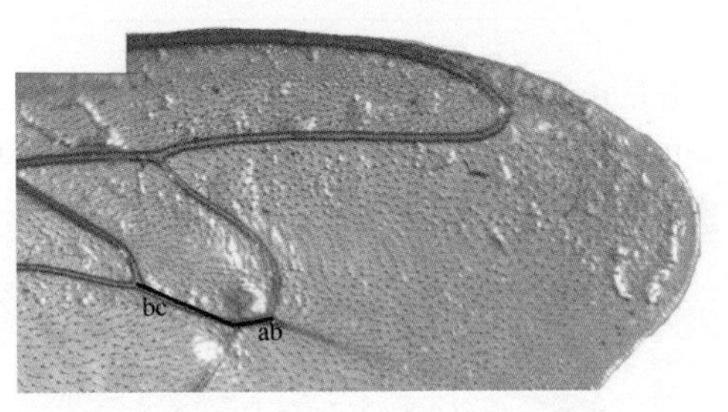

Figure 1.105 The measurements between the branch points – this is cubital index 2.8

Nervous system

Brain and ganglia

The brain is well developed in the emerged larva on day 3. In the honey bee the nervous system is arranged with a central complex within the head (brain) and a series of ventral nerve centres (ganglia), one for each segment including one in the brain.

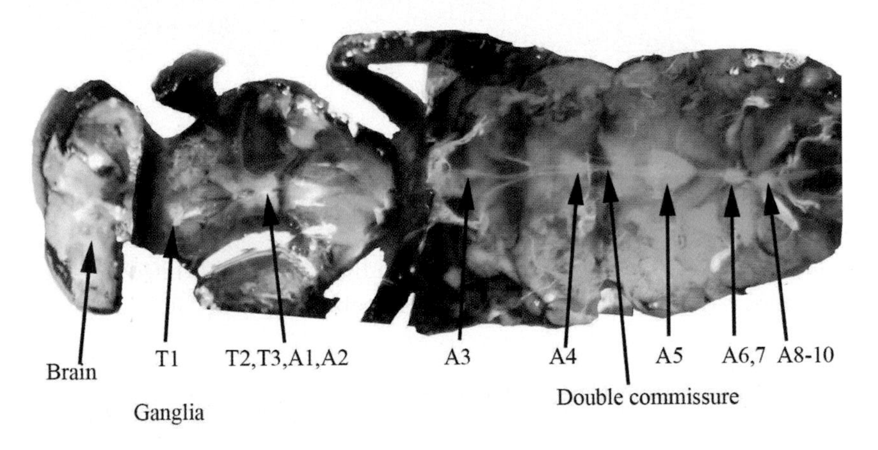

Figure 1.106 Gross dissection in the frontal plane dorsal view of the worker honey bee sternum revealing the neurological system

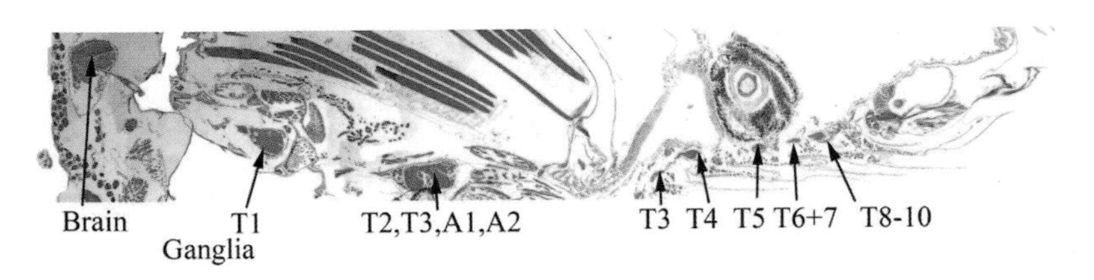

Figure 1.107 Midline sagittal section of the worker honey bee revealing the neurological system

Brain

While the brain is complex, the bee having more neurons per m^3 than the human brain, it can be divided into regions, which can be readily identified. Note the appearance of the brain differs between the castes and also with the experience of the workers.

In the middle of the worker brain four cup structures are obvious, two on either side of the midline. Honey bees are interesting as they have four mushroom bodies, most insects only have two. The compound eyes are controlled by two very large bodies medial to the compound eyes – the medulla and the lobula. The antennae also have a large lobe, which helps to analyse information. The antenna lobe has multiple large bodies called glomeruli in the lobe.

Ganglia

There is a ganglia for each segment including the brain. In the brain the ganglia lies underneath the oesophagus and is directly attached to the brain. This is called the suboesophageal ganglia. Running from this ganglia are two neurological commissures, which run to each of the segmental ganglia. The different segmental ganglia are clearly seen in the larval stages. In the adult, the various segmental ganglia may have fused into five ganglia.

The double ventral commissure is clearly visible in transverse sections of the bee.

Combined with the circular heart (dorsal) the two commissures (ventral) these structures can be used to orientate transverse sections of the abdomen.

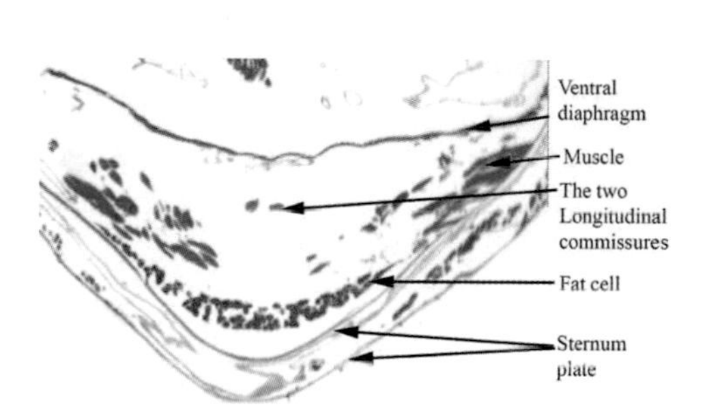

Figure 1.108 A transverse section of the abdomen of a worker honey bee showing the two longitudinal commissures.

Note the ventral diaphragm dorsal to the commissures

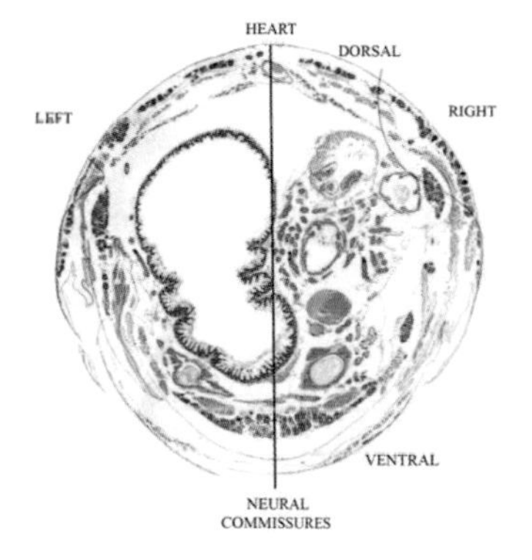

Figure 1.109 The dorsal circular heart and ventral commissures orientate transverse sections of the abdomen.

Antenna

The honey bee has a filiform antenna. The antenna is composed of four basic structures; the base, scape, pedicel and a flagellum. The female (worker and queen) antenna is divided into 11 annuli. In the male the antenna has 12 annuli (annuli #2 is divided into two). The divisions in the antenna are not 'segments' as there are no muscular attachments. The antenna is filled with sensory equipment and provides the bee with information on temperature, humidity, chemical environment (smell etc.) and location. Note that bees inside the hive live in total darkness, thus touch and chemical communications are extremely important.

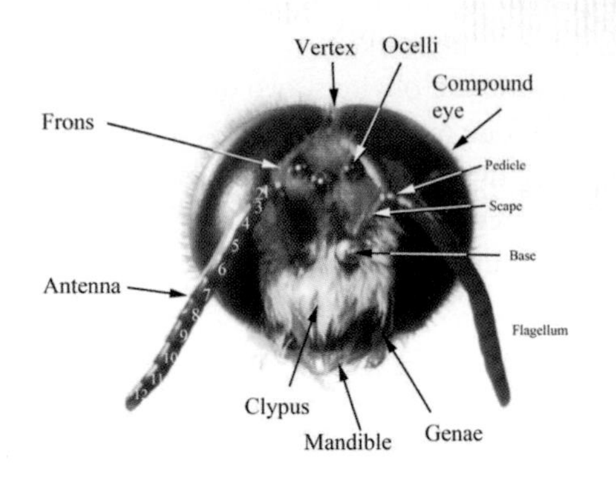

Figure 1.110 The face of the drone with detail of the anatomy of the antenna

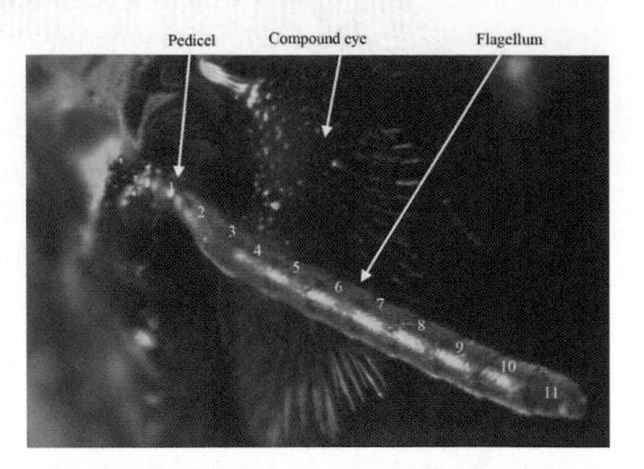

Figure 1.111 The antenna of the queen honey bee

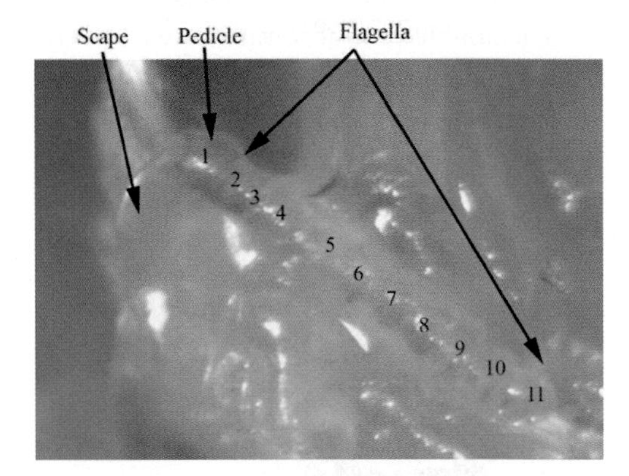

Figure 1.112 The antenna is clearly visible in the pupal stages. The photograph shows the antenna of a pink eye (16 day) worker pupa

Compound eye

The honey bee has two large compound eyes. The drone compound eye is particularly large and occupies most of the head.

The compound eye is composed of a series of units called the ommatidium. These units collect light and focus it onto the retina where the signals are the transmitted to the brain to the medulla and the lobula through the inner and outer chiasma. The worker and queen bees have around 4000 ommatidium in each compound eye, the drone 7000 or more in each eye.

The ommatidium is composed of a clear lens, crystalline core, eight or nine retinula cells and surrounding pigment cells which stop light leaving the ommatidium and scattering around the eye. The compound eye is protected by numerous hairs. The bee's compound eye/brain is 10x faster at forming an image than the human eye/brain but the brain probably does not form as clear an individual picture. The eyes are designed to detection motion, vital in flight. Note it is believed that bees are able to recognise each other by facial appearance.

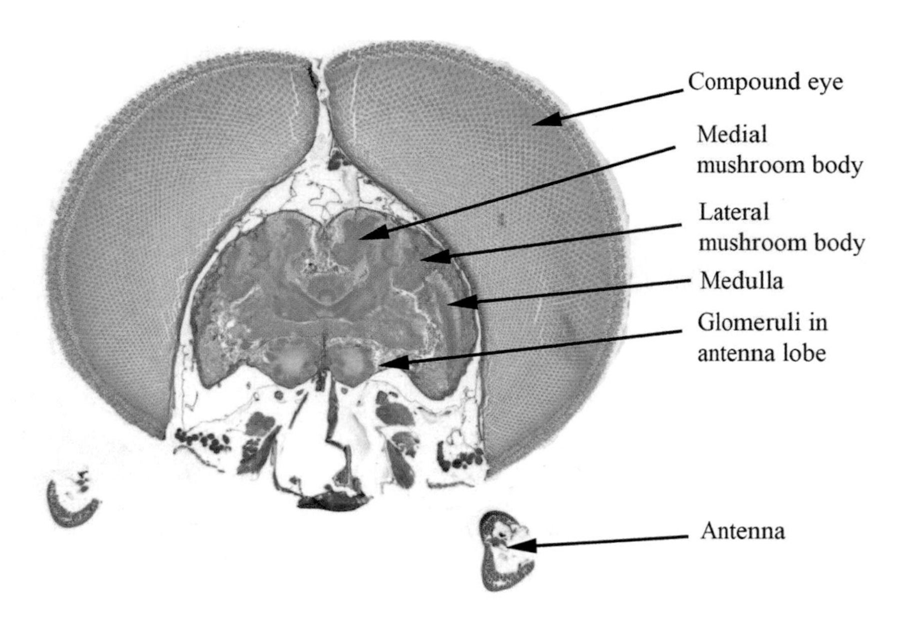

Figure 1.113 Transverse section of the head of a drone showing the large compound eyes

The eyes are kept clear of pollen or dust by the actions of the prothoracic leg.

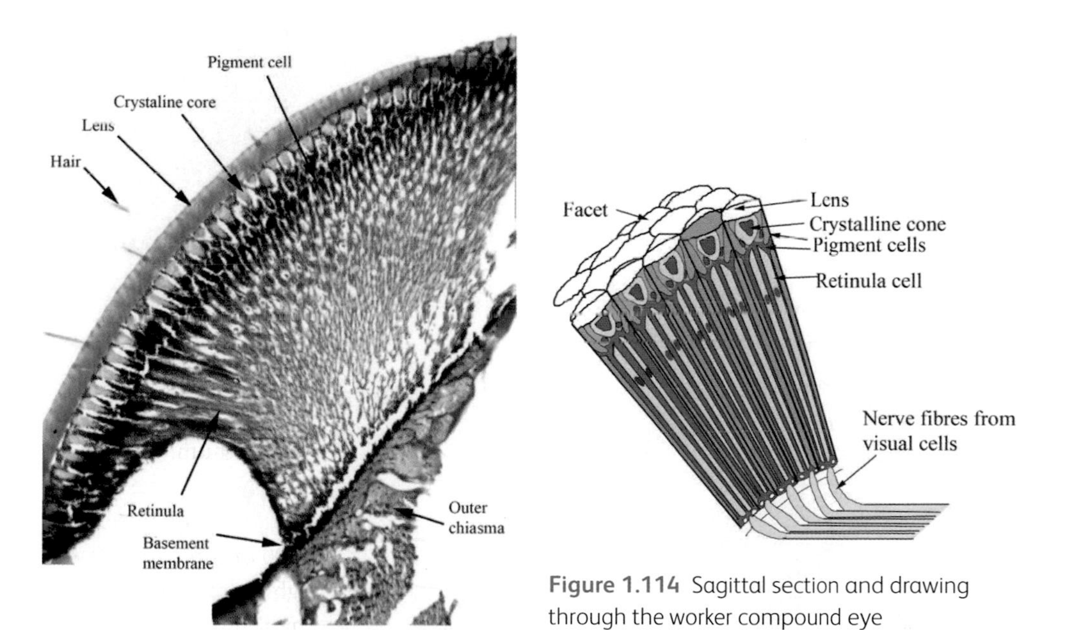

Figure 1.114 Sagittal section and drawing through the worker compound eye

The structure of the single ommatidium can be seen in transverse sections.

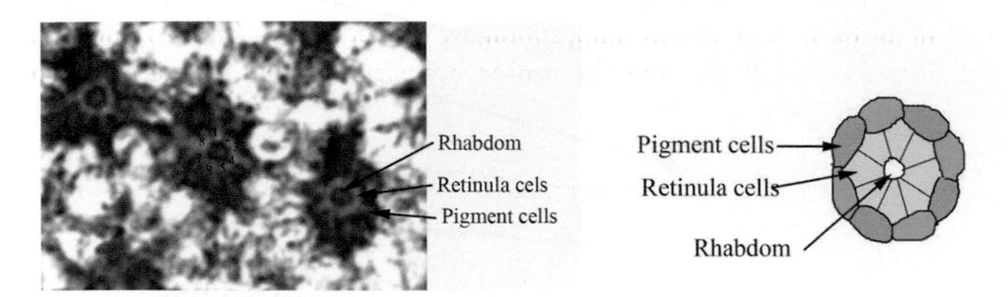

Figure 1.115 Transverse section and drawing of an ommatidium in the compound eye

Ocelli (simple eye)

These do not form a proper image and are important in determining the position of the sun. There are three ocelli located at the dorsal (top) of the head in the female and in the middle of the face in the drone. The drone's ocelli are in the middle because the large compound eyes nearly meet in the middle of the top of the drone's head. The three ocelli are located in a triangular arrangement. In these eyes there is a single large lens with the retinula cells underneath. Hairs are arranged to control the angle of light entering the ocelli.

Figure 1.116 Ocelli shielded by surrounding long hairs in the drone

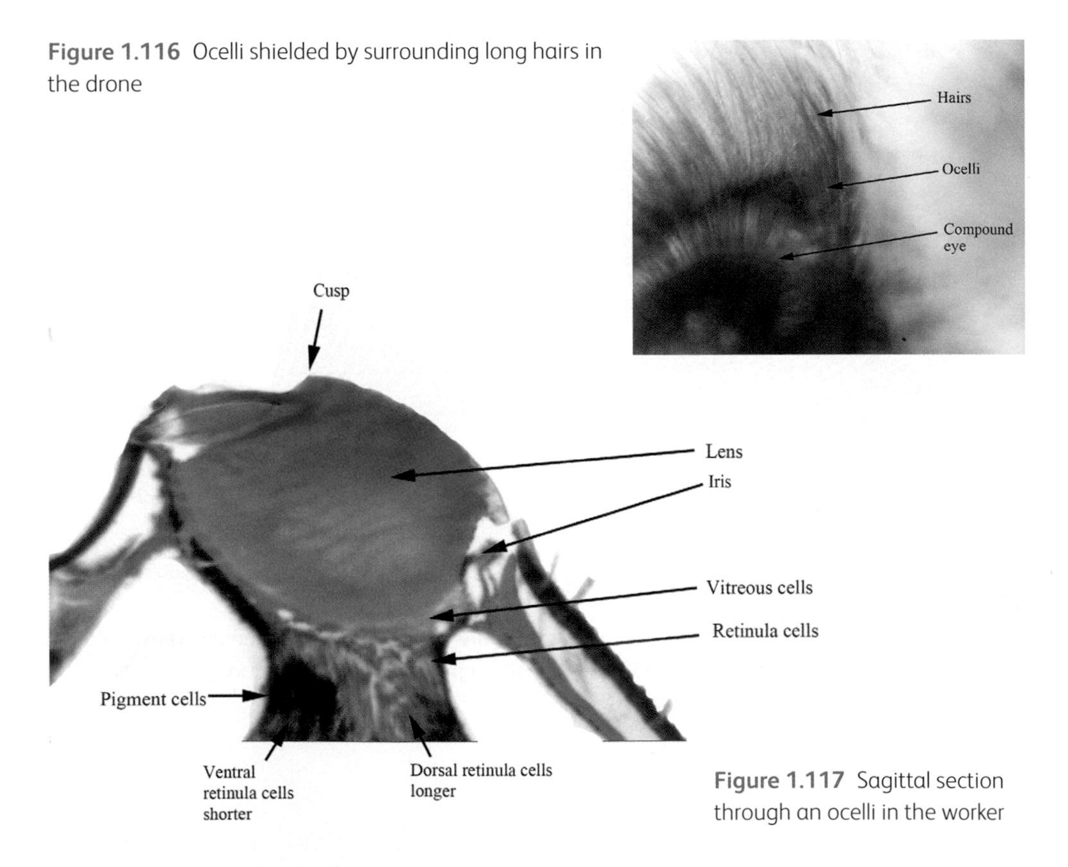

Figure 1.117 Sagittal section through an ocelli in the worker

Visualising light

It is interesting to note that within the hive there is no light and therefore the queen will spend most of her time in the dark. Prior to becoming a foraging worker the developing honey bee has spent all its life in the dark. Within the hive, sensory transfer of information is by touch and smell. Despite spending so much time in the dark the bee does not fly at night.

Sensory plates and cells

The bee needs to know where its legs, wings, head, thorax and abdomen are located in three dimensions. These are all manipulated under exquisite control during flight. In order to achieve this degree of control, the bee is provided with nerve plates and hairs.

For example, in the cervix (neck), there is a complex nerve plate on the cranial edge of the thorax which contacts with the caudal aspect of the head.

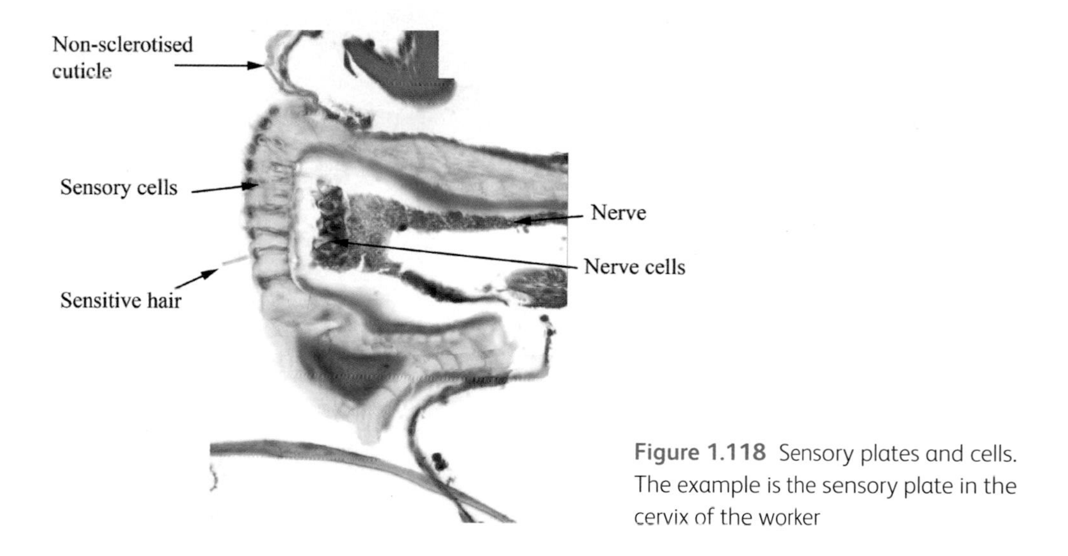

Figure 1.118 Sensory plates and cells. The example is the sensory plate in the cervix of the worker

Reproductive system

Female – queen reproductive organs

The female reproductive tract starts with two massive ovaries, which are located in the proximal (front) portion of the abdomen. The ovaries contain 400 ovarioles, which start with the primary oogonia and develop into full sized eggs as the cells develop distally (backwards) through the ovary. The developed egg passes through bilateral oviducts (seen as two circles under the ventriculus). These oviducts fuse into a common oviduct. The common oviduct communicates with the vagina. A large sac, the spermatheca, stores semen and is connected with the vagina by a short duct. If the queen determines that the future bee is to be female, a small amount of sperm is added to the passing egg. The spermatheca can hold around 5 million sperms. The egg is fertilised through the micropyle.

The worker female

The worker female has a much reduced (although in some potentially viable) reproductive tract with less than 30 ovarioles. The ovaries can be so indistinct that they can be difficult to visualise in gross dissections.

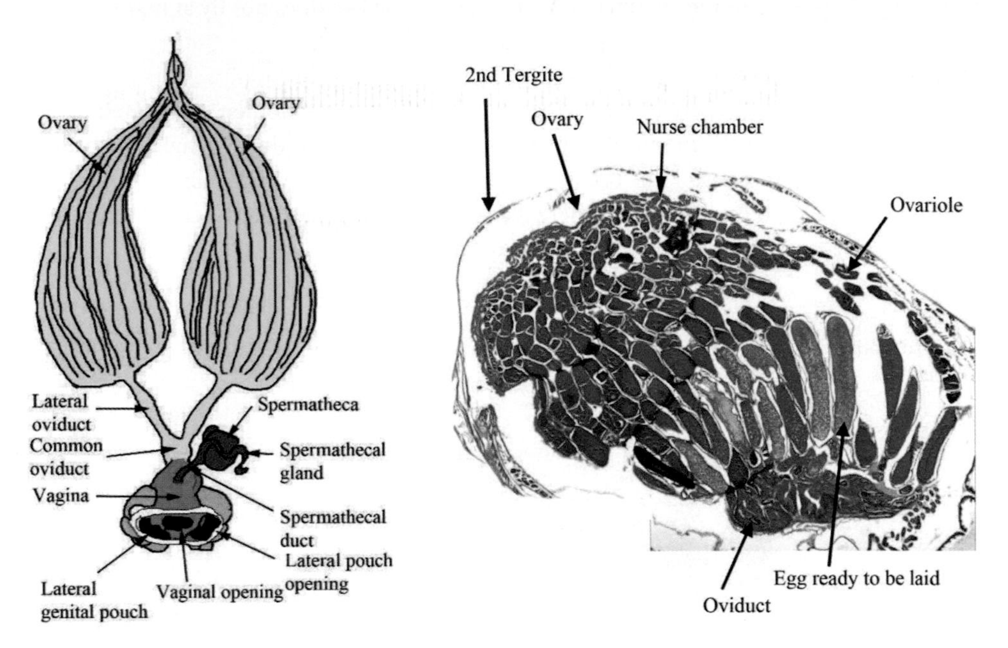

Figure 1.119 Drawing of the queen reproductive tract coloured for clarity

Figure 1.120 Sagittal section through one ovary

Segments VIII and IX are drawn into the abdomen and form the chamber that contains the sting. The sting is described later.

Male reproductive tract

The male reproductive system starts with the large testes 5mm in length in the developing bee. In the mature drone (12 days post emergence) the testes will be small flat, triangular and empty of semen. From the testes there is a tube, the vas deferens, leading into the ejaculatory duct. Mucus from large mucus glands is added to the sperm. The two vas deferens and mucus glands unite distally into the single ejaculatory duct. The ejaculatory duct leads to the large and complex penis. Each ejaculation is about 11 million sperm.

Mating

The drones fly to the drone congregation areas and await a suitable queen to arrive. The courting arena is about 10 to 20 metres in the sky and about 50 to 200 metres in diameter. In *A. mellifera* the drone congregation area is in the open. In other *Apis* species it is more within forest areas. There will be about 100 drones in the drone congregation areas. Males come to the arena from around 12km. The drones are able to remain in the drone congregation areas

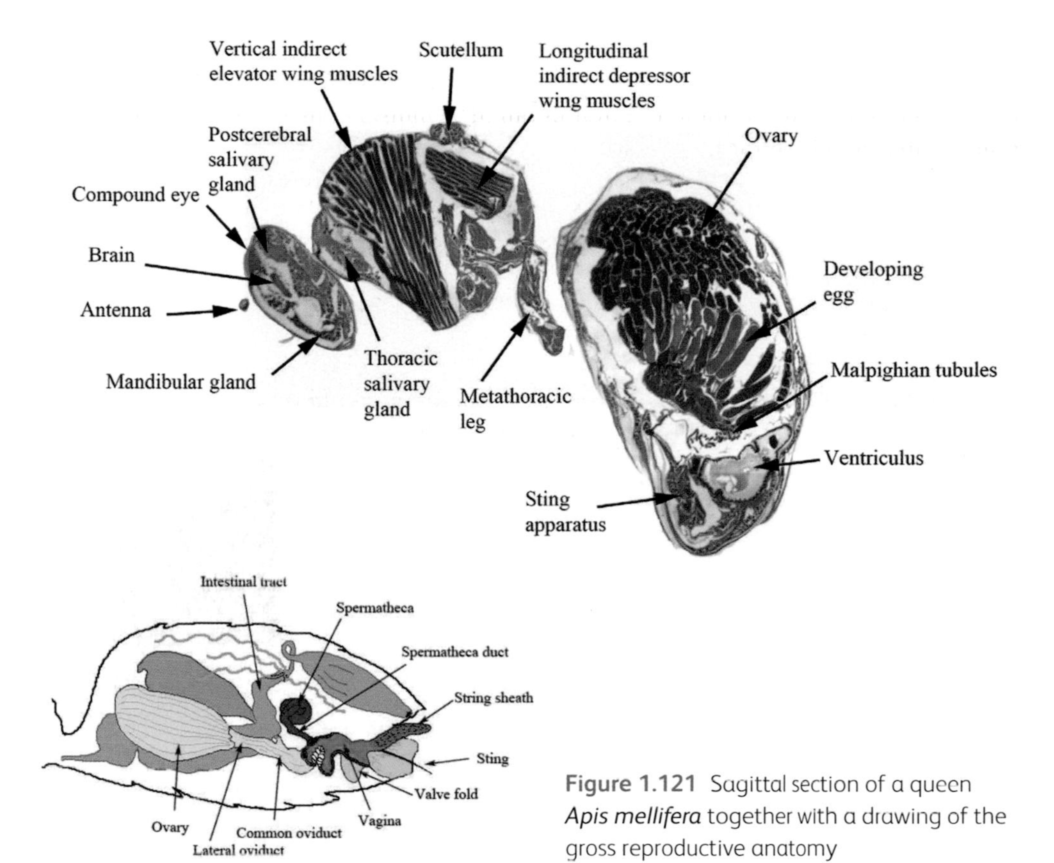

Figure 1.121 Sagittal section of a queen *Apis mellifera* together with a drawing of the gross reproductive anatomy

for about 30 minutes before they must return to the hive for a honey refill. It is possible that the drones locate the mating area using magnetite in their abdomen – which increases in drones more than six days postemergence. If mating occurs, the sperm mass stored in the bulb of penis is discharged by eversion into the queen's vagina using hydraulic pressure from the haemocoele. Mating takes around two seconds. After mating the queen separates from the male, with the bulb of penis remaining in her genital tract. The male reproductive organs tear at the penis neck. The drone subsequently bleeds to death. The spermatozoa are discharged in the distended lateral oviducts. The spermatozoa are then moved into the vagina and then the spermatheca gland where they remain for the productive life of the queen – up to five years. Once back at the hive, the workers remove the penis bulb (mating sign) from the queen. The 'virgin' queen will mate with 20–30 males on her mating flights, after which she will not mate again. After 8–10 days after emergence and a successful mating flight, the queen starts laying eggs. If the queen 'runs out of semen' she will only lay unfertilised drone eggs.

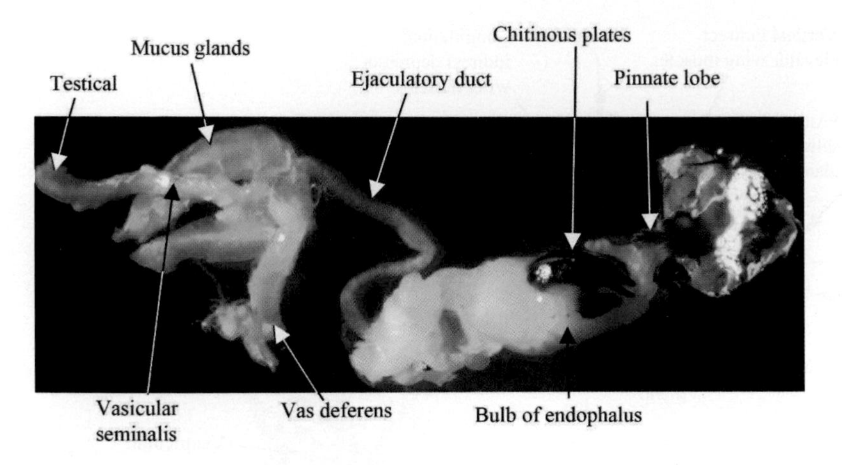

Figure 1.122 Gross dissection of a mature drone reproductive tract

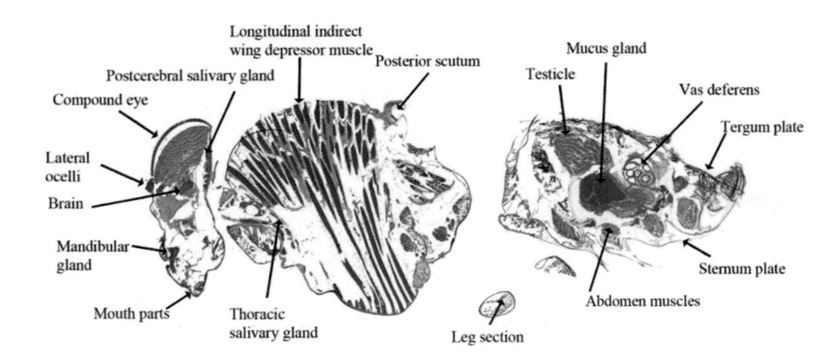

Figure 1.123 Sagittal section of a drone honey bee

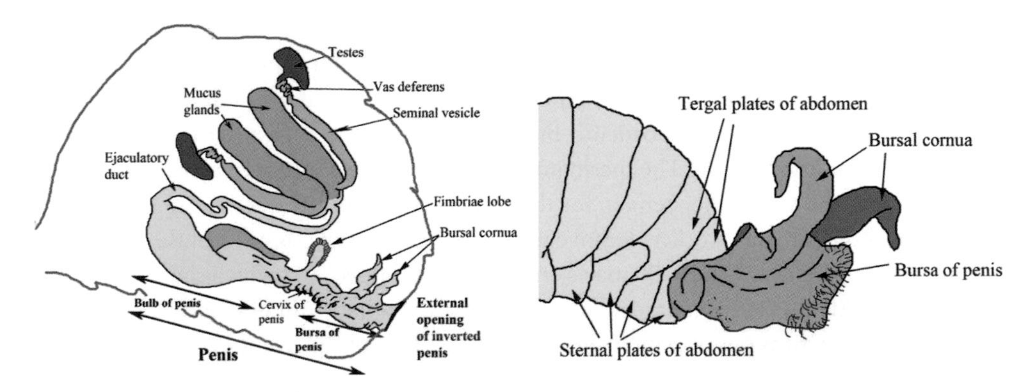

Figure 1.124 Drawing of the male sexual organs *in situ*

Figure 1.125 Drawing of the everted male penis

Figure 1.126 Drawing of the sequence of events in mating as the drone leans backwards his endopenis separates in his abdomen and remains in the queen as a mating sign

Table 1.6 Time of day of mating (various species of *Apis* in Borneo and *A. mellifera* in Europe)

Species	Start	End
A. nuluensis	10:45	13:15
A. andreniformis	12:00	14:00
A. cerana	14:15	15:30
A. koschevnikovi	17:00	Dusk
A. dorsata	19:00	Dark
A. mellifera (Europe)	14:00	17:00

The time of the mating flight and the position of the drone congregation areas help define the various species of bees.

In *A. andreniformis* there is an additional sexual organ, a 'thumb' on the hind leg of the drone, which he uses to hook to the hind leg of the queen. There is also very little mucus with the semen.

Worker sting and reproductive organs

The female honey bee does not use an egg laying organ, the ovipositor, which is present in many hymenoptera. This egg laying organ, however, is modified further into a stinging organ.

There are five components to getting stung: the needle, the driving mechanism, the poison, the petiole, the attractants to other bees. All of these are necessary for the bee to defend the hive from intruders.

The piercing needle

The piercing needle is formed from two lancets and a stylet. The stylet is a large bulb organ, which supports the two lancets on a monorail. The two lancets work alternately, as one advances the other is withdrawn. However, the lancets are barbed and these act as an anchor point, so as one advances it locks in place and as the other lancet moves forward the entire sting mechanism is then drawn forward thus pushing the forward-moving lancet deeper into the victim. The poison from the poison gland enters the cranial portion of the stylus and passes between the stylus and the lancets.

The queen's lancets are not so heavily barbed and can be withdrawn from the victim without injury to the queen. The queen's sting is generally used to kill other rival queens.

The size of the barbs varies with the species of *Apis*.

Table 1.7 Length of sting

Species of Apis	Length of the sting
Apis dorsata	More than 3mm (longer than clothing)
Apis mellifera	2mm long
Apis andreniformis	Less than 2mm

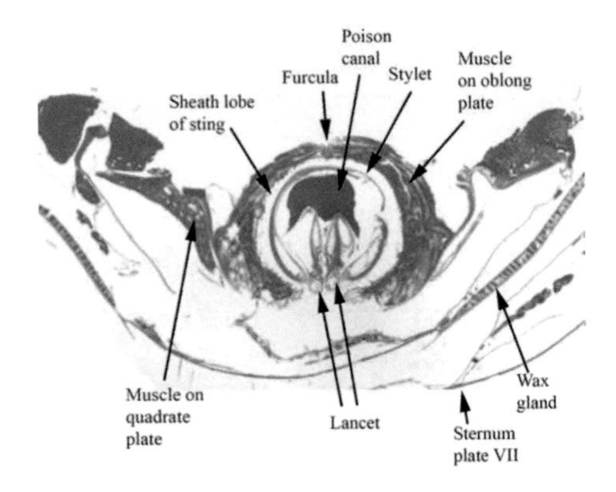

Figure 1.127 Cranial view of a transverse section through the sting apparatus of a worker honey bee

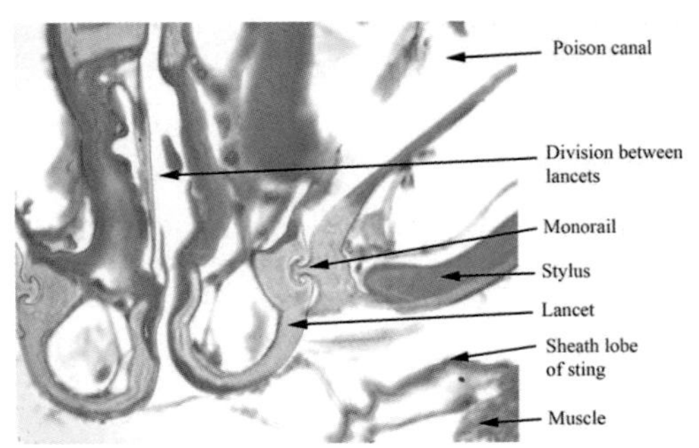

Figure 1.128 Detail of the lancet attachment to the stylus of a worker honey bee

Figure 1.129 Detail of the tips of the two heavily barbed sting lancets with the sheath in the background. Worker honey bee

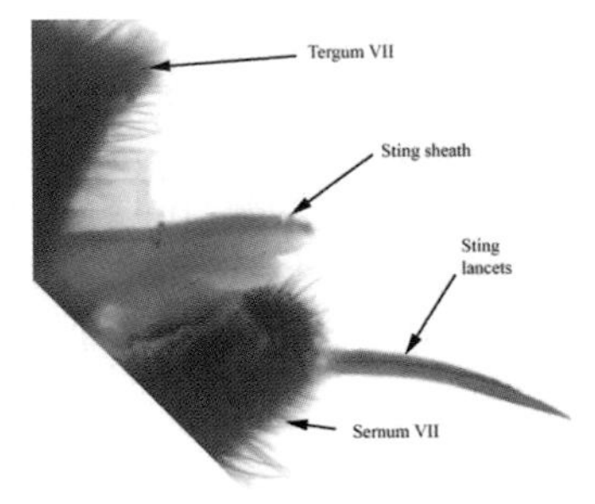

Figure 1.130 The sting of the queen honey bee. Note the lancets are only poorly barbed

The driving mechanism

The needle lancets are moved by the action of three plates called the quadrate place (segment IX), the triangular plate and the oblong plate.

These three plates are connected to produce a rotation motion moving the lancet. The protractor muscle connects the quadrate plate to the oblong plate. Contraction results in the quadrate plate swinging outwards, causing rotation of the smaller triangular plate. The triangular plate continues into the first ramus, which ultimately becomes the lancet. Thus the lancet moves forward a small amount with each contraction and when it pierces a victim's skin the barbs lock and each muscle contraction causes the lancet to penetrate deeper. In the final moments the whole sting apparatus is torn from the abdomen rupturing the rectum and possibly the heart. The wound is fatal to the worker bee. Note the sting is not torn from the abdomen when the bees attack other insects, only when they attack vertebrates such as humans.

The poison gland

The poison is produced by a poison gland, easily recognised in gross dissections. The poison is then stored in the poison sac (filled with colourless liquid in a gross dissection) ready for use. The queen's poison sac can be particularly large. On H&E histological sections the poison gland will be easily recognised as a large sac coloured red. Each sting contains about 150μg of poison venom.

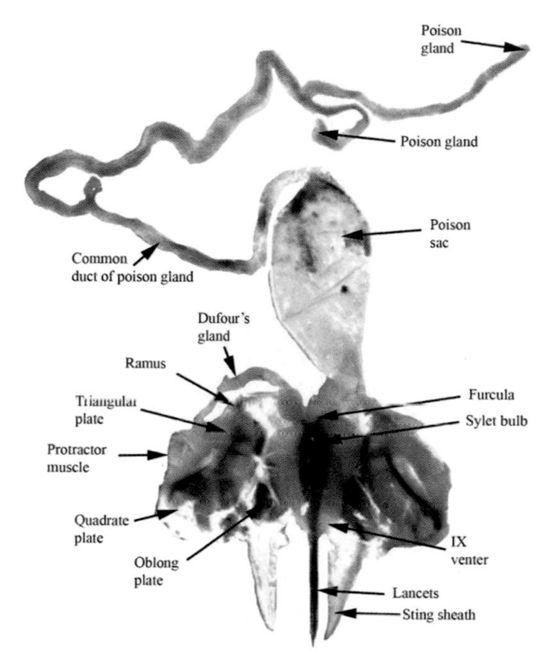

Figure 1.131 The dorsal anatomy of the sting apparatus, slightly flattened

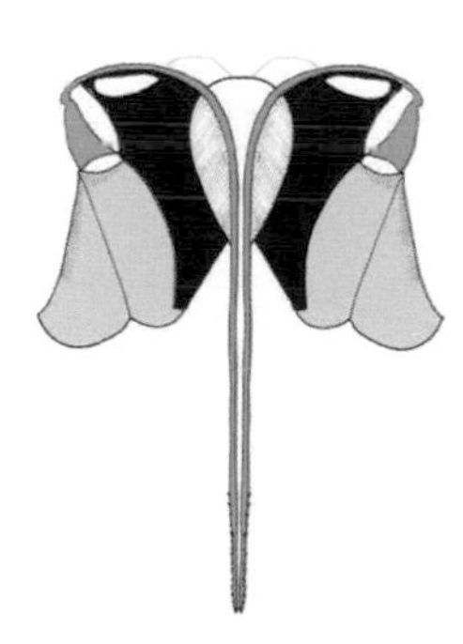

At rest.
Yellow is the quadrate plate
Blue is the oblong plate
Green is the triangular plate with the attached lancet.

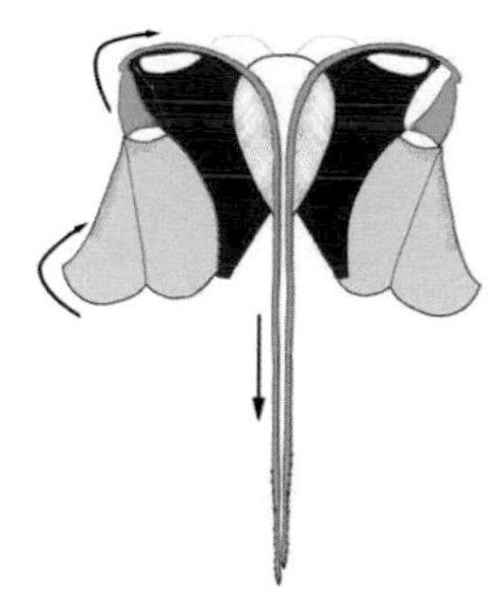

The protractor muscle on the quadrate plate contracts, pivoting the triangular plate around the oblong and thus the lancet moves forward.

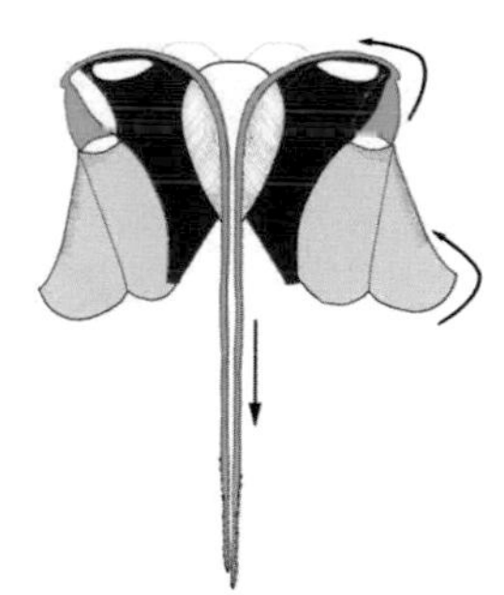

The action is repeated on the other side, the lancets move forward. The victim's skin holds the lancet's forward movement in place.

Figure 1.132 Drawing of the mechanism of stinging in the honey bee

The petiole

The petiole (waist) is a major part of the sting apparatus. The bee is able to rapidly manipulate its abdomen around to sting its victim. The fulcrum to this action is the narrow petiole, in the honey bee between segments A1 and A2.

The attractant glands

The Dufour gland (the alkaline gland) may appear yellow in a gross dissection and in the worker connects to the ventral aspect of the sting. The other gland is the Koschevnikov gland, which produces the 'alarm' hormone. This is a collection of cells in the middle of the quadrate plate. These glands produce very volatile chemicals, which stimulate other bees to attack the same area. They aim to ultimately drive the victim away from the hive through the worker bees self-sacrifice.

Note the secretion of the alarm hormone can occur without the need to sting.

The ganglia 8–10

Once the sting is torn from the body, the ganglia 8-10 is no longer controlled by the central nervous system and continues to act to stimulate the muscles of the oblong plate, driving the barbed sting deeper into the victim.

Respiratory system

The respiratory system of the honey bee transports, warms, moistens and cleans the air from outside the bee to each individual cell.

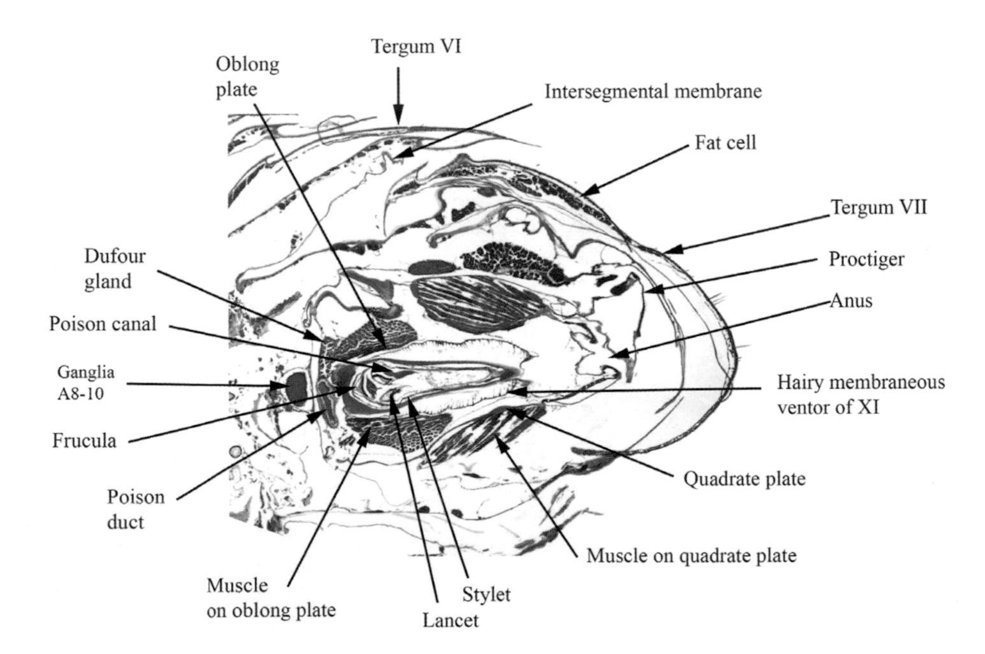

Figure 1.133 Frontal section through the sting apparatus of a worker

Spiracles

Air enters the adult bee through the nine visible spiracles. There are actually 10, but the tenth is hidden inside the string apparatus on internal tergum plate AVIII.

There is no spiracle in the head. Each of the thoracic segments has its own spiracle. The first eight abdominal segments have their own spiracle. Segments 9 and 10 do not have a spiracle.

Apart from the first spiracle on the prothorax, the anatomy of each spiracle is very similar.

First spiracle on the prothorax ST1

The first spiracle is protected by a large cover, the spiracle lobe. The internal tracheal system leads directly to the spiracle. The first tracheal system is extremely wide and easily seen when the prothorax is removed.

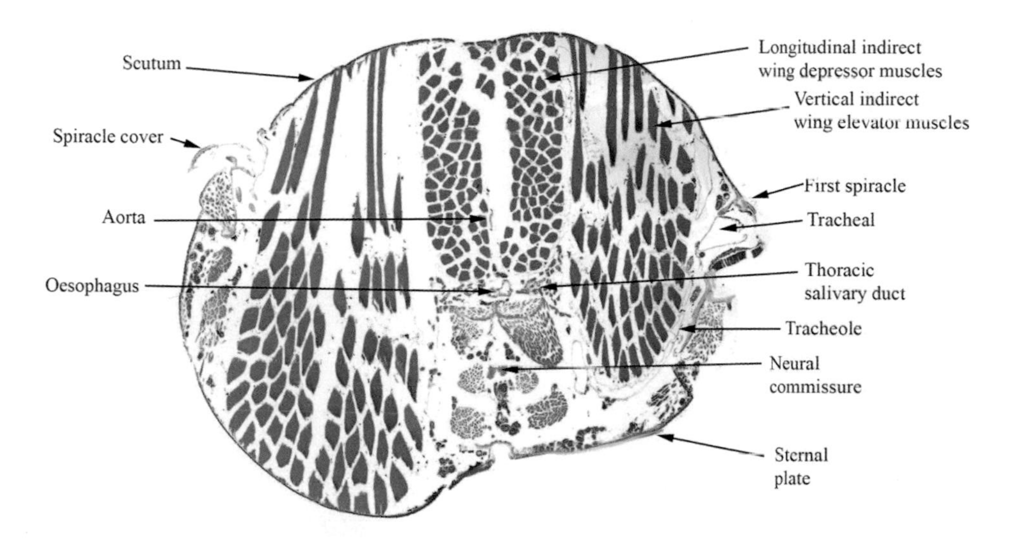

Figure 1.134 Transverse section of the first tracheal spiracle

Other spiracles (excluding hidden spiracle SA8)

The anatomy of the other spiracles is very similar. The spiracles are guarded by a cover with hairs. The spiracle is a narrow slit about 0.06mm long. Just inside the bee there is a chamber called the atrium. The entrance to the tracheal system is protected by a valve with more cleaning and protection hairs.

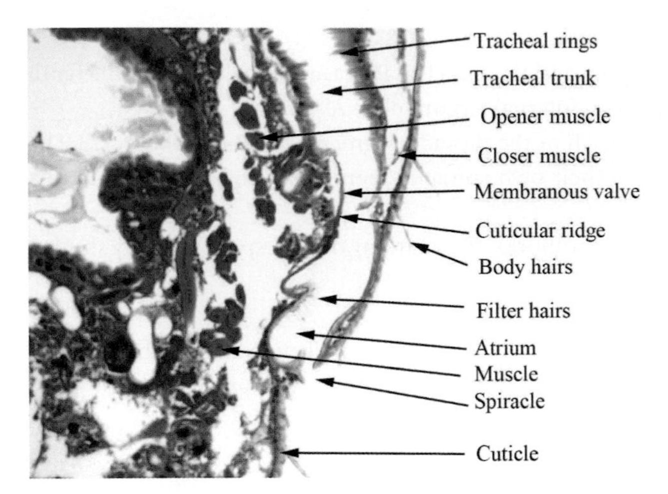

Figure 1.135 Frontal section of the abdominal spiracle of the honey bee

Tracheal system

After the air enters the spiracle it is transferred around the bee through the tracheal system. The largest trachea leads from the first thoracic spiracle. If the head and prothorax tergum is dissected from the bee the first tracheal system is easily recognised.

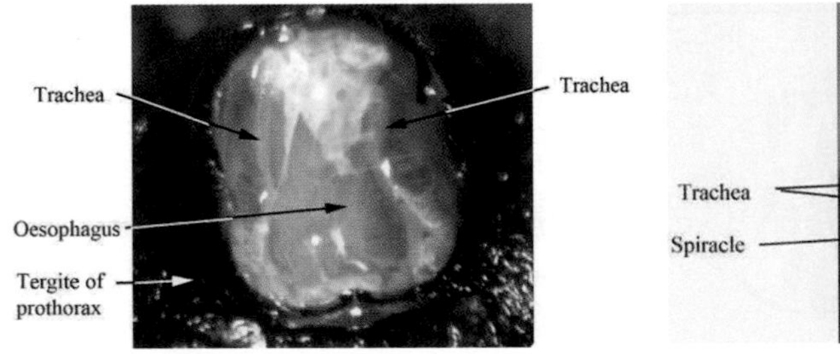

Figure 1.136 The gross appearance of the first tracheal system in the thorax with the prothoracic tergum removed

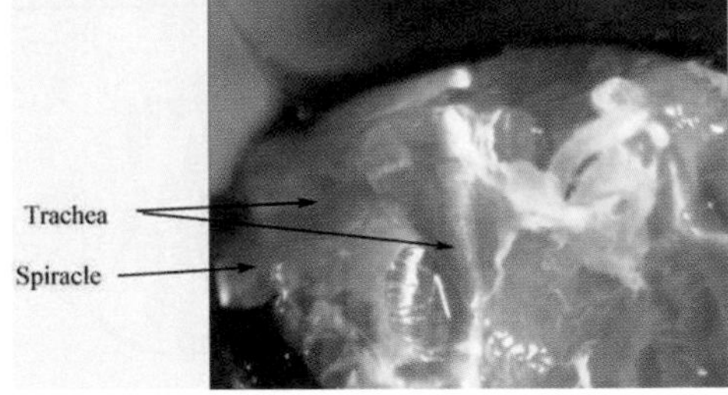

Figure 1.137 Detail of the first tracheal system

Air sac

The tracheal system from the spiracles enters the air sacs. The arrangement of the air sacs anatomy varies from bee to bee. The airs sacs extend throughout different segments although there is often an air sac associated with each segment. They are easily seen on dissection of the abdomen, especially under isopropanol alcohol when they will be seen as large white inflated thin walled bags. On sections they may be difficult to recognise as they have a very thin wall, and lack taenidia. The function of the air sac may be primarily buoyancy.

Trachea and tracheoles

The fine trachea emerge from the air sacs. In gross dissections these can be seen as white lines, often branching. The trachea and tracheole wall is supported by a spiral ring structure called the taenidia.

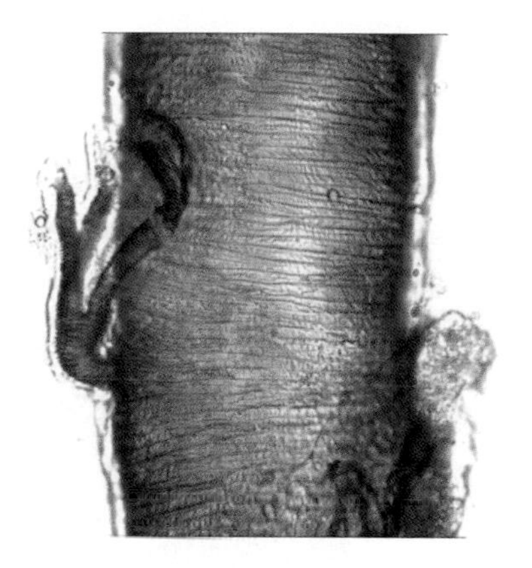

Figure 1.138 Gross examination of a trachea with branching tracheole

Figure 1.139 Histological examination of a trachea with the characteristic taenidia rings.

The tergum and sternal plate movement system

Note the whole of the bee's body is involved with respiratory movement with the tergum and sternal plates playing a vital role in expanding and contracting the abdomen. This is an interplay with the various muscles and intersegmental membrane system.

Urinary and waste product removal

Malpighian tubules

The Malpighian tubules are responsible for removing nitrogenous waste from the haemolymph. They originate from the epithelium but are not covered by cuticle. In the larval stages there are four large Malpighian tubules, which may be distinguished in dissection of a L5 larva. They are attached physically to the pylorus but their lumen does not connect with the small intestine. When the L5 larva is capped and the ventriculus and small intestine connection open, the four Malpighian tubules also connect and empty their contents into the small intestine. These Malpighian tubules then are reabsorbed. Numerous adult Malpighian tubules develop from the pylorus and 'float' free in the haemocoele.

The Malpighian tubules filter the body 200 times a day (compared with 12 times a day in the kidneys of humans).

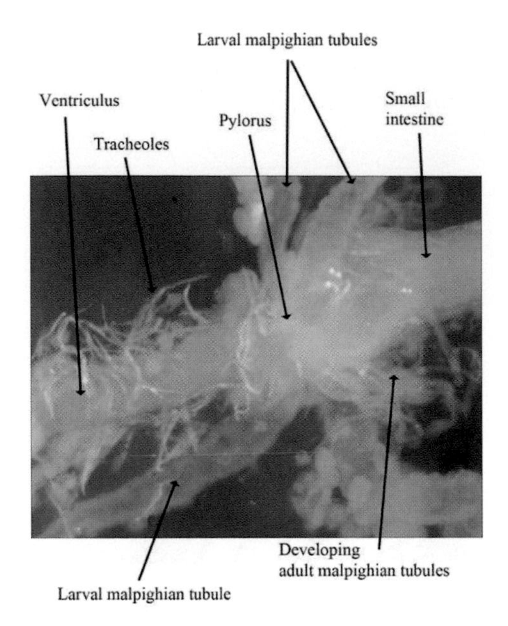

Figure 1.140 Gross dissection of the L5 larvae demonstrating the larval Malpighian tubules and the imago developing Malpighian tubules of the future adult

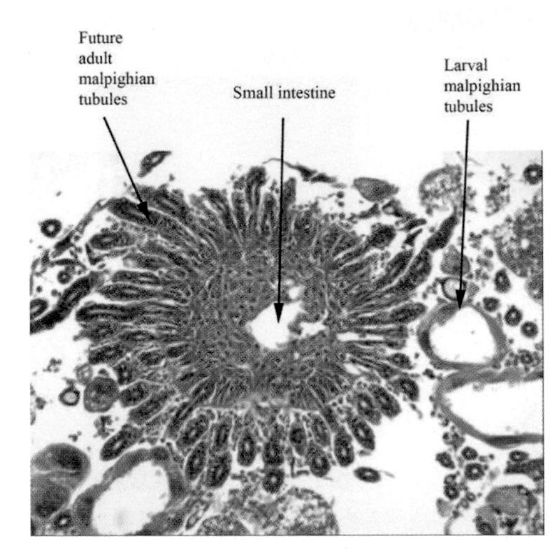

Figure 1.141 Transverse section of the pylorus of an L5 larvae revealing the imago Malpighian tubules around the pylorus

Rectal pad

The rectal pad is designed to conserve salts, being particularly responsible for chlorine metabolism and maintaining the osmotic pressure of the bee's haemolymph.

Dorsal diaphragm

The dorsal diaphragm will remove particulate matter from the haemolymph.

Water conservation

The honey bee does not lose water through its impervious cuticle. The Malpighian tubules and rectum are integral components of water conservation of the honey bee. Water is easily lost through the respiratory system and the spiracle coverings play a vital role in conserving water.

Other bees

The basic body plan described is followed by the *Apis* family. The various organs may differ in size and sometimes in function depending on the bee's feed source and environment. These can be used to differentiate between bee species. Note *Apis mellifera* may be confused with *Andrena florea* but *A. florea* has a short tongue and a different wing structure. Following the honey bee layout, organs and structures in other bees can be identified and their function realised.

Bumblebee

This longitudinal section of a tree bumblebee (*Bombus hypnorum*) shows the various anatomical features that can be recognised and similarities and differences realised. For example the small intestine in this species is longer than in the *Apis mellifera*. The anatomical differences between species highlight different ecological roles and lifestyle adaptations.

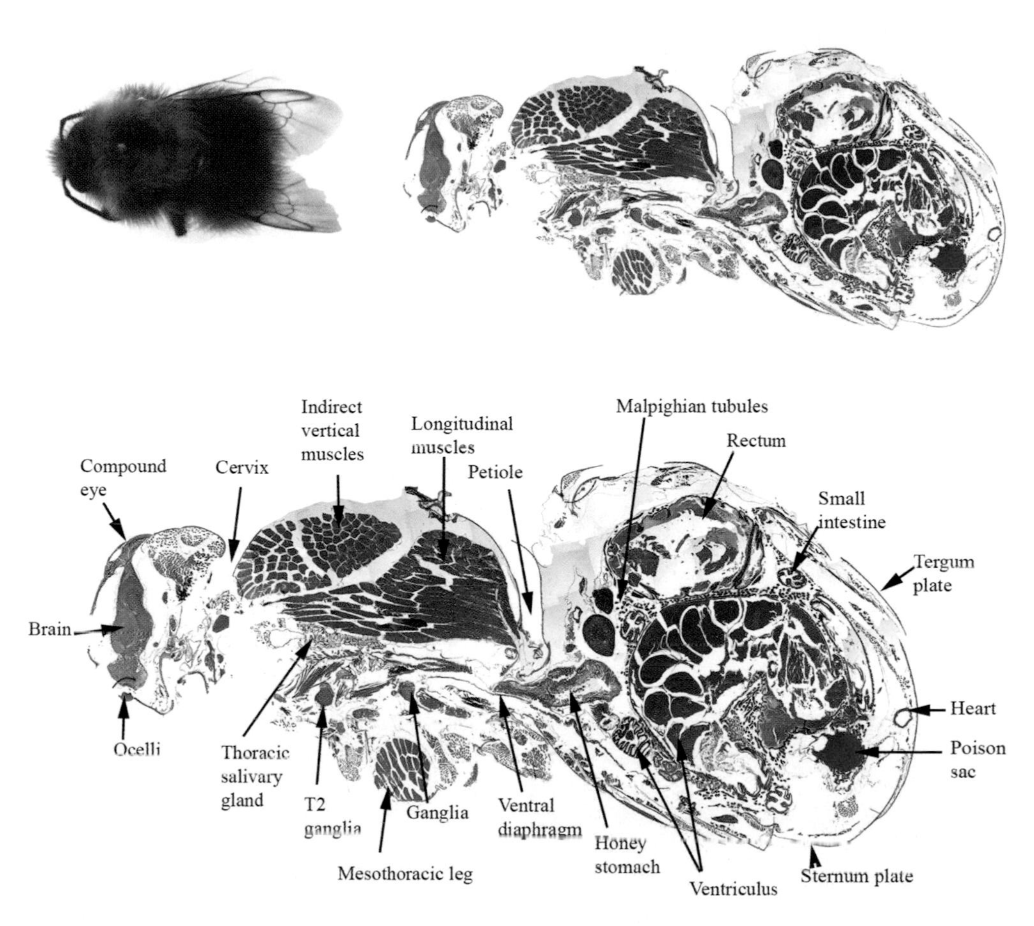

Figure 1.142 Sagittal section of *Bombus hynorum* (tree bumblebee) labelled and unlabelled. The similarities with *Apis mellifera* are obvious. But also spot the differences

2

Managing the health of bees – nutrition

To provide sufficient nutrition how much space is required for each hive?

This clearly depends on the amount of food available to the hive. A healthy hungry worker bee may fly 7–9km to find food and then has to fly back. But this also takes a lot of energy to find the food. Normally the bees will work an area around 3km from the hive. This is still a 6km trip to the shops! As a guide it is advised to have one hive per 4–12 hectares (1 hectare is 2.5 acres).

There are other considerations. Bees will collect only one type of pollen in a flight. The worker will then tell her sisters where the food source is situated. Therefore, having multiple plants of the same species is much more appreciated than solitary plants. Encourage your friends to have multiple planting.

Figure 2.1 Single flowering plant is not very attractive to the hive

Figure 2.2 A whole larder of the same plant type all flowering as a group is very attractive to the bees

The bee also requires the environment to be correct. Bees will generally not fly below 12°C and ideally need the temperature to be above 19°C.

They will not work during the rain nor at night (unless forced).

Major components of nutrition

There are three major components of nutrition which the bee has to satisfy:

- Water
- Protein
- Energy

Water

A bee, similar to most life on this planet, is mainly composed of water. Water holds the various parts of the body together, it provides a solvent so that nutrients, waste and oxygen can move around the body and cell.

Water is the forgotten nutrient and the ready availability of clean fresh water is essential. Insufficient attention is given to clean water supplies. It is useful to consider the role that water plays in all the normal metabolic functions of the bee.

- It helps to maintain and control body temperature, through both during intake by cooling and exhalation when the heat is dissipated from the bee. It is lost primarily in two ways, either by respiration or in the faeces.
- An imbalance between water intake and loss results in dehydration. This is serious in some parts of the world subject to droughts.
- It is responsible for transporting food and waste products throughout the body. Waste products are eliminated via water through the malpighian tubules and the rectal pads.
- Hormones and chemicals are transported around the body through the haemolymph.
- Water regulates the acid/alkali balance in the body through the controls exerted by the rectal pads.
- Water is used in protein synthesis. The digestive process will not function without adequate water.

However, apart from the water within the stored food sources – such as honey – the bee does not actively store water in the hive. Honey is also quite a dry substance and needs water to be metabolised. Water must be obtained frequently. Obviously in the winter the bee cannot go outside the hive to obtain water supplies and must rely on the stored honey reserves. In cool months, a honey bee hive requires about 150g of water per day to survive. In the hot months this increases to 1 kg of water per day per hive. In other parts of the world with wet and dry seasons, during the wet season there is plenty of water, but the bees do not fly.

It is essential to place the hive near a good water supply – ideally a pond or river. In the summer months providing additional fresh clean water in a basin may be very beneficial. Place a leaf or brick in the water so the bees can land and drink. This will increase your honey yield. Add a little salt to the water initially to encourage the bees to access the water. Bees like other insects are attracted to urine for its salt content.

Figure 2.3 Place the hives near a good fresh water source

Figure 2.4 Support the hive in the hot summer months by additional water but add landing materials

Simple arrangements of gravel, moss or peat filled dishes topped up with water are ideal. Place several around the hive and garden. This is especially important in hot dry climates such as Australia.

Clean water

It is vital however to ensure that the water you supply is clean and not contaminated.

Figure 2.5 This waterer while well designed was heavily contaminated with soil materials

Figure 2.6 Newly established top box with plenty of easy access to water

Water metabolism

Water is taken in directly through the mouth or via food stuffs. Water is lost through the faeces and the respiratory system. The first spiracle has a large mobile cover which plays an important role in regulating water loss through the tracheal system.

Water and salt levels within the bee are regulated through the Malpighian tubules which lie in the haemolymph. In the adult there are many hundreds of Malpighian tubules; in the larva there are four. The Malpighian tubules take water from the haemolymph and pass it into the pylorus and then the small intestine. The four larval Malpighian tubules only empty into the pylorus at the L5 larval stage. Urine then passes into the small intestine and rectum and out of the body. Water balance is regulated by most of the water being reabsorbed through the rectal wall. The chemical composition of the haemolymph is regulated by the six rectal pads at the proximal end of the rectum. Their primary responsibility is to regulate and conserve chlorides. Particular impunities are filtered out of the haemolymph by the dorsal diaphragm.

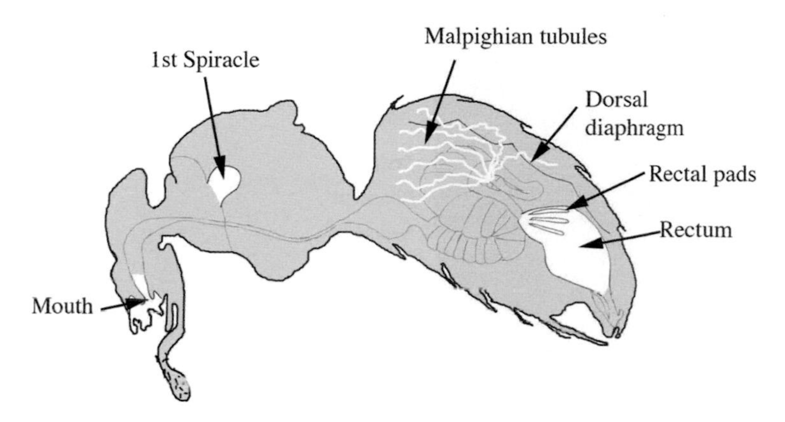

Figure 2.7 Drawing of the main water and mineral control points in the bee: mouth, spiracle, Malpighian tubules, dorsal diaphragm, rectal pad and rectum

Water in the hive

Water is used in the hive to maintain the hive's humidity and to dilute thick honey.

How much water do bees require?

In cool months a hive requires about 150g of water per day to survive. In the hot months this increases to 1kg water per day per hive.

Protein and energy

While not exclusive, a major source of protein (pollen) and energy (nectar) is provided to the bee via flowering plants. To understand this symbiotic mechanism where the evoluation of both the flowering plants and various insect families has developed with mutual benefit, it is essential to have an understanding of the basic anatomy of the flower.

General anatomy of a flower

Flowers are fascinatingly complex structures themselves. They first evolved around 160 million years ago but only became the dominant type of plant, replacing conifers, less than 100 million years ago. This was mainly after the demise of the dinosaurs.

There are many ways of describing flowers which is beyond the scope of this book. To describe flowers from a bee perspective they can be considered simple – as in a single flower or compound (complex) where multiple flowers are arranged to look like a single flower.

There are two major components to a flower, the vegetative and the reproductive.

The vegetative is made of two rings: the outer septal ring called the calyx and the petal ring called the corolla. Inside these two rings of leaves, there are the reproductive organs: the stamens of the male and the stigma of the female.

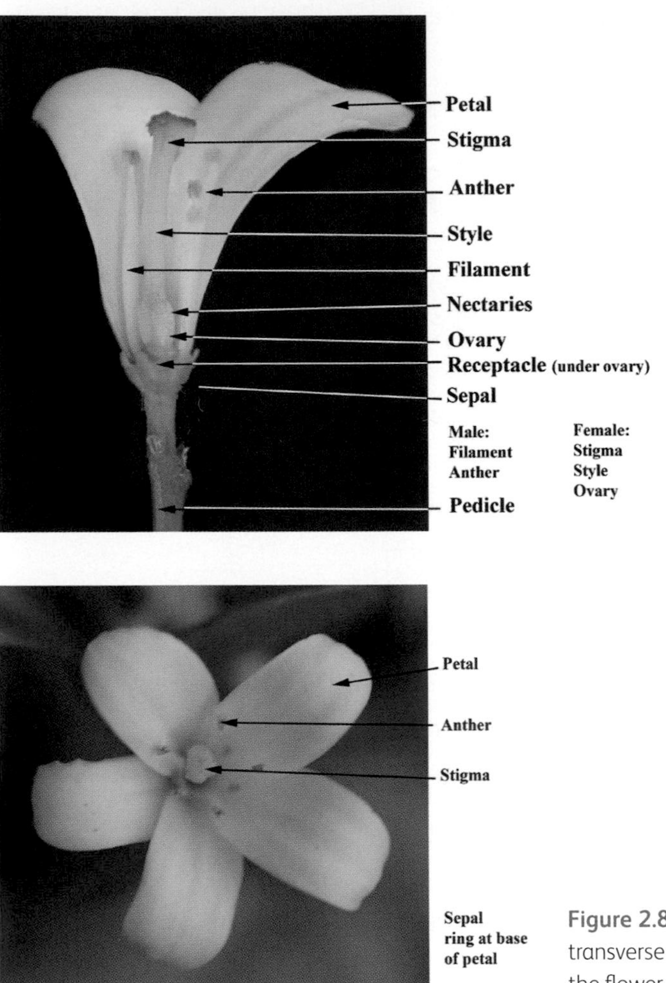

Figure 2.8 The major parts of a flower in transverse section and a frontal view into the flower

If there are many flowers on a single stem, this is called an inflorescence. These may be very closely packed together to make a compound (complex) flower where the simple flowers are collected into a group and only appear as a whole flower.

Figure 2.9 Compound flower main anatomical features

Figure 2.10 Detail of the anatomy of a stamen from amarantus

Figure 2.11 Bee collecting pollen: note the number of hairs covering the bee to which the pollen sticks

Protein

The major source of protein to the colony is pollen. On an average day around 15-30% of the foragers may be collecting pollen. A colony requires about 30–50kg of protein per year, or about 50–100g protein a day and as pollen is about 20% protein, this is about 250–500g of pollen per day. However, different plants produce different amounts of protein and the quality of pollen can vary tremendously. Thus if only poor protein content flowers are available, this demand increases to 340–600g of pollen per day.

There are three major means whereby the flowering plants (angiosperms) are pollinated:

- Self pollination when the flowers may not even open, Salvia species for example.
- By animals, including bees.
- By the environment mainly air. The grass family – Wheat, Barley, Maize are examples of environmental, (air) pollination plants.

It is important to realise that a large field of maize (corn) (*Maize zea*) may offer no food for the bees. These plants have no need to attract insects or other pollinators, and so do not produce nectar and their pollen will be low in protein (15%) and will not be collected by bees.

Mankind then acts as guardians to the maize crop by removing all weeds (which may have flowers for the bees) or by deliberately eliminating all other plants using herbicides which maize is genetically engineered to be resistant to. Conifers are another example of wind pollination. However, the bees may collect the pine pollen but it is very low in protein. On the other hand plants relying on insect pollination will often have high protein content in their pollen – white clover (*Trifolium repens*) at 26% is an example.

When a plant uses insects to pollinate, this is referred to as entomophily.

Bees are able to recognise pollen with higher protein and fats and will select those preferentially. Note there are also species differences within families of plants. Also different hives, even side by side, will collect different types of pollen and therefore may have radically different food stores. This can explain why one hive is strong but next door the hive is weak.

Table 2.1 The quality of the pollen from different plant species. Bees require pollen to have more than 20% protein

Plant species	Water content %	Protein %
Cytisus	8	29
Erica	6	16
Eucalyptus	10	27
Halimium	5	14
Quercus	5	18
Raphanus	5	25
Rubus	9	21
Sunflower		17

Note: Green have good pollen, Red poor quality pollen.

Amount of pollen produced

The amount of pollen produced by each flower can also vary tremendously.

Pollen is only produced by the flowering plant while it is preparing to disperse its genes and reproduce. In many cases this may only be for a few days. The bee therefore has to find and harvest the food resource quickly, while it is present.

Pollen selection

Bees can also show a preference for some types of pollen, Elderberry (*Sambucus nigra*) is one example. This can be taken into account when planting for bees. Fat and protein content of the pollen encourages bee preferences.

Pollen calendar

A pollen calendar records when particular plants in your area are likely to be producing pollen and thus protein for your bees. The following pollen calendar is for typical flowering plants in the UK. Create your own pollen calendar for your locality.

Table 2.2 Pollen calendar for the UK, sorted by the month when pollen production starts

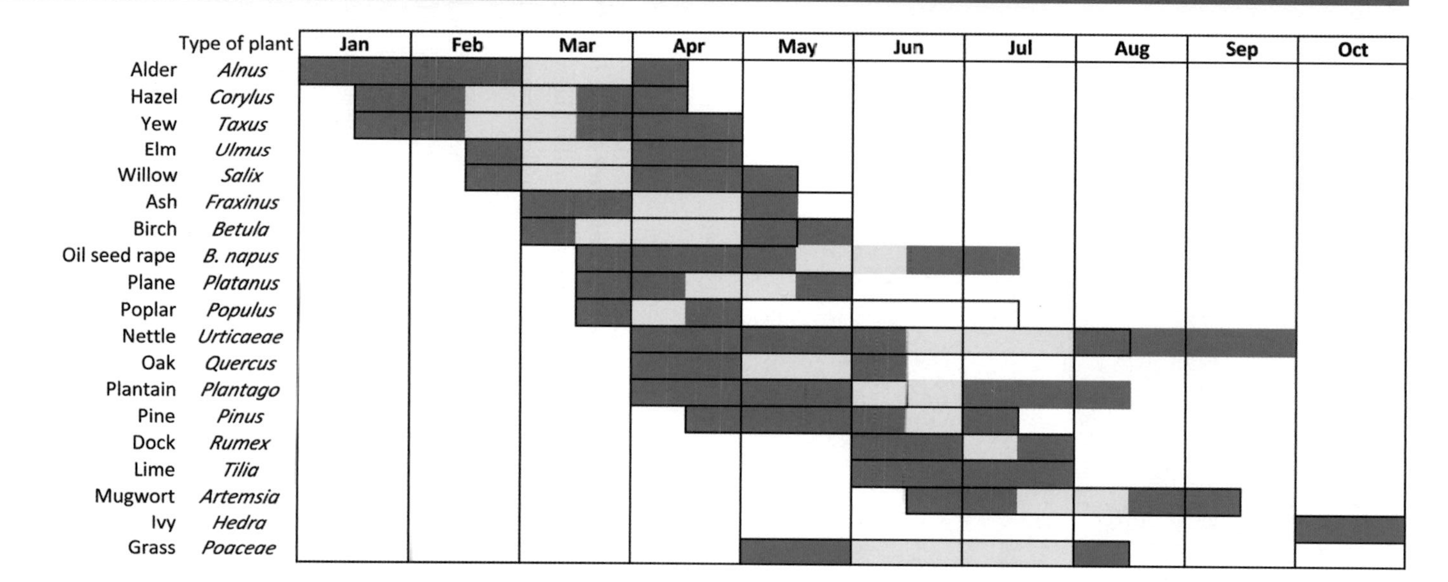

Types of pollen

Each plant species produces pollen with a specific character – often colour, shape and size of pollen grain. This can be used to classify the source of the pollen. This can be very useful to determine where the bees are working. It is possible to calculate the % of each type of plant the hive is working.

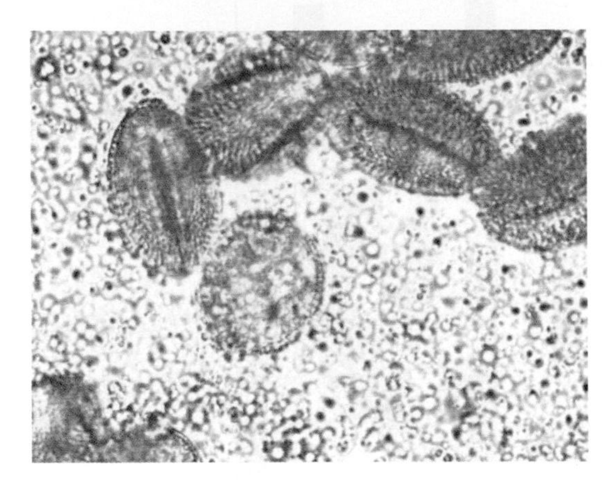

Figure 2.12 Pollen down the light microscope.
To monitor the type of pollen and identify the food sources can be useful in planning your hives food source.

How does a honey bee collect and move pollen back to the hive?

The honey bee has a specialised piece of anatomy, the corbicula or pollen basket, on the tibia (see Chapter 1). The bee is covered by hairs to trap pollen grains, including over the compound eyes. The worker bee uses its tongue to moisten the forelegs (prothoraxic legs) with some nectar. These are then used to brush pollen from the body of the bee. The inside of the metathoracic (hind leg) basitarsus has nine series of hairs called the pollen brush. The pollen is removed from the pollen brush by the pollen comb. It is transferred through the pollen press where it is compacted. The compacted pollen is transferred to the smooth concave cavity surrounded by a fringe of hairs on the outside of the tibia, the corbicula. The pollen ball is held in place by a single hair in the middle of the ball of pollen.

A typical honey bee (0.1g) can carry about half its bodyweight in pollen – thus 0.05g per flight. To collect 50kg of pollen requires a million flights.

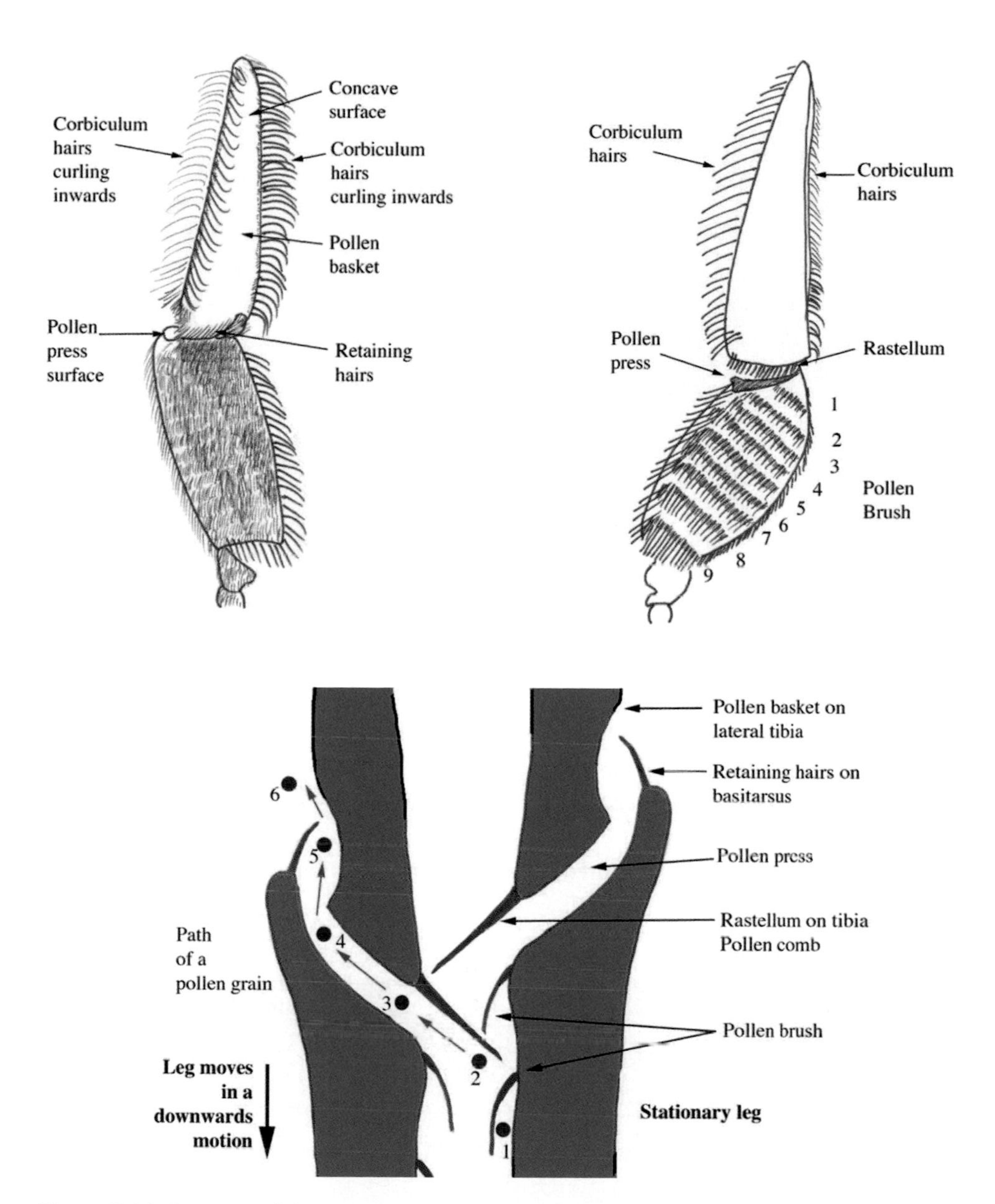

Figure 2.13 Drawings of the important anatomical features of the metathoracic leg responsible for the collection and storage of pollen. Left is the lateral, outer, surface and right is the medial, inner, surface. Also shown is a drawing of the production of the pollen ball, illustrating the movement of pollen from the pollen comb to the pollen basket

Figure 2.14 Yellow pollen being delivered back home

Figure 2.15 Bees with three different pollen types coming home

How does a bee store pollen in the hive?

Pollen is stored in cells towards the top of the hive. This can be a problem in the winter time. If the weather outside is too cold, the feed may become too far for the bees to obtain. With supers it is possible to turn the super during the winter to make it easier for the bees to access their feed sources.

Pollen from one type of plant is stored in one cell. This is an amazing feat considering this storage is all done in total darkness.

Fig 2.16 Protein stored in cells: note the different colours but also each pollen cell only contains the pollen from one type of plant

Where is pollen stored in the hive?

Dissection of the hive demonstrates the relative position of the honey, pollen and brood-nest. The queen cells are produced at the very bottom of the brood-nest where the wax is newest.

The same pattern is seen in the typical 'frame'. The honey is stored at the top of the frame and the brood in the warm centre. In Figure 2.18 the honey is high (1), the brood lower (3) and at (2) there is a narrow band of pollen. Note the cappings are uniform in colour and are convex (higher in the centre than at the edges).

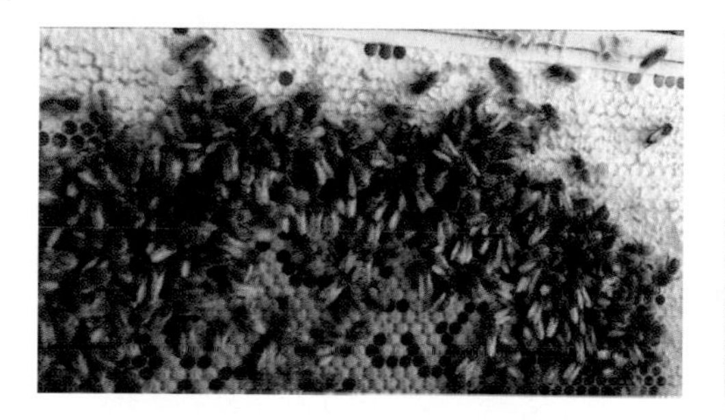

Figure 2.17 The frame with honey bridge

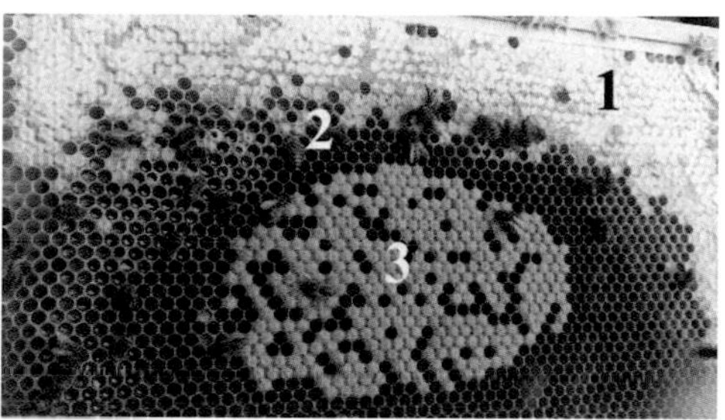

Figure 2.18 Bees shaken off revealing the various zones

Amino acids

These are specific molecules, which when combined in different combinations (as determined by DNA and RNA), form different proteins. There are approximately 22 amino acids and whilst the bee can synthesise the majority of these, there are a number it cannot and these are described as essential for normal health and metabolic processes.

The major essential amino acids required by the honey bee are:

- Arginine
- Histidine
- Iso-leucine
- Leucine
- Lysine
- Methinine
- Phenylalanine
- Threonine
- Tryptophan
- Valine

In the honey bee the limiting aminoacid is leucine. This is important to realise when designing supportive nutritional rations. This means that in order to use other amino acids the leucine must be correct first, otherwise the other amino acids are surplus to requirements and are wasted or turned into fats.

Amino acids required by bees:

Leucine	4.5%
Valine and Iso-leucine	4%
Threonine, Lysine and Arginine	3%
Phenylalanine	2.5%
Methinine and Histidine	1.5%
Tryptophan	1%

Protein stores within the bee

Protein is stored within the bee primarily in the muscles. The flight muscles in particular are a large source of protein.

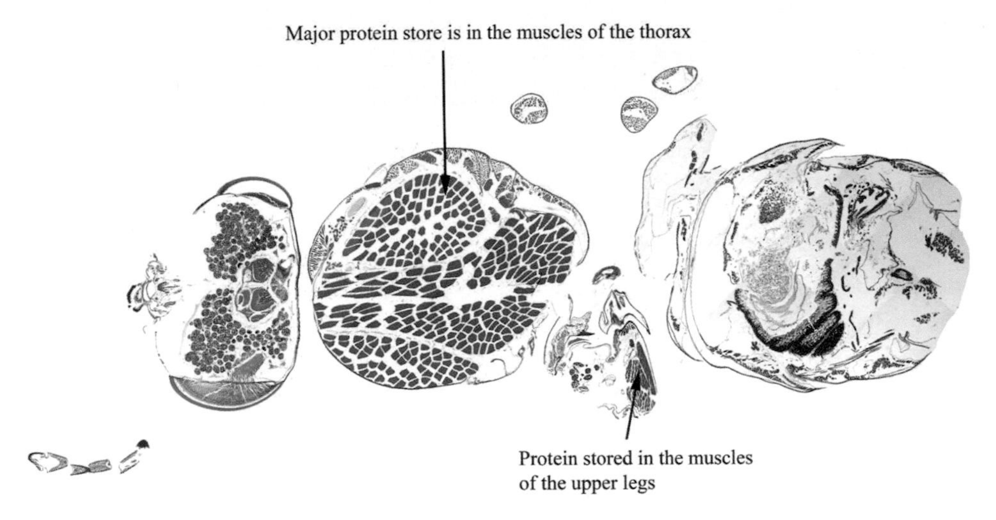

Figure 2.19 The major protein stores in the bee are in the muscle masses which stain red in an H&E stained frontal section of an *Apis mellifera* worker

Total protein percentage of the bee as an indication of its general health

A measure of the health of the bee can be used to monitor its total protein content. A normal healthy bee can have a total protein content of 60% dry matter. Bees going into winter should have a total protein concentration of more than 45%. If the total protein is below 30%, the bees are starving and it is likely the hive will not survive winter.

Problems with protein sources

When working certain plants the protein levels can be poor and this results in stress on the hive. If the hive becomes deprived of protein the bees will continue to try and raise the brood using their own protein sources but eventually the hive will stop producing brood.

*White box stress (*Eucalyptus albens*)*

White box has poor protein content of its pollen at 17% and the iso-leucine content is only 3.8% reducing the digestibile protein content further. The plant flowers in the winter months. Bees are therefore cold stressed and protein stressed and this can lead to hive collapse.

*Alfalfa/lucerne stress (*Medicago sativa*)*

While the lucerne plant has plenty of protein – 20–24% – the aminoacid ratio is incorrect for bees being short of iso-leucine. The bees require 4% iso-leucine whereas lucerne has 3%. Therefore, if the bees only are working lucerne, the hive increasingly comes under nutritional distress.

*Sunflower (*Helianthus annuus*)*

Sunflowers have a low protein content and thus when bees are used to help pollinate large fields of sunflower without some supplemental protein support the hive can actually starve.

Pollen trapping

Pollen can be an important by-product of the hive and have great nutritional value for human consumption. However, only collect pollen when the hive is strong enough to be able to lose the pollen. Many pollen traps may also trap drones, so ensure the trap has a drone escape mechanism present.

Pollen trapping may also be a benefit to the bees as you can collect pollen at peak times of the year, when it would be surplus to the hive's requirement, and then store in a freezer and feed back to the hive in times of stress. Note stored pollen can be subjected to mould and spoilage.

Bee pollen or bee bread

This is made by mixing nectar or honey with pollen by the worker bees. Bee bread is fed to workers and drones after day 6 post-lay – thus to the L3, L4 and L5 larva.

Energy

Energy in the diet is measured either by megajoules (MJ) or in the older kilocalories (kcal) as used in the USA and Canada. To convert kilocalories to megajoules multiply by 4.184.

The most common nutritional deficiency in bees is that of energy and the amount available in the diet is usually measured either as, net energy (NE), digestible energy (DE) or metabolisable energy (ME) (ME = 0.96DE).

The bee could not function without a ready supply of energy because it is the fuel that supports maintenance and drives the whole of the metabolic process. A major use of energy is flying. If the outside temperature is lower than 14°C bees cannot fly, as they cannot create enough energy resources.

The major source of energy is from sugars – ultimately glucose – and from fats obtained from the environment.

In addition, energy can be obtained by the metabolism of protein resources such as muscle, but this is not sustainable for a long period.

Sources of sugars

To make 1 litre of honey takes about 75,000km of flying to collect.

One litre of honey weighs 1.36kg.

Nectar

Flowering plants produce a unique compound called nectar by specific cells called nectaries. Different plants place their nectaries in different positions, depending on where and who they want to attract to the plant. Flowers, which are of interest to the bee, often have their nectaries deep within the flower by the ovaries.

Nectar is extremely variable in quality depending on species and may contain only 5 to over 80% sugar in water plus small amounts of protein etc. The main sugars are fructose, glucose and sucrose. Nectar provides around 50 times the energy expended in its collection so is an extremely valuable resource. Depending on the location of the hive and the severity of the winter, a hive requires about 25kg of honey to overwinter. The normal water content of nectar is about 80% whereas honey is less than 19% water thus a lot of water needs to be driven out of the nectar to make it into honey.

Plants and nectar production

Nectar is only produced by the plant when it requires assistance. It is not constantly produced. It is not made to benefit the bee! Thus many flowers will produce nectar for only 4–5 days while it is pollinating the specific flower. Nectar production in some plants – the Brassica family for example – demonstrates a daily rhythm (diurnal pattern) with maximum production 4–6 hours after dawn. Lime trees also produce more nectar in the morning.

Some plants also only produce nectar when the air temperature is high enough. For instance with Lime to optimise the production of nectar the temperature needs to be above 20°C.

Nectaries

Nectaries are often placed towards the base of the ovary to attract the pollinator (bee) to ensure that the bee places its body well into the flower to allow pollination – transfer of male pollen to the female stigma.

Figure 2.20 The nectar lines and colour markers indicating to the pollinator the path required to take to get to the nectaries. Note it has recently rained

Nectar lines

Note the nectar lines the bee sees may be very different from the 'colours' humans perceive. The colour vision of bees and humans is illustrated in Figure 2.21. For more detail regarding the function of the bee's eye, see Chapter 1.

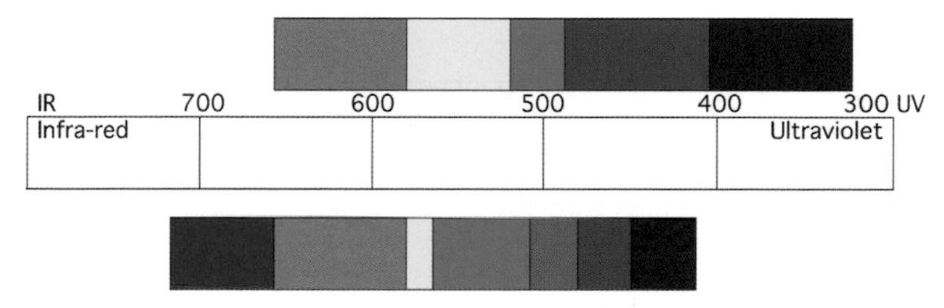

Figure 2.21 The colour perception of bees and humans

Note how the bee 'sees' less red than man, but much more blue, especially into the ultraviolent range. So pretty flowers to man, may be unappealing to a bee. This needs to be taken into account when designing your bee garden.

Colour of flowers

Figure 2.22 The flower's appearance to the human eye

Figure 2.23 The flower's appearance to the bee's vision under ultra-violet

Composition of nectar

The composition of nectar varies between different species of plants.

Brassica produces nectar with a glucose and fructose ratio of 1:1 with almost no sucrose. On the other hand, with Cerinthe 93% of the sugars in its nectar are sucrose. In Hellebore the sugar contents are sucrose 97%, fructose 2% and glucose 0.4%.

Honey bees do not seem to select flowers for their sugar composition. However, bumble-bees, *Bombus pratorum* and *Bombus terrestris*, can display a preference for nectar with specific sugar compositions.

Note the composition of nectar can also change depending on the quality of soil.

The nectar can also become more concentrated during the day due to evaporation from heat. After the rains, the nectar can be washed away or at least diluted.

Honey collection

An average honey bee will collect around 1 to 0.4g honey in its foraging lifetime. The average hive produces around 25kg of honey a year.

Honey crystalisation/candies

Different nectar sources will also crystallise over different periods of time. For example asters' nectar will crystallise rapidly whereas Echium nectar may take 9–15 months to crystallise. Rape/canola (*Brassica napus*) nectar rapidly crystallises and it may be extremely difficult for the bee and beekeeper to remove this honey from the super.

Honeydew

There are a number of other sources of sugar. Honeydew is a sugary compound produced by animals and fungi. Many sap-sucking insects produce honeydew as a bioproduct of their metabolism. Of the animals which produce honeydew the aphid is probably best known. Ants will actually 'farm' aphids to collect their honeydew products.

Honey bees will also collect the honeydew deposited on leaves by aphids. In some hives this can become a very important source of energy and produces a particular type of honey.

Note honeydew may also contain viruses which can be transmitted between aphids and bees.

Figure 2.24 Honeydew being deposited by Aphids

Figure 2.25 Detail of honeydew production by aphid

Honeydew is also produced by a variety of fungi – notably ergot (*Claviceps purpurea*) a common fungus of wheat.

Figure 2.26 Ergot (*Claviceps purpurea*) sclerotium (black) in wheat seeds

Composition of honeydew

The composition of honeydew varies with the plants the sap-sucking insect is living on. For example:

Plant	Total sugar content %
Citrus limon	2
Ficus benjamina	5
Nephrolepis biserrata	2

The sugar composition also varies and tends to match the sugar content of the phloem (sap) of the infested plant. It is interesting however, that many plants will change their sugar composition of their sap between being un-infested and infested with sap-sucking insects.

Plant	Glucose %	Fructose %	Sucrose %	Abrabinose %
Citrus limon	55	8	40	0
Ficus benjamina	50	2	35	18
Nephrolepis biserrata	85	0.5	15	0

Problems with honeydew honey

If the bee has been collecting honeydew as a major energy resource, it is advised to remove this honey before winter. The honeydew honey has a much larger proportion of indigestibles than light floral nectar honey. It is believed that this may result in digestive problems and dysentery. Bees collecting this resource also have to be fed balanced protein supplements, as honeydew lacks the protein-rich pollen accompaniment gathered from flowers.

Pollen

Pollen is also a source of carbohydrates although much lower than that available in nectar or honeydew. The ethereal extract % indicates the amount of fat in the pollen.

Plant species	Glucose %	Fructose %	Ethereal %
Cytisus	11	21	3
Erica	19	24	3
Eucalyptus	13	18	2
Halimium	14	18	5
Quercus	18	23	7
Raphanus	9	20	9
Rubus	21	23	3

Collection of nectar/water and honeydew from plants and leaves

Nectar is collected from the flower or as honeydew from a leaf using the bee's tongue and the associated mouthparts forming a tube, which sucks up the nectar and other fluids.

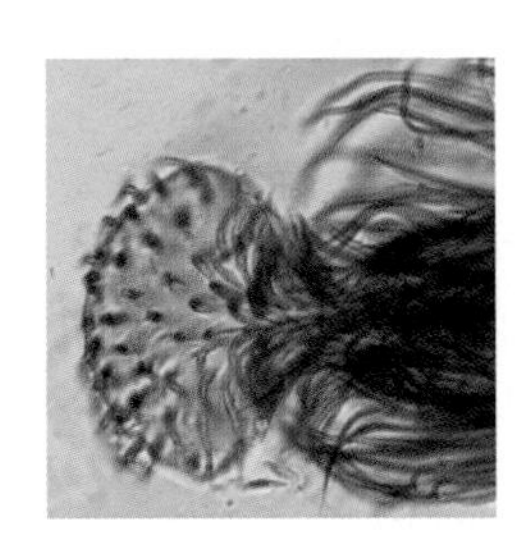

Figure 2.27 The labellum

The tongue has a spoon, the labellum, at the end which is covered in hairs. This attracts the fluids of the nectar and is then sucked up into the bee's mouth.

The maxilla and labium mouthparts fit together to form a single unit, the proboscis (tube or straw). The whole maxilla and labium swings around the pivot provided by the cardio, to move from the resting position, tucked under the head, to a forward vertical position in contact with the front of the mouth.

The following figures show the proboscis mouthparts. Note the other mouthpart, the mandible, is for biting and chewing.

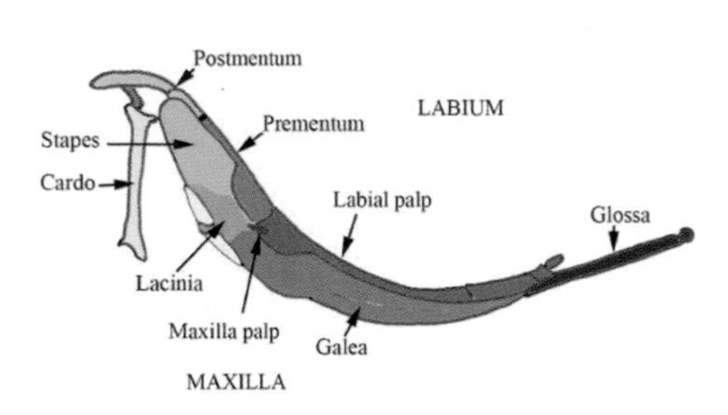

Figure 2.28 Drawing of the side view. The maxilla and labium lie parallel and embrace each other to form a close-fitting tube

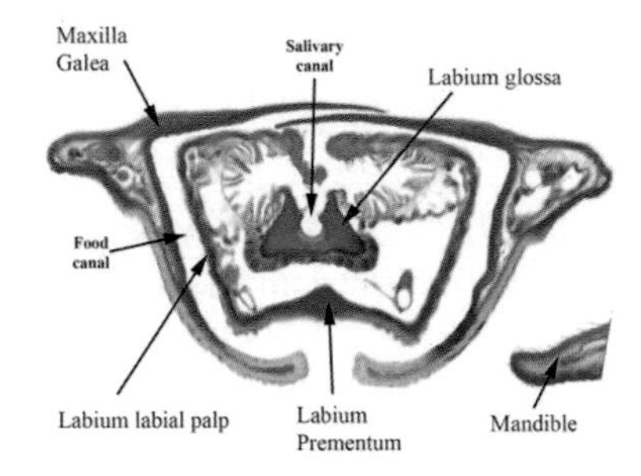

Figure 2.29 Transverse section of the posterior mouthparts demonstrating the embrace of the maxilla and labium to form a tube, by which liquids can be sucked into the mouth

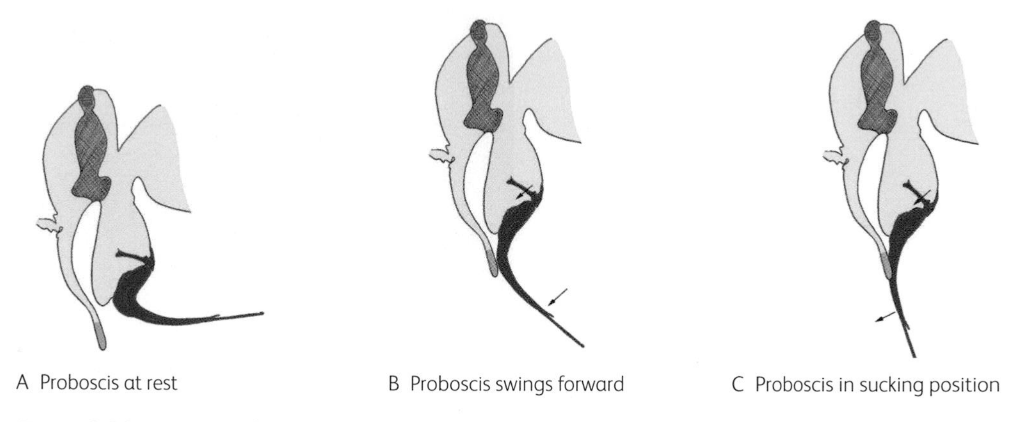

A Proboscis at rest B Proboscis swings forward C Proboscis in sucking position

Figure 2.30 Drawing of the maxilla and labium (proboscis) forming the sucking mouthparts

Transporting the water/nectar to the hive

The nectar is stored in the honey stomach or crop. This is a large diverticulum of the oesophagus and is situated immediately after the oesophagus passes through the petiole (see Chapter 1).

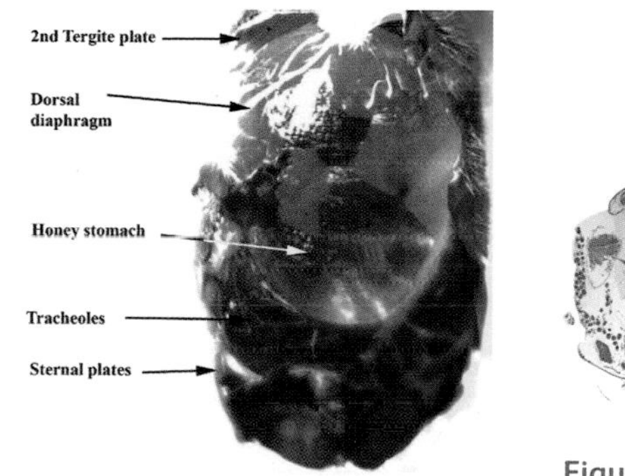

Figure 2.31 The honey stomach *in situ*

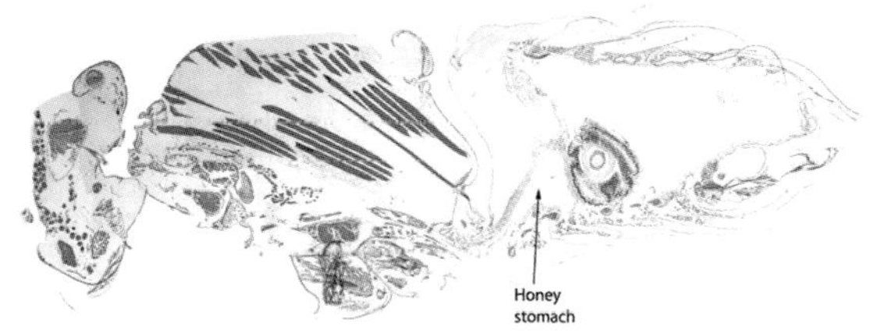

Figure 2.32 The honey stomach in transverse section H&E stain

Where is energy stored in the hive?

The major energy resource is stored as modified and dehydrated nectar called honey.

Where is energy stored in the bee?

Energy resources are stored in the bee's fat pads. There are four major fat pads in the bee's abdomen.

- Dorsal and sub tergum fat pad
- Lateral fat pad
- Ventral fat pad
- Visceral fat pad – surrounding various organs

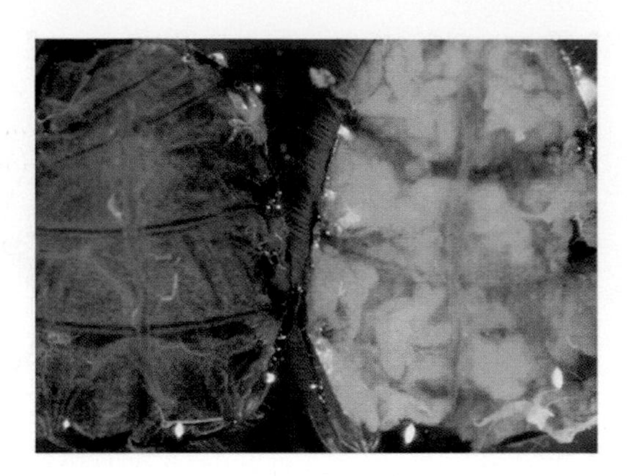

Figure 2.33 Starving left and healthy right showing clearly the difference in fat content

In the bumblebee the quality of the fat pad determines the success of the mated queen to over winter by hibernation.

Turning nectar into honey

Nectar has a water content of 80% whereas bees require honey to have a water content of less than 20% to stop the natural yeasts fermenting the sugars. The house bees expose the nectar to the warmth of the hive to create evaporation concentrating the sugars. The house bees also covert the sucrose in the nectar to fructose and glucose by the addition of the enzyme invertase, produced in the salivary glands. Once dried the honey is protected by a seal of wax over the cell. Note there is no air gap between the surface of the honey and the wax seal and so the cap is white.

Properties of honey

Honey from plant nectar

Honey from plant nectar consists of:

Water	< 18 % (or fermentation starts)
pH	3.5–5.5
Enzymes	Saccharase, Amylase (Diastase), Glucose oxidase
Water insoluble solids	The lower content greater clarity
Osmotic pressure	2000 mOsmols/kg
Refractive index	1.55 at 13% water
	1.49 at 18% water
Energy content	2.6 MJ/kg (1380 cal/lb)
Sugars	Glucose and fructose

Honey from honeydew

Honeydew honey is very dark brown in colour, with a rich fragrance of stewed fruit or jam, and is not as sweet as nectar honey. Honeydew honey is popular in some areas, but in other areas beekeepers may have difficulty selling the stronger flavoured product. In Greece, pine honey (a type of honeydew honey), constitutes 60–65% of the country's annual honey production. The composition of honeydew can vary tremendously depending on the plant species the honeydew was obtained from. However, it tends to be higher in minerals and amino acids than nectar honey. There may be more longer-chained sugars such as melezitose and raffinose. Because honeydew is lower in glucose and fructose the honeydew honey tends to stay liquid longer and resists crystallisation longer than nectar honey.

Measuring the quality of honey: from a human perspective

To measure the quality of the honey there are a variety of methods. One of the most reliable is to use a refractometer. To store honey, the water content needs to be less than 20% and ideally less than 18%. The lower water content controls the natural yeasts preventing it from fermenting the sugars. Note the refractometer used is actually inverted and actually measures water in sugar content.

Minerals, vitamins and lipids (fats)

Pollen, nectar and honeydew all contribute to provide the essential minerals, fats (lipids) and vitamins essential to keep a hive and its resident bees healthy.

B vitamins are essential to most insects. Vitamin A and K are linked to the development of the hypopharyngeal glands. Pollen is an excellent source of vitamins but many types are unstable.

Minerals are also essential especially potassium, phosphate and magnesium. However, sodium chloride and calcium in high concentrations may be toxic. Pollen is an excellent source of minerals.

Figure 2.34 A jar of honey

Finding food resources

There are several communication dances, indicating object(s) of interest such as food, water and future home resources.

The interesting object is fairly close – less than 100 metres

Circular or round dance:

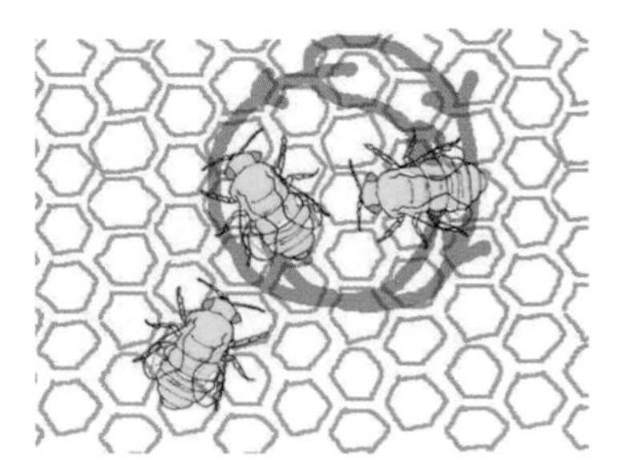
Figure 2.35 The round dance movements

The worker engages in a circular pattern. The other workers pick up the odour information and then will look in the immediate vicinity of the hive.

The object of interest is 150 metres or more away

The waggle dance

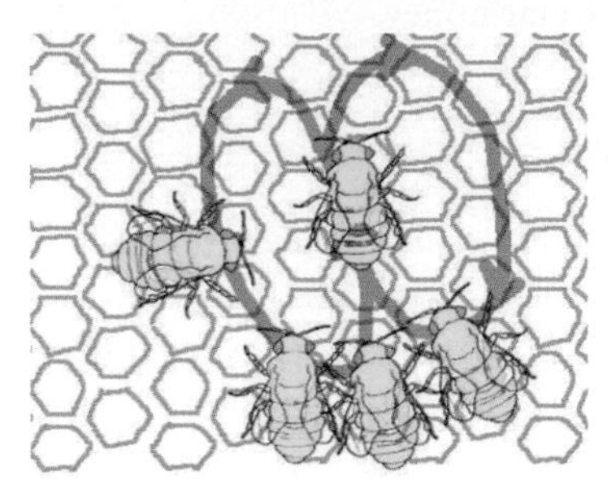

Figure 2.36 The waggle dance choreograph

The dance is composed of:

- A straight run – the direction of which conveys information about the direction of the resource.
- The speed – how often the dance is repeated indicates how far away the resource.
- The duration (correlated also with the distance) – the further the waggle run, the further the resource.

In the lifetime of a worker bee she might travel 1000km going back and forth between different resources!

The pattern of waggle dances can be illustrated in Figure 2.37:

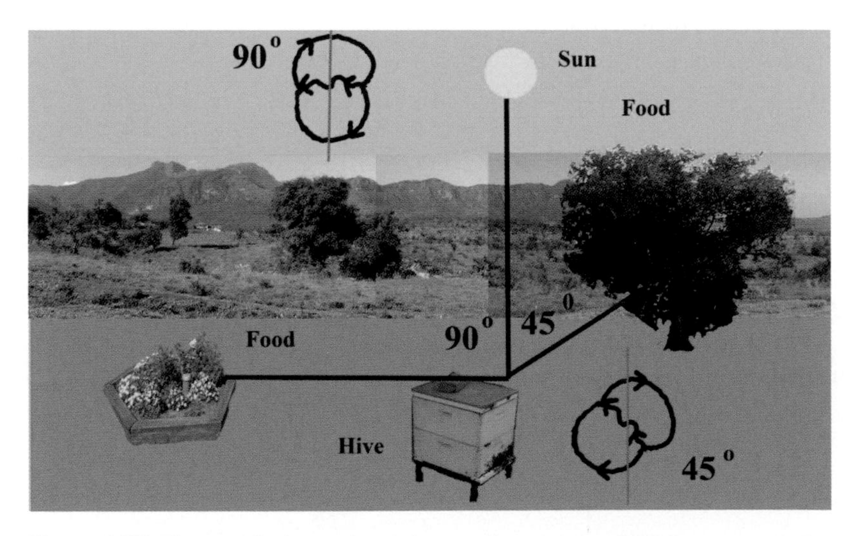

Figure 2.37 The waggle dance depending on the position of the sun

Feeding the offspring

The feeding schedule required differs between the different castes. The quantity and quality of the feed is a major determinate in the development of the larva and especially with the queen, her future as a mother and egg producer. Royal jelly is the protein rich product of the hypopharyngeal glands of the worker honey bee.

Queen (future)

Days	Stage	Approx weight mg	Food source
1	egg	0.1	yolk
2	egg	0.1	yolk
3	egg	0.1	yolk
4	larva	0.3	Royal jelly
5	larva	1.5	Royal jelly
6	larva	5	Royal jelly
7	larva	35	Royal jelly
8	larva	150	Royal jelly
9	larva	200	Royal jelly
16	Emergence	180 to 300, depending on nutrition level	

Worker

Days	Stage	Approx weight mg	Food source
1	egg	0.1	yolk
2	egg	0.1	yolk
3	egg	0.1	yolk
4	larva	0.3	Royal jelly
5	larva	1.5	Royal jelly
6	larva	5	Royal jelly/honey and pollen (bee bread)
7	larva	28	honey and pollen (bee bread)
8	larva	135	honey and pollen (bee bread)
9	larva	155	honey and pollen (bee bread)
21	Emergence	120	

Drone

Days	Stage	Approx weight mg	Food source
1	egg	0.1	yolk
2	egg	0.1	yolk
3	egg	0.1	yolk
4	larva	0.3	Royal jelly
5	larva	2	Royal jelly
6	larva	7	Royal jelly/honey and pollen (bee bread)
7	larva	35	honey and pollen (bee bread)
8	larva	50	honey and pollen (bee bread)
9	larva	220	honey and pollen (bee bread)
10	larva	300	honey and pollen (bee bread)
24	Emergence	290	

This feeding difference can be seen in light microscope sections of larva and pupa. Whereas digested pollen can be seen in the ventriculus of the L3, L4 and l5 worker and drone larva, it is not seen in the queen bee larva.

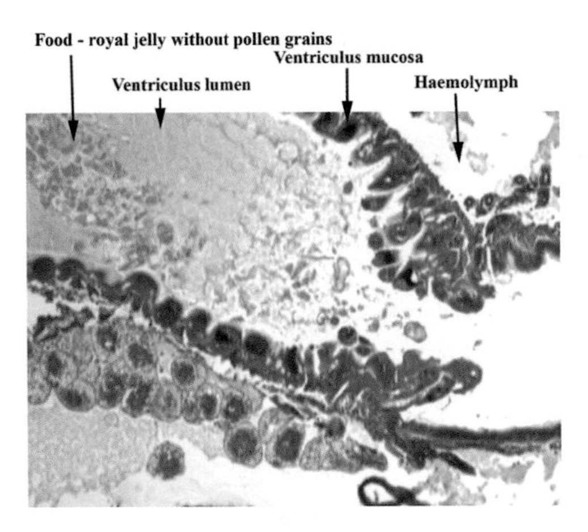

Figure 2.38 Ventriculus of a queen L4 larva without pollen grains

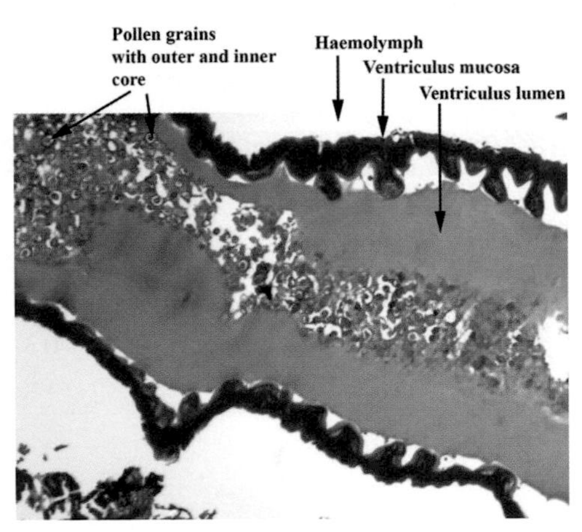

Figure 2.39 Ventriculus of a drone L4 larva with clear evidence of digested pollen grains

Adults

Adult drones obtain food from young workers as a mixture of glandular sections, pollen and honey.

After emergence and for the first 5–6 days the new worker bees consume lots of pollen to complete their growth and development, in particular the development and function of the hypopharyngeal glands. If this does not happen properly they will be unable to make royal jelly and brood development will decrease.

After day 10 the worker becomes more of a forager and the requirement is more energy from honey and nectar.

Monitoring hive health and food resources

The standard method of measuring hive food resources is by estimating the weight of the hive and physical examination of the frames. Note, however, examining the frames at time of nutrition stress (winter for example) only adds to the bee's problems.

Talking to other beekeepers and newsletters are a great resource to assist in determining if you need to produce nutritional support for your hives.

At the end of the day – ask the bees. Like all animals, they can be lazy and will go for the easiest resource. But if they are not taking the additional sugars in the spring, perhaps they do not need the extra.

Note nutrition stress can occur at any time of the year not just winter. Prolonged periods of rain wash away nectar resources. This is important in countries with wet and dry seasons, rather than cold and warm. Drought may make water resources difficult for the hive. The use of monocrop farming can put a lot of stress on the local environment including the bees.

Weighing the hive

There are a number of methods of weighing the hive. To be accurate the scales may need to be expensive and temperature effects need to be taken into account. Weigh the hive at about the same time each day to minimise the diurnal effect. This can be combined with the internet to allow a beekeeper to constantly monitor the weight of their hives.

Manually feeding the hive

To ensure that the bee survives through periods of poor food supply and still provides mankind with a source of bee products, it is required to supplement the hive's nutrition.

This is generally in the form of sugar products but it should be remembered that protein will also be required if brood development is also required.

In Northern Europe a hive requires about 15–17kg of stores as a minimum. In North America and Russia this could increase to 20kg to survive the winter. Therefore, when we supplement feed we need to provide sufficient feed to keep the stores above 15kg. The stores in a hive may vary over the winter months (in Northern Europe for example).

In Northern Europe, if it is assumed that the hive loses 10kg over winter, then a hive weighing more than 15kg in October should be able to overwinter. If the hive is less than 15kg in October, provide supplemental feeding. Provide the feed early rather than late, for example in September (northern hemisphere). Do not trust to the late nectar flow from ivy, for example. The bees also have to have time to convert the nectar and sugar solution into storable honey and drying can be difficult in the wet autumn months. The ivy flow should be considered a bonus not a saviour. In North America, winter can be very severe and the storage requirements by a hive can be over 20kg. Note opening the colony to inspect the bees during the cold weather can result in even more stress and reduction in stores. The use of a slatted crown board can significantly reduce the temperature loss by the brood area.

Note that each beekeeper needs to understand the climate zone in which the hive is present and be able to assess the normal hive requirements. If the spring is wet and cold, additional feeding will obviously be required. The bees will use their winter stores most efficiently at 7°C.

When selecting materials to supplement the feeding of your bees with natural bee products ensure that the honey or pollen used is free of any pathogens – especially European and American Foulbrood.

Figure 2.40 The ivy flow of Northern Europe

Figure 2.41 Detail of ivy flower

Feeding energy or water in a drought

Energy can be provided in a liquid form or a solid/fondant.

Liquid form – sugar syrup

This is made from granulated sugar (sucrose) and should be presented in a concentration of more than 50% sugar so the bees can readily use the sugar. It is vital that the concentration is not too strong or the sugar will start to crystallise out. Note liquid sugar syrup should not be used in very cold climates, here only use fondant.

There are two major concentrations, which would be commonly used.

Autumn feeding – a 61.7% sugar solution.

This is made by adding 2.5kg of boiling water to 4kg of granulated sugar. This will make 5 litres of sugar solution. The boiling water will help the sugar dissolve. Stir until the sugar is totally dissolved. Do not reheat the solution to get the sugar to dissolve quicker or you will create toffee. Add one drop of thymol solution (see Chapter 12) per litre of fluid to help reduce fungal growth.

If there is no handy weigh scale available this concentration can be made by half filling any container with water and then filling to the top with sugar.

Spring feeding – a 50% sugar solution

This is made by adding 1kg of boiling water to 1kg of granulated sugar. This will make 1.57 litres of sugar solution. The boiling water will help the sugar dissolve. Stir until the sugar is totally dissolved. Do not reheat the solution or you will create toffee. Add one drop of thymol solution per litre of fluid to help reduce fungal growth (see Chapter 12).

Administration of sugar solution

The winter solution is often presented to the bees in a rapid feeder placed over the feed hole in the crown board; a super frame is used to provide additional head room. If there are two holes in the crown board cover one to prevent bees using it.

A frame feeder can also be used, especially when there is limited headroom, in a nucleus for example. Note add some pieces of bamboo or fresh straw to float on the top of the liquid to reduce the number of bees, which drown in the liquid. A frame feeder holds about 3 litres of sugar solution.

Place the sugar solution in the hive in the late evening. If you have multiple colonies, feed all colonies at the same time. Reduce the number of entrances, ideally to even an individual bee entrance. This will reduce robbing.

Figure 2.42 Contact feeder

Figure 2.43 Frame feeder

Solid sugar feeding

Fondant feeding can be used in mid-winter. The fondant is placed over the feed hole in the crown board.

Place the fondant in the hive in the late evening and repeat procedure for sugar solution feeding.

Feeding protein

If there is a problem with protein sources it might be better to reduce the number of hives and the amount of brood to feed. Limit queen movement within the hive using a queen excluder and consider combining weaker hives to make a strong hive (see Chapter 12). Pollen should be stored frozen (–15°C) ready to use. Note bees may make use of the pollen substitute just to get at the sugar contained, which is a very expensive method of providing energy.

Figure 2.44 Feeding commercially made fondant. Note: not often recommended as it may contain AFB spores of EFB bacteria.

Figure 2.45 Robbing in the spring

Consider feeding supportive protein when:

- The bees' flying is restricted – for instance, during cool weather.
- When the major pollen available is low in protein – for example, sunflower flow.

Figure 2.46 Protein bee bread offered to bees

Pollen bee bread

Bee bread can be made to make an even more ideal supportive material for the bees.

Ingredient	Makes 1402g
Pollen	1000g
Honey	150g
Water clean	250
Whey	2

Artificial protein source

This is never as successful as natural collected pollen. Most artificial substitutes are actually low in protein. Soya products are generally used. Filter the products to get the particle size below 0.5mm. Once made, the final product can be rolled and cut into dry biscuits or cakes and stored frozen.

Recipe

Ingredient	Amount
Pollen	10–25%
Soya Flour	20–100%
Yeast	20–25%
Sugar/honey/water	20–50%

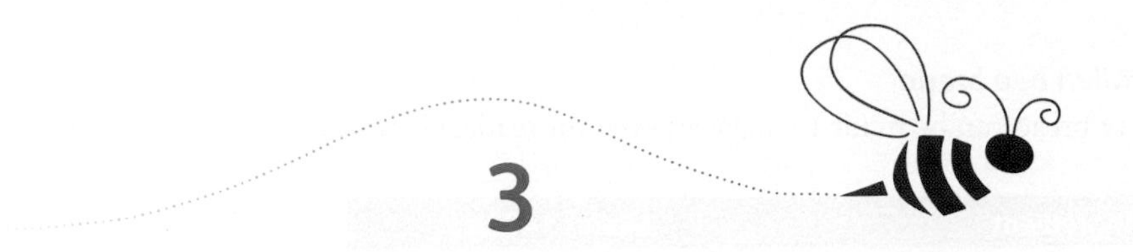

3

Understanding health and disease

Definition of health

The term 'health' means different things to different people. It is a state of physical and psychological well-being that allows the bee to express its genetic potential for maximising productivity, reproductive performance, honey, wax, royal jelly, pollen and other products.

The most important factor in maximising health, production and preventing disease, is good husbandry.

The term 'disease' means an unhealthy disorder of body and mind, sometimes with pain and unease that is likely to prevent the bee from exploiting its genetic potential resulting in lowered productivity.

The level of clinical disease is described by the term morbidity. Disease can be clinical (the affected bee shows clinical signs) or sub-clinical (the affected bee shows no obvious clinical signs). Sub-clinical disease can also have an adverse effect on productivity. It is important to distinguish between sub-clinical disease and sub-clinical infection. Every healthy hive, without exception, carries a multitude of potentially pathogenic organisms, mainly in the gut but also in the respiratory tract, mouth, skin and reproductive tract, which are not causing disease either clinical or sub-clinical. Bees also produce detritus, hive debris and this is host to a massive number of organisms. Bumblebee nests are an environment all on their own, where many organisms live their entire lives in the nest.

The absence of all 'pathogenic' organisms is not possible. In fact it is not possible for the bee to survive without the presence of bacteria and other organisms within its intestines. It is an interesting concept that in fact 90% of a bee's dry matter bodyweight may actually be bacterial! The concept of metagenomics looks at the total genetics of the bee – the bee and its resident microflora. This concept will become increasingly important in diagnostics and treatment regimes. It has already been used to attempt to identify the causal agent(s) of Colony Collapse Disorder and has led to the discovery of numerous natural viruses of bees.

There is a delicate balance between these potential pathogens and the bees' immunity to them. Any physical or psychological disturbance of this immunity may render the bee susceptible.

Good husbandry, including good stockmanship aims to avoid such disturbances, provided the more virulent pathogens are absent (e.g. *Paenibacillus larvae* American foulbrood or *Melissococcus plutonius* European foulbrood). Good husbandry means good housing, good nutrition and good management. Good stockmanship means care and attention to the bees' health and welfare.

This delicate balance between potential pathogens and the bees' immunity becomes even more precarious when looking at the entire hive or collection of hives in an apiary. By causing disease in a small group of bees as a result of poor husbandry, the pathogenic organism multiplies up to a concentration that may overcome the more resistant bees. The concentration again builds up and threatens to overwhelm the collective immunity of the hive (i.e. hive immunity). It is interesting that various bee behaviours, and the recognition of their full and half-sisters, may affect pathogen transfer within the hive.

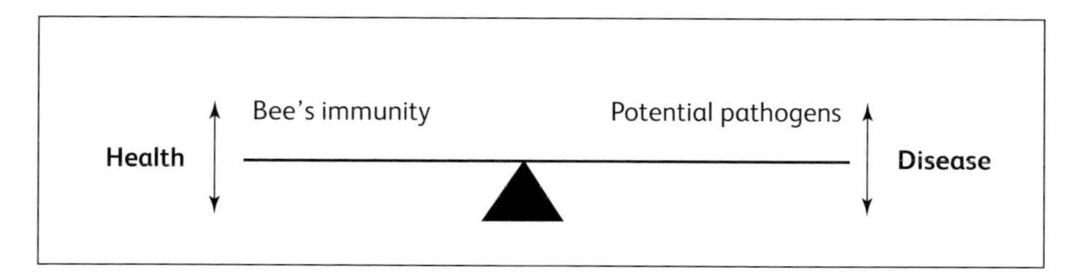

Figure 3.1 The balance of health and disease

The causes of disease

It is likely that when considering the causes of disease you think first of infectious pathogens. You would be right to do so in that infectious disease plays a much bigger role in bee hives, particularly large bee hives, than in animals kept individually such as dogs and cats. In most cases you would also be wrong, however, because clinical disease in bee hives usually results from the interplay between a number of predisposing, primary and contributory causes. You should bear this in mind when you are thinking how to suppress clinical signs and effects of disease. Most clinical diseases in bee hives have multiple causes.

Causes of disease are considered here under six infectious and five non-infectious headings:

Infectious agents

Prions
Viruses
Bacteria
Fungi
Protozoan parasites
Metazoan parasites

Table 3.1 Relative sizes of the infectious agents

Infection	Can be seen	Approximate size
Viruses	Electron microscope only	20–300nm
Spiroplasma	High power microscope	0.5μm
Bacteria	High power microscope	0.5–30μm
Fungi	Low power microscope	5–80μm
Protozoa	Low power microscope	6–12μm
Metazoan parasites	Low power microscope	5–20mm
Mites	Low power microscope	0. 5mm
Varroa	By eye	4mm

Note: 1 metre = 100cm = 1000mm
10^{-2} or 1/100 metre = 1cm
10^{-3} or 1/1000 metre = 1 millimetre (mm)
10^{-6} or 1/1,000,000 metre = 1 micrometre (μm or mcm)
10^{-9} or 1 billionth* of a metre = 1 nanometre (nm)

This gives you an idea of the actual size.

* Billion here = 1000,000,000 = 10^{9}

Non-infectious agents

Trauma
Hereditary and congenital defects (developmental abnormalities)
Nutritional deficiencies and excesses
Toxic agents (poisons)
Environmental stress

Infectious agents

Prions

Prion diseases include bovine spongiform encephalitis (BSE) in cattle and scrapie in sheep. Prion diseases have not been reported in the bee.

Viruses

There are many viruses of bees and many more to be found. Most will do little or nothing to the bee and do not affect their health. However, others, especially in combination with other stress factors, such as the *Varroa* mite, may end the viability of the whole hive.

What are viruses?

Viruses are pathogenic organisms that infect bees and can be listed in order of decreasing size and complexity. Viruses are the smallest of the infectious agents (Prions are still not classified). They cannot be seen by light microscopes that are used for looking at bacteria. They can only be seen through an electron microscope. They so small they are filterable – pass through filters – this is what the word virus means in Greek.

At their simplest, viruses consist only of nucleic acid (i.e. their genes) and proteins which are arranged around the nucleic acid in a geometrical design and which protect the gene. Some larger ones also have a loose outer coat (the envelope) which may contain lipids (fats) and carbohydrates which are derived from the hosts cell.

Viruses contain their genetic code in the nucleic acid but they do not contain the full mechanisms for their own multiplication. They have no energy-generating systems and lack chemicals such as enzymes required for their own reproduction. They depend entirely on those of the host.

- Viruses can only multiply inside the host cells.
- Viruses cannot multiply outside the host.

Viruses cannot multiply outside the bee. Outside the host viruses are inert. They have no metabolic activity. Inside the host's cell they behave like pirates. Their genes, using the host's nucleus, take over control of part or all of the cells' mechanisms. They control these mechanisms to make many more viruses exactly like, or similar to, the ones that invaded. This activity usually damages or destroys the cells.

How do they get into cells?

The proteins on their surface stick specifically to receptors on the cells' surfaces. The cells then engulf them as if they were taking in particles of food, but the particles are destructive invaders. They quickly lose their outer envelopes and the proteins covering their genes. These released genes then take over from the cell's nucleus and direct the cell to make more viruses. Some viruses, RNA viruses, replicate in the cytoplasm and some DNA viruses, in the nucleus.

The classification of a virus

All animal, plant and bacterial genes are made of a nucleic acid called DNA (deoxyribonucleic acid) which is found in the central nucleus of the cells and a further one RNA (ribonucleic acid) found in the cytoplasm. Virus genes only contain one or the other.

The first broad classification is therefore into DNA viruses and RNA viruses. These are then classified into families based on their shape, size and structure.

Many viruses are recognised in the bee but do not appear to have a pathogenic action against the bee. This is the normal state of most of the viruses that are to be discovered. We must not assume that any virus found causes anything. Bees have been around for a long time and have evolved with these viruses. Viruses found in many families of bees – *Apis* and *Bombus* – are likely to be very old and well adapted. If the bee is healthy it reduces the chance that the virus will cause clinical disease.

Table 3.2 Viruses associated with diseases in bees

Virus name	Family	Comment
DNA viruses		
Invertebrate iridescent virus IIV	Iridiovirus	*Apis mellifera*
RNA viruses		
Acute bee paralysis virus	Dicistroviridae	*Apidae*
Bee virus X	Unclassified	*Apis mellifera*
Bee virus Y	Unclassified	*Apis mellifera*
Black queen cell virus	Dicistroviridae	Many species of *Apis*
Chronic bee paralysis virus	Unclassified RNA	I and II *Apis mellifera*
Cloudy wing virus	Unclassified	*Apis mellifera*
Deformed wing virus	Iflavirus	Many species
Israel acute paralysis virus	Discistroviridae	Affects many species of *Apis mellifera*
Kakugo virus	Iflavirus	*Apis mellifera*
Kashmir bee virus	Dicistroviridae	Many species
Morator aetatulas – Sacbrood	Picornovirus	*Apis mellifera* and *cerana*
Slow paralysis virus	Unclassified	*Apis mellifera*
Thai sacbrood virus		*Apis cerana*

Table 3.3 Viruses which are believed to cause no clinical disease

Virus name	Family	Comment
Aphid lethal paralysis virus	Dicistroviridae	Through honeydew
Arkansas bee virus	Picornovirus	Many species of *Apis*
Berkeley bee virus		*Apis mellifera*
Big Sioux River virus		Through honeydew
Chronic bee paralysis virus satellite	RNA	Sometimes seen with Chronic Bee Paralysis virus
Egypt bee virus	RNA	Non pathogenic
Filamentous virus	DNA	*Apidae*
Lake Sinai 1		*Apis mellifera*
Lake Sinai 2		*Apis mellifera*
Macula-like virus		*Apis mellifera*
Varroa-destructor virus I	Iflavirus	*Apis mellifera*

RNA viruses generally are very fragile and survive only a matter of days whereas others, often DNA viruses, can persist for months. In general, viruses survive long periods if frozen. A virus can survive fairly long periods in damp overcast cold weather, but only short periods in hot sunny dry weather. The action of UV light from the sun is particularly damaging to viruses. This is why viral diseases are more common in winter than summer. This information clearly is important in their control. Most, but not all, viruses are destroyed by strong acid or alkaline solutions and those that have a fat (lipid) coating are quickly inactivated by fat solvents/ detergents. This is why it is vital for beekeepers to understand the nature of bee viruses.

Virus may be shed in saliva, faeces, expired air, semen, eggs or from the skin. The transmission may be by direct bee to bee contact, or indirect contact on machinery, trucks, or by vectors such as the wind, flies, bumblebees, pests or discarded human bee products.

Clinical signs of disease

The major clinical signs of the known virus are neurological as this is easily seen. The different manifestations of clinical disease provide a framework for determining the significance of the laboratory result. Many viruses, even pathogenic, cause no clinical sign in an individual bee. When examining hives consider the following as major clinical signs:

Hive entrance

- Large number of dead bees
- Trembling bees
- Bees crawling around aimlessly
- Black and or hairless bees
- Bees being rejected by the guard bees
- Deformed wing bees
- Bees gripping onto local grass
- Bees form a sunburst patten
- Bearding
- Reduce flight activity
- No bees

Inside the hive

- No bees
- Low numbers of adult to brood
- Varroa on back of bees
- Deformed wing bees
- Bees with head in brood cells
- Open brood unhealthy
- Capped brood unhealthy
- Scattered brood
- Irregularly coloured capping
- Dead discoloured larvae
- Chalk or stone brood

Table 3.4 Virus pathogens of the bee and associated clinical signs

	Nothing	Neurological	Deformities	Death	Infertility	Miscellaneous**
Acute bee paralysis virus		✓	✓	✓		
Aphid lethal paralysis virus	✓					
Arkansas bee virus	✓					
Bee virus X						✓
Bee virus Y						✓
Berkeley bee virus	✓					
Big Sioux River virus	✓					
Black queen cell virus					✓	✓
Chronic bee paralysis virus		✓	✓	✓		
Chronic bee paralysis virus satellite	✓					
Cloudy wing virus			✓			
Deformed wing virus			✓			
Egypt bee virus	✓					
Filamentous virus	✓					
Invertebrate iridescent virus IIV						✓
Israel acute paralysis virus		✓		✓		
KaKugo virus						✓
Kashmir bee virus				✓		
Lake Sinai 1	✓					
Lake Sinai 2	✓					
Macula-like virus	✓					
Morator aetatulas – Sacbrood			✓	✓		
Slow paralysis virus		✓		✓		
Thai sacbrood virus			✓	✓		
Varroa-destructor virus I	✓					

Diagnosis of viral infections

It is more difficult and expensive to grow viruses than bacteria in the laboratory. They have to be grown in living cell-cultures artificially in tubes or bottles. Some cannot be grown by either of these methods. Modern techniques have speeded up and simplified laboratory diagnosis because the virus does not need to be cultured. Fluorescent-antibody techniques use antibodies labelled with fluorescent dyes. The Lateral flow test is also commonly used in many diseases (e.g. American foulbrood bacteria) to provide a rapid accurate diagnosis. Techniques such as polymerise chain reactions (PCRs) have been developed to demonstrate the virus genetic material (i.e. the DNA or RNA) in samples. Their advantage is that they can accurately detect tiny amounts of the DNA or RNA. Unfortunately they are expensive if done in small numbers. They also do not indicate if the virus is still viable, only that the genetic material is present. Immunohistochemistry, where the presence of a pathogen can be demonstrated in the tissues, can be particularly useful.

In addition the genetic sequence of a virus can now be obtained and allows for detailed epidemiology to be carried out.

Interpretation of results

It is important to understand that no medical test is perfect. Sometimes test results can be positive even though the animal does not have the pathogen, a phenomenon called a false positive. Other times, the results can be negative when in fact they do have the disease – a false negative.

Very few tests are able to detect all positive cases, and sensitivity and specificity are used as measures of accuracy.

- The sensitivity of a test refers to how many cases of a disease a particular test can find.
- The specificity of a test refers to how accurately it diagnoses a particular disease without giving a false-positive result.

The higher the sensitivity and specificity of any given test the more accurate the result (see Chapter 12).

Treatment of a virus

Some newer medicines are available for the treatment of a few viral infections in human beings but they are expensive and not available for use in bee diseases. Since viruses have no cell wall and no metabolism of their own, antibiotics will not destroy them, although they may help by preventing secondary bacterial infection.

Remember, antibiotics have no effect on viruses only on any secondary bacteria. But with regard to the bee this knowledge is still rudimentary.

Bacteria

These can be readily seen under the microscope, particularly when they are stained. They are recognised by family group according to their ability to take stains, shape, size, biochemical characteristics, antigenic characteristics and recently by identification of their DNA using PCR.

Most bacteria can be grown in nutrient liquids and on solid media containing nutrients set in agar gel. Most grow profusely within 48 hours. The bacteria of bees are unusual in insects as they need to be cultured at 35–36°C the normal temperature of the hive. Bacteria tend to form little colonies consisting of billions of organisms and the shape and colour of these may be characteristic for a particular family or specific bacteria. A tiny smear from a colony spread on a glass slide and stained will give both the shape and the staining reaction of that particular bacterium. The stain commonly used is called a Gram stain, and bacteria either stain positive (purple) or negative (red). This stain is of great help in carrying out primary observations.

There are relatively few pathogenic bacteria recognised in bees. The major ones are the bacteria associated with American and European foulbrood. These are recognised because of the devastation they do to the hive.

Table 3.5 Bacteria which may cause disease in bees

Bacterial name	Disease	Type of bacteria
Paenibacillus larvae	American foulbrood	Gram positive rod
Paenibacillus pulvifaciens		Gram positive rod
Melissococcus plutonius	European foulbrood	Gram positive cocci
Pseudomonas apiseptica	Pseudomonas	Gram negative rod
Spiroplasma apis	Spiroplasmosis	Bacteria without a cell wall
Spiroplasma melliferum	Spiroplasmosis	Bacteria without a cell wall

Many bacteria coexist with the bee and may actually help the bee to survive by providing essential vitamins and digestive capabilities. Others help protect the bee's surfaces – the pharynx for example – by providing a microbiota stopping pathogenic bacteria from colonising the surface.

Table 3.6 Bacteria that coexist with bees

Bacterial name	Type of bacteria
Alpha 1	
Alpha 2.1	
Alpha 2.2	
Bifidobacterium	Gram positive rod
Frischella perrara	Gram positive rod
Gilliamella apicola	Gram negative rod
Lactobacillus	Gram positive rod
Snodgrassella alvi	Gram negative rod

Each bacterium has a number of specific characteristics that are peculiar to itself. Some, for example *Paenibacillus larvae*, form spores which can survive outside the bee for many years. Like viruses, the survival of bacteria is also dependent upon the material surrounding them (faeces, honey, wax, soil), temperature, moisture and exposure to ultra-violet light. Again, like viruses, they can survive indefinitely when frozen and for long periods in cold damp dark weather but only for a short time in very sunny weather. Such knowledge is important in controlling the spread of the disease. Freezing prolongs the survival of infectious agents and sunlight and drying kills them.

Fungi

Fungi (moulds and yeasts) are found in damp conditions and therefore, tend to occur in the wet spring or autumn months. Pollen can become soiled by fungi and it is possible that they would then produce toxins. If the hive is stressed the bees might be willing to eat this soiled pollen and become weakened.

Fungi are a common problem in bees and are associated with a number of disease conditions. But again many fungi will live in the hive debris and are vital as part of the natural cleaning process. Nosema belongs to a family of fungi called microsporidia.

Table 3.7 Fungi which may cause diseases in bees

Fungi name	Disease
Apicystis bombi	Nosema bumblebee
Ascosphaera apis	Chalkbrood *Apis*
Aspergillus flavus	Stonebrood
Aspergillus fumigatus	Stonebrood
Aspergillus niger	Stonebrood
Nosema apis	Nosema/dysentery
Nosema caeanae	Nosema/dysentery

Table 3.8 Fungi which appear non-pathogenic

Fungi name
Ascosphaera major
Ascosphaera aggregata

These two *Ascosphaera* fungi are sometimes found in cases of Chalkbrood in honey bees, but *Ascosphaera Apis* is the pathogenic version.

Protozoa

These are single-celled organisms without cell walls more typical of animals. They have their DNA enclosed within a nucleus, unlike bacteria.

Table 3.9 Pathogenic protozoa

Protozoan	Family/description	Bee
Malpighamoeba millificae	Amoeba	*Apis mellifera*
Crithidia bombi	Typanosome	Bumblebee
Crithidia expki	Typanosome	Bumblebee

Many protozoans are also found in the gut of bees and they appear to have no pathogenic features. Many may actually assist digestion.

Table 3.10 Non-pathogenic protozoa

Protozoan	Family/description
Monocia apis	Gregarines
Aprigregarina stammeri	Gregarines
Acuta rousseui	Gregarines
Leidyana apis	Gregarines

Metazoans

Metazoans are multicellular organisms. The pathogens here are parasites that are organisms that either live in the body (internal parasites or endoparasites) or externally on the skin (ectoparasites). They invade and live in the lining of the small intestine.

Note this list does not include animals and plants that may capture and eat the bee, for example wasps and hornets.

Parasites unlike bacteria have a life cycle which is the process of development from the egg through larval stages and finally to the adult. Some parasites require an intermediate host. This type of cycle is called an indirect one. A knowledge of the life cycle is important in preventing parasitic diseases. The most effective and cheapest way of controlling parasites is to break the cycle either by good hygiene or by removing the intermediate host if there is one.

Because they are easy to see there are a number of metazoan parasites recognised in the bee.

Table 3.11 Metazoan parasites

Parasite	Real name	Site
Beetle	*Trichodes ornatus*	Skin
Conopid	*Apocephalus borealis*	Abdomen
Conopid	*Conops quadrifasciatus*	Abdomen
Conopid	*Physocephala paralleliventrius*	Abdomen
Conopid	*Physocephala rufipes*	Abdomen
Conopid	*Sicus ferrugineus*	Abdomen
Fly	*Brachicoma devia*	Abdomen
Fly	*Brachicoma sarcophagina*	Abdomen
Fly	*Senotaina tricuspis*	Abdomen
Hive beetle large	*Hoplostomus fuligineus*	Skin
Hive beetle small	*Aethina tumida*	Skin
Hymenoptera	*Mutilla europaea*	Abdomen
Hymenoptera	*Syntretus splendidus*	Abdomen
Sphaerularia	*Sphaerularia bombi*	Abdomen
Tracheal mite	*Acarapis woodi*	Respiratory
Tracheal mite	*Locustacarus buchneri*	Respiratory

Parasite	Real name	Site
Tropilaelaps	*Tropilaelaps clareae*	Skin
Tropilaelaps	*Tropilaelaps koenigerum*	Skin
Tropilaelaps	*Tropilaelaps mercedesae*	Skin
Tropilaelaps	*Tropilaelaps thali*	Skin
Varroosis	*Varroa destructor*	Skin
Varroosis	*Varroa jacobsoni*	Skin
Varroosis	*Varroa rindereri*	Skin
Varroosis	*Varroa underwoodi*	Skin
Wasp	*Melittobia*	Abdomen
Wasp	*Philanthus bicinctus*	Abdomen
Wasp	*Syntretus spp.*	Abdomen
Wasp	*Mutilla europaea*	Abdomen
Wax moth	*Achrocia grisella*	Hive
Wax moth	*Galleria mellonella*	Hive

There are also many large organisms which live with bees and may look like parasites but have not been demonstrated to actually parasitise the bee.

Table 3.12 Metazoans which are not parasitic

Parasite	Real name
Bee louse	*Braula coeca*
Beetle	*Antherophagus nigricornis*
Euvarroa	*Euvarroa adreniformis*
Euvarroa	*Euvarroa sinhai*
Euvarroa	*Euvarroa wonsirii*
Fly	*Volucella bombylans*
Mites	*Acarapis dorsalis*
Mites	*Acarapis vagans*
Mites	*Neocypholaelaps indica*
Mites	*Parasitellus fucorum*

Note, just because an organism is called parasitic does not mean it actually is. For example, *Patasitellus fucorum*, which lives on bumblebees, may have its own parasitic mite.

Non-infectious agents

Trauma

Good management reduces traumatic diseases. The most common form of trauma occurs when the hive is examined. Bees can become trapped, crushed, injured and killed between the different portions of the hive.

Hereditary and congenital defects (developmental abnormalities)

Considering the number of bees in a hive of up to 80,000 bees, it is amazing how few bees demonstrate congenital or hereditary defects. When the beekeeper looks closely however, there are a number of deformities which can be seen. Check the wings and the wing position. Head and eye deformities are also commonly reported. Close examination may reveal hermaphrodite features (male and female characteristics) for example workers with large drone like eyes. As long as these deformities are unusual/rare do not worry. If they become commonplace, re-queening will be required.

Nutritional deficiencies and excesses

The nutritional requirements of the honey bee still need to be explored in more detail and this is examined in Chapter 2 of this book. Nutritional stress is briefly discussed here.

Shortage of water

During droughts and summer the hive may become dehydrated. This can lead to overheating of the hive.

Shortage of energy

If there is a shortage of flowers there will be a general shortage of nectar and therefore energy. Bees will travel a long way to get nectar but it costs energy to collect it.

If the weather is generally wet this dilutes the nectar in the flower and may even wash the nectar away completely.

If the weather is cold the bees will not be able to fly and therefore will be unable to collect nectar.

If the weather is very humid/wet it makes it more difficult for the bees to dehydrate the nectar. Rain can wash away nectar. If honey is too wet, the yeast will ferment and soil the sugars. This can be important to provide time for the bees to convert artificial sugars into honey.

Shortage of protein

If there is a shortage of flowers there will be a general shortage of pollen and therefore protein and energy.

The crop which is being worked by the bees may have a protein content less than 20% protein. This will lead to a progressive starvation in the hive.

The protein aminoacid mixture may not be suitable for bees, especially if short of leucine.

Toxic agents – poisons

Poisoning by a variety of agents, many man made, is common in bees. Many substances if taken at excessive levels become toxic and cause disease. Poisoning can occur in an individual bee or in a group or even affect a whole hive. In the latter two cases a number of animals will be affected at the same time all showing similar clinical signs. A study of the history may indicate a common exposure to the poison by contact or ingestion.

Medicines

Many medicines are highly toxic if used above their therapeutic levels. It is a common fault with people when treating animals to assume that twice the dose will act twice as well. This is a fallacy. Overdosing may well have the opposite effect. Medicines and treatments are discussed more in Chapter 4.

Terrorised hive

If the hive is being terrorised by hornets or wasps the worker bees may be unwilling to forage. This will lead to a progressive state of starvation.

Colony Collapse Disorder

This condition still needs a better description. The lack of foraging bees leads to a progressive failure of the hive to look after the brood.

Environmental stress

Stress is a condition which occurs in all bees when confronted with adverse management and environments. Better management of the environment has a beneficial effect on the health and the biological efficiency of the bee.

Environmental stress can be summarised into five major categories.

Water shortage

Bees need water to survive and do not actively store water in the hive. In periods of drought the bee hive may become severely stressed over the lack of water. Protein and honey consumption require water to be digested.

Water excess

Bees do not fly when it is raining. Rain also dilutes or even washes away the nectar store within plants. Many days of raining can lead to acute energy shortage for the hive.

If the rain persists, flooding of the hive can also become a problem and this needs to be considered when placing the hive. An attractive point near a bubbling brook may become a disaster when it turns into a raging torrent.

Figure 3.2 Parched landscape with little or no fresh water available

Figure 3.3 After a month of rain, the stream next to the hive started to burst its banks. Fortunately the hive survived

Figure 3.4 Rain dilutes and can even wash out the nectar in the flower

Cold weather

Worker bees do not fly below 10°C. Their stores are most proficiently used at 7°C. While bees are capable of surviving down to –40°C, for short periods, they must be given sufficient stores, within easy reach within the hive, to survive the winter months or a cold snap. Note too great a gap between the cluster of bees and the stores can result in a hive collapsing.

Table 3.13 Effect of temperature on bee behaviour

Temperature °C	Activity
40° or more outside hive	Bearding and cooling hive
38° Outside hive	Workers actively forage for water
33–35° inside hive	Normal brood development
	Wax secretion in hive
29° inside hive	Brood cannot be made
20° outside hive	Queen bee does not attempt to mate
16° outside hive	Drone bees do not fly
10° outside hive	Worker bees do not fly

Do not check the bees when the weather is very cold. Opening the hive disturbs the hive cluster allowing cold air in which may kill the queen. The advent of remote temperature monitoring may allow the cluster's health to be assessed through the Internet.

Figure 3.5 Hives covered in snow

Having a slatted crown board/inner cover allows visualisation of different frames without disturbing the rest of the hive. Alternatively use a towel to cover the open box while you examine individual frames. In general do the minimum you need to do to check the bees.

Figure 3.6 Solid crown board/inner cover

Figure 3.7 Solid crown board/inner cover needs to be removed exposing the brood box

Figure 3.8 A slatted crown board allows individual frames to be examined without exposing the whole set of frames

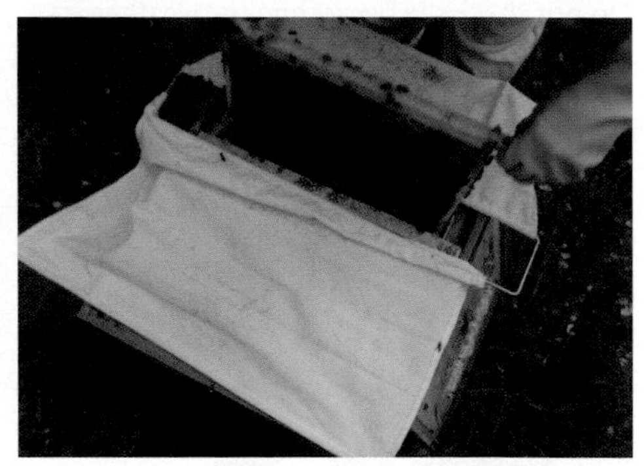

Figure 3.9 A towel can be used to partially close the hive. This significantly reduces the number of flying bees which also makes examination much easier

Ensure that the hive is well maintained. A hive damaged by woodpeckers or accidents with large cracks and wood rot is not going to provide a suitable environment for the bees over winter.

Figure 3.10 Old and worn hive which has been damaged and would offer little insulation protection for the bees

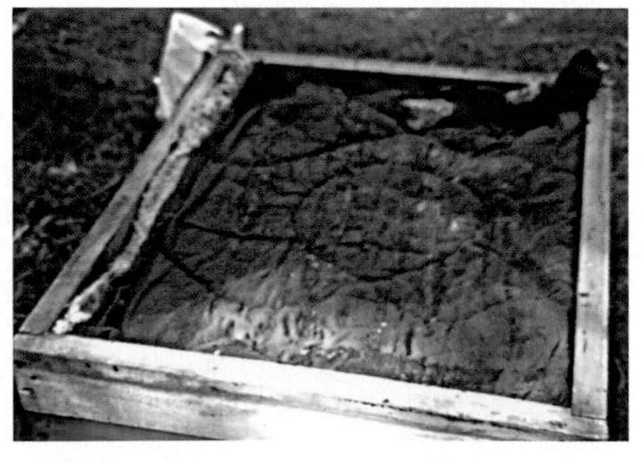

Figure 3.11 Hive insulation between the super and topbox space provided by an eke

Hot weather

Bees can tolerate quite high temperatures but the brood area must stay around 36°C. With 80,000 bees in an active hive, the hive can quickly become too hot for the health of the brood to be maintained. Bees must then cool the hive. Bees will collect water and place drops of water around the inside of the hive to increase evaporative cooling. Note if the weather is hot and humid this effect can be marginal. Bees will then stand in a line on the entrance and fan their wings to create a cooling breeze inside the hive. The nurse bees help to keep the inner temperature correct by fanning the frames and the cells.

Figure 3.12 Bees fanning the entrance

Bees will leave the hive to help reduce the temperature and this may be seen as large number of bees loitering around the front of the hive – but not fighting. This may even look like a swarming activity. This is called 'bearding' and generally starts with the air temperature around 38°C. Thus, this behaviour is normally in the summer while swarming is in the spring. Also bearding tends to be seen in the late afternoon while swarming generally occurs between 10:00 and 14:00.

Figure 3.13 Bearding of a hive getting too hot

Figure 3.14 Bees moving outside the hive to avoid over heating the hive

If your bees are getting too hot

- Reduce the number of bees by splitting the colony. Remove a frame and replace with a dummy frame thus providing the bees with a little more room.
- Paint the hive box white or some other reflective colour.
- Consider where you place the hive. It might need the morning sun but provide shade in the afternoon.
- Provide 2 openings.
- Provide an inner insulation area as provided in the WBC hive for example.
- Ensure the frames are placed perpendicular to the opening to increase ventilation.
- Ensure a good water source is within 0.5km of the hive or even provide local water sources.
- Provide water in water bottles next to the hive.

Food shortage

The hive may start starving very rapidly. At the peak there may be 80,000 mouths to feed. Possibly starvation and shortage may be of a more chronic insidious nature.

- Food can be physically short which quite often occurs in the middle of summer, as there is a general shortage of flowers.
- Food can be unavailable because the nectar is washed away or is dilute through too much rainfall. Or flower production fails because of a drought.
- Food can crystallise too quickly as honey. This may happen with Rape nectar.
- Food can be poor quality but in large quantities. This can occur when certain plants pollen is low in protein or the amino acid ratio is inadequate, as may happen with Sunflower or Lucerne.
- Food may be short because the stores are too far for the bees to travel which may be in the hive during a hard winter. The stores in the outer reaches of the hive become unobtainable because of the cold.
- Food may be short because the pollen has soiled with mould or the honey was not dehydrated enough allowing for yeast to start fermentation.
- Food may be short because of inadequate numbers of foraging workers as may happen in a queen-less hive, a drone laying queen, a poisoned hive or in cases of Colony Collapse Disorder.

How infectious agents are spread

To control infectious disease it is helpful to understand how pathogens are disseminated and gain access to the bee. Each pathogenic organism has individual properties that determine how long it will survive outside the bee, how infective it is, and how easily it is transmitted.

Of these, the two greatest risks of contamination to your hive are neighbouring infected bee hives (robbing and drifting) and the introduction of pathogens through the purchased queen.

How infectious agents are spread:

- Locality – the location of your apiary and any neighbouring properties.
- Airborne transmission in aerosol droplets.
- Direct contact between infected bees, including newly purchased bees, wild bees and infected pollen, honey and wax and semen (artificial insemination).
- Mechanical spread by vehicles and equipment, particularly bee transporters.
- Spread by people, boots and clothing.
- Spread by birds, rats, mice, insects, dogs, cats and wildlife (e.g. bumblebees).
- Contaminated food (nectar and pollen) or water.
- Medicines and medicine containers.

Problems and realities of biosecurity

Bees are probably their own worst enemies. The hive is a self-contained unit, however, older bees may forget where they live and tired bees will land in unfamiliar hives and if laden with food are allowed access by the guard bees. Thus movement of pathogens between hives is natural.

Bees also practise robbing, stealing the honey from other hives. This inadvertently also 'steals' pathogens between hives. Note wasps may also rob hives particularly as the beekeeper collects the honey supers on a number of hives.

Virgin queen bees are still naturally mated in large congregation areas where hundreds of drones meet and the queen flies into the arena. She will mate with perhaps 30 different drones to ensure her spermatheca is filled with sperm. The male organs remain in the queen (the mating sign) until being removed in the hive by her attendants. Therefore 30 different hive pathogens can enter the hive with the newly mated queen. Artificial insemination is still infrequent but even this is not without its risks.

Nectar, honeydew and pollen are natural products, they are not sterile. Many organisms will enter the hive through these products. A bee from a particular hive will not be the only bee to have found the hoard of nectar; bees from several competitive hives may have visited and left pathogens behind to be 'collected' with the food resource.

Apis mellifera is not a new species and has several related species – notably *Apis cerana* and will share many of the pathogens of other species of *Apis. Bombus* (bumblebees) are even more common with many thousands of very closely related species and many of the pathogens are not species specific.

Many pathogens will simply hitch a ride on related animals – *Varroa* for example, will ride a bumblebee and be deposited close to a honey bee hive and the cycle starts again. This is called phoretic transfer.

Many would argue what is the point of biosecurity?

Practical biosecurity

While there are innate issues with having a biosecure bee hive it is still useful to attempt a degree of biosecurity to protect your investment. Even reducing the level of infestation or infection increases your hive productivity. Avoiding European foulbrood or American foulbrood could save your hives and investment from total destruction by the local government control scheme.

Locality

Mapping programmes such as Google Earth are extremely useful to evaluate the risk of pathogen spread. Note all the apiaries within your location. Ideally one hive should work 4 to 12 hectares, but it all depends on the amount of flowers in the locality. Note if you take the hive to pollinate certain crops: heather, almonds, sunflower or rape for example, you may not be the only beekeeper doing so. Register your hive with the local bee inspector so they can monitor the spread of pathogens within an area.

Bee hives should be placed more than three metres apart to reduce drifting and/or face in opposite directions.

Older bees will be more forgetful and have a habit of drifting to the first hive available. The older bee may also be carrying more pathogenic organisms including *Varroa* mites.

When you have multiple hives make it as easy as possible for the bee to recognise their own hive. Have the entrance different coloured. Point the hives in different directions so the bees have to orientate themselves differently before entering the hive.

Bee to bee contact

Infectious agents enter and leave the bee as shown below.

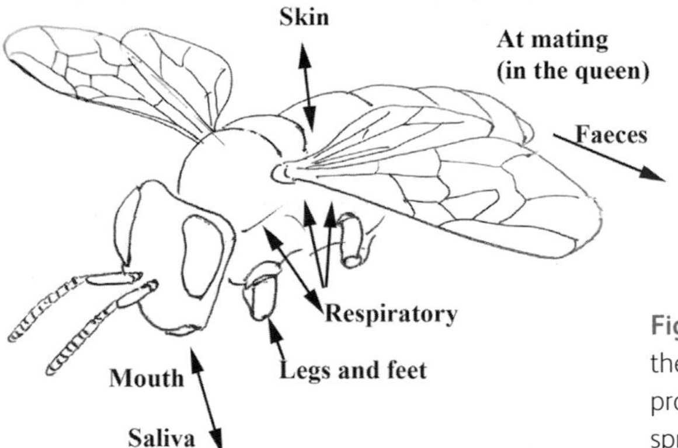

Figure 3.15 Organisms may enter or leave the bee by the arrows shown. The arrows provide general direction of pathogen spread.

The common methods are by mouth (ingestion) or by inhalation/respiration. In general inhalation requires smaller doses of the organism than ingestion to set up an infection and produce disease because the action of the ventriculus and the normal bacterial flora of the intestines inhibit the multiplication of the organisms, or microbiota. Organisms can gain access to the bee through the skin straight into the haemolymph via secondary agents such as *Varroa*.

Whether clinical disease develops or not is dependent upon a number of factors including:

- The resistance of the bee to the pathogen.
- The survival of the organism outside the bee.

- The virulence of the organism.
- The numbers of the organisms to which the bee is exposed.

The queen and her entourage is the obvious deliberate new bee introduction. Queen bees are transported all over the world with only minimal understanding of bee pathogens.

Semen/artificial insemination (AI)

Viruses could be transferred by semen during the mating flight or even accidentally through AI. A queen can mate with 30 drones during her maiden flight and they all could have come from different hives. Beekeepers selling queens need to be particularly vigilant regarding the potential pathogen transfer risk.

Spread of pathogen may be within or between hives and ultimately between countries

The ways in which pathogens spread within hives in a single apiary are similar to those by which they spread between apiaries except that the relative importance of each is different. The spread of pathogens within a single apiary: equipment (hive tools), robbing and drifting, rodents, insects, and airborne dust and aerosol are more important.

Biosecurity at a country and state level is taken very seriously, especially by island nations. New Zealand is free of European foulbrood. Australia is free of *Varroa*. Most of the world is free of Large Hive Beetles. The control of foreign bee species is also an active programme. Australia wishes to remain free of *Apis cerana* and *Braula coeca*. Northern Europe is trying to control the spread of the Asian Wasp (hornet).

Wild bees

Not all bees are kept by beekeepers: *Apis mellifera* is quite capable of living on its own in the "wild". Many pathogenic organisms can infect other bee species. Some may only use the

Figure 3.16 Empty monitoring hive in Australian docks

Figure 3.17 *Apis cerana* in a cupboard

other species as a carrier. For example *Varroa* mites may have a ride on bumblebees. This is called phoretic spread. In countries where other *Apis* species live there may be cross-transfer of pathogens. It is even possible for *Apis cerana* and *Apis mellifera* to live in the same hive and actually assist each other's brood rearing. Although this normally only happens in an existing weakened *Apis mellifera* hive. A rare example where two queens can live in the same hive!

Bee products

Victoria in Australia and the whole of North America became infested with the small hive beetle presumably through the accidental introduction of infested bee products.

All bee products should be considered with suspicion. The massive reuse of wax as foundation material is a particular concern regarding the spread of viruses as the temperature to melt wax is not sufficient to kill viruses. Perhaps wax should be pasteurised. Bringing honey from other sources and then allowing the bees to have access to this is a classic method of pathogen transfer. Note the honey jar proudly brought in from abroad, left sitting on the kitchen window can be 'robbed' by native bees.

Bee products which should be viewed with extreme caution would be wax, honey, royal jelly and pollen. Note that wax is often sold straight from the hive, with little preparation. It is vital that such products are kept out of the reach of your own bees. Wax is often only heated to a low heat, not sufficient to kill viruses.

Figure 3.18 New super frame made from purchased bees wax

Figure 3.19 Whole comb being shown in a country show

Do not be tempted by cheap, second hand frames or even free ones from friends. Even moving frames between hives within an apiary to re-invigorate a hive is going to move all the pathogens from one hive to the hive already in trouble.

Old wax and hive debris should not just be thrown on the ground around the beehive. This attracts wax moths, wasps and other bees to your hive. Remove all materials and locally burn. Leaving wax around the hive attracts wax moths to the hive's location.

A serious risk of spreading pathogens is with abandoned bee equipment. This should be carefully and thoughtfully disposed. Even if you have lost interest in your bees, you should still care about their overall well-being.

Figure 3.20 Older comb with honey going to be moved into a new hive – this should be avoided

Figure 3.21 Wax removed while examining the hive just thrown on the ground by the hive

Figure 3.22 Hive debris should be carefully disposed of and not just thrown into the local hedge. Note the chalkbrood

Figure 3.23 Ideally burn all hive debris. This will kill any *Varroa* which have fallen or been pushed out of the hive. Use a small incinerator

Figure 3.24 Equipment which is no longer going to be used should be carefully disposed and not left to rot

Figure 3.25 A controlled fire is an excellent method of disposing of old hives and frames

Mechanical Spread by Vehicles and Equipment

Figure 3.26 Truck loaded with bee hives ready to move to pastures new

Figure 3.27 Bees working almond flowers

Thousands of hives are moved all over the world to assist in the pollination of some of the major feed crops on this planet. In the USA it takes 1.5 million hives to pollinate the annual almond crop, that's some 200,000 million bees. These bees come from all over the USA and then go back home again – after sharing all the pathogens in the USA. It is perhaps fortunate that *Apis mellifera* is not a native species of the USA.

Cleaning the truck

As a guide:

- Remove all equipment and partitions from the vehicle.
- Completely remove all animal materials including wax or dead bees.
- Soak the internal surfaces in water and detergent for at least 1/2 hour at a temperature of 50°C.
- Pressure wash with hot water or use a steam cleaner.
- Pressure wash the exterior of the vehicle.
- Examine the efficiency of the above procedures. Repeat if there is visual contamination.
- Finally spray both internal and external surfaces with a non corrosive rapidly acting approved disinfectant at recommended levels.
- Document the cleaning procedure.
- If the vehicle is to transport nucleus bees it must stand empty for 12 hours at least after disinfection.

Problems can arise in countries during the winter when temperatures remain below freezing for long periods. Under such circumstances facilities must be made available to carry out efficient washing and disinfectant procedures and holding the vehicle in an equitable environ ment not subjected to freezing.

Equipment

Each hive should be numbered and ideally coloured.

Each hive should have its own equipment and this should be clearly marked. This is impractical in many large apiaries where hive tools and capping tools are moved from one hive to the next. As a basic minimum there should be several tool sets and between hives these should be cleaned and placed in alcohol to attempt to sterilise/disinfect their surfaces. Tools

Figure 3.28 Hives should be numbered and coloured

Figure 3.29 Each hive should have its own hive tools

Figure 3.30 Each hive should have its own capping fork as this is particularly likely to become contaminated by viruses and bacteria

Figure 3.31 Combining old frames into other hives should be avoided. Adding new frames provides more room but does not reduce the pathogen load on the hive

should be wiped clean of all hive materials including propolis before being used on another hive.

At an absolute minimum, hive tools should not be shared between apiaries. This should include training days.

Spread by people, boots and clothing

The role of people as a mechanical means of transmitting disease has evoked considerable debate over the years.

Recommendations for the control of people:

- No visitors should visit the apiary at any time unless wearing boots and coveralls provided by the beekeeper.
- Keep outside footwear outside the hive area.

Figure 3.32 Protective suits should be regularly washed ideally boil washed.

Figure 3.33 Boots should be cleaned before each visit. A foot bath can be advised – but note that the smell of the disinfectant may upset the bees

- No clothing used near the bees should be used off the apiary.
- Foot dips should be used for entry to the apiary and be properly maintained with a rapid acting disinfectant.
- No equipment having had contact with other bees should be allowed on the apiary unless cleaned, disinfected, frozen and/or fumigated followed by a gap of at least seven days.
- Staff working on commercial apiaries should not own other bees nor visit other bee apiaries.
- No wax products should be brought onto the apiary without significant forethought.
- No other bee products should be brought onto the apiary.
- Hands should always be washed before and after meals (to control zoonotic diseases) and after going to the toilet.

Figure 3.34 Well-worn clothing covered in bee faeces

Transmission by mice

Mice will try and get into the hive to steal the honey and other food stuffs. They can wreck the hive, eating the wax comb and its contents. The mice may also be killed by the bees but they will be unable to get the mouse out of the hive. Thus it is found later as a mummified decaying mass. Mice will visit multiple hives and therefore will act as a carrier of pathogens to the hive.

Reduce the entrance by the use of mouse guard or to reduce robbing reduce further even to a single bee entry space so the guard bees have more chance to monitor who is returning to the hive.

Transmission by flies and other insects

Fortunately most pathogens are fairly specific in their host species. Viruses from Aphids may be transmitted to honey bees through honeydew nectar but most currently recognised aphid viruses have created no specific clinical signs. However, the potential is there and the hive should be made difficult for other insects to enter. Bees do a great job of keeping most insects out by sealing holes using propolis but the beekeeper can still help by making it more difficult for other insects to gain entrance. This is primarily by managing the hive stand. Leaves and grass should not be allowed to grow up around the hive stand and provide easy routes for insects, particularly ants into the hive.

Figure 3.35 Easy access to the hive for ants using leaves

Figure 3.36 Even when the stand area is cleaned; dried plant stems can undo all the great work

Figure 3.37 Hives partially abandoned in the summer surrounded by vegetation

Figure 3.38 Well loved beehives with carefully manicured hive stands

Domesticated/farm animals

There are few pathogens that domestic animals are going to transmit to bees. However they may find the hive an interesting scratching post. This will disturb the bees and may damage or even destroy the hive.

Understanding cleaning

Many beehives and bee equipment are made from wood. While this material is fantastic in many environmental aspects of raising bees, it is extremely difficult to effectively clean. Polystyrene and plastic hives are being developed but bees have to like them as well.

Disinfectants are substances that kill both harmless and disease producing organisms. They act either as bacterial poisons, coagulate bacterial protein or act as oxidising or reducing agents. Note many disinfectants have a particular smell and you need to observe the bees behaviour with the disinfectant. If the bees exhibit an unusually aggressive behaviour you may need to change brand or type of disinfectant.

Disinfectants have two prime functions. First to prevent infectious agents gaining access to the apiary and second equally important to control those organisms already on the apiary and that persist in large numbers in the environment. The process of disinfection can be considered in three stages.

- The removal of the gross contamination within the building – that is, the dried faeces, slurry and dust – by pressure washing, preferably with hot water.
- The use of detergents to assist in the final removal of the organic material. Only after completion of these two should a disinfectant be applied.
- The use of the disinfectant.

Remember that the cheapest methods of disinfection are the physical removal of contaminated material and final removal by water.

Water + detergent is >90% efficient in removing pathogens from a hive.

Detergents

These are cleansing agents which have good wetting powers and the properties of penetrating surfaces. Detergents can either be acid, alkaline or neutral. The neutral ones tend to be those such as soaps and liquids that are mainly used to remove soiled materials. They are often combined with the disinfectant to provide a dual action.

Properties of a good detergent

- It should be efficient in removing organic material.
- Acts well in pitted surfaces.
- Has a good degreasing action.
- Good penetration.
- Quick acting.
- Works in the presence of the disinfectant.
- No residues are left.
- Active with hard water.
- Non toxic.
- Does not make surfaces slippery.
- No toxic residues.

Disinfectants

There is a bewildering array of disinfectants available and these are generally underused by beekeepers. Each of the disinfectants comes with attractive claims made as to their effectiveness. To make the best cost effective decision a number of questions should be asked.

Key points to consider when selecting a disinfectant

- Always use a disinfectant that has been independently proven and has been shown to be effective against a wide range of pathogens but particularly those that are present on your apiary. In many countries there are lists of approved disinfectants. Ask your veterinarian or supplier.
- Dilution rate – always read carefully the instructions as to use and in particular the amounts to be added to water for general disinfectant purposes. Check also the amounts required for the highly infectious diseases such as American or European foulbrood.
- From the cost of the concentrated disinfectant work out the cost of the diluted chemical. A more expensive disinfectant with a dilution rate of 1:300 may be a better buy than a lower priced disinfectant with a dilution rate of 1:100.
- Time to act – there is always a minimum time before the disinfectant has killed microorganisms. Since most disinfectants are used at low temperatures always look at the killing time relative to this.
- Effectiveness in the presence of organic matter – this is important when using disinfectants on the apiary, because invariably under such conditions they are going to come into contact with large amounts of organic matter. Some disinfectants such as chlorine are very quickly neutralised in the present of such materials.
- Penetration – it is very important that the disinfectant has the ability to penetrate organic matter (detergency). In most cases however it is preferable to apply a detergent cleaner prior to the use of the disinfectant for more effective use.

Each group of disinfectants have their own special properties and an understanding of these will help you in your selection.

Table 3.14 The characteristics of the different disinfectant chemicals

	Chlorine based	Peroxygen compounds	Phenols unchlorinated	Phenols chlorinated	Iodophors	QAC compounds
Can be used in aerosols	A Few	Yes	No	A Few	Yes	Yes
Corrosive to metal/rubber	No	No	Yes	Yes	No	No
Detergent action	No	Yes	No	Some	Yes	Yes
Effectiveness in presence of organic matter	Moderate	Yes	Yes	Yes	Moderate	No
Good action against bacteria	Moderate	Yes	Yes	Yes	Yes	Moderate
Good action against viruses	Yes	Yes	Poor	Poor	Yes	No

	Chlorine based	Peroxygen compounds	Phenols unchlorinated	Phenols chlorinated	Iodophors	QAC compounds
Persistent residues	No	No	Yes	No	Poor	Yes
Speed of action	Quick	Quick	Moderate	Moderate	Quick	Moderate
Staining	Some	No	Yes	Yes	Some	No
Suitable for foot baths	No	Yes	Yes	No	Yes	No
Toxic or irritant	Yes	No	Yes	Yes	Some	No

The ideal disinfectant should be:

- Active quickly against a wide range of viruses, bacteria and fungi.
- Safe to handle.
- Active in the presence of dust or organic matter.
- Have a long period of activity.
- Non-irritant, non-staining, non-toxic and non-corrosive.
- Combined with a detergent or have such properties.
- Capable of use as an aerosol.
- Safe and effective when used in water systems.
- Capable of use through pressure washers.
- Coloured.
- Leave no toxic residue for the bees.
- Leave no residual smell which might discourage the bees.
- Leave no residue which may enter the honey.

There are six classes of chemicals used for disinfection:

- Phenols
- Chlorine based compounds
- Iodine based compounds
- Quaternary ammonia substances (QACs)
- Aldehydes
- Peroxygen formulations.

Phenols

These are organic compounds that may or may not be combined with chlorine. They are usually effective in the presence of organic matter but do not normally have a high detergent action. They are not however corrosive to metal but they can cause damage to plastic and rubber compounds. Their action is moderately slow. Tar acids may be combined with other organic acids such as acetic and sulphuric acids to increase efficiency to improve their effects against viruses.

Key facts about phenolic based disinfectants include:

- They are active in the present of organic matter.
- Their activity persists for a long period of time.
- They are ideal for vehicle dips and concrete floors.
- They have no detergent activity.
- Their rate of activity is slow: two to twelve hours.
- They can be toxic and damage tissues.
- They are usually very effective against bacteria but not so good against viruses or spore producing bacteria.
- They are usually quite cheap.
- Some phenols such as chlorxylenols contain chlorine which adds properties of quick action.
- They taint honey.

Chlorine based compounds

The chlorine based compounds can be considered in two groups – those without organic compounds such as the hypochlorites which depend on the liberation of chlorine for their disinfection action and those that contain organic substances. Chlorine disinfectants have a very quick action but are very quickly neutralised in the presence of dirt or organic matter.

Key facts about the chlorine based disinfectants are:

- They can be very corrosive.
- They have a very quick action.
- They are inactivated by organic matter and hard waters.
- They do not persist for very long periods of time.
- They have no detergent activity.
- They are very active against viruses and bacteria.
- They may cause taint.
- They may be ozone unfriendly.
- They are very cheap.

Iodine based compounds

This group of disinfectants includes substances called iodophors, where the iodine is dissolved in a surface active agent and then phosphoric acid is added. Iodine substances are very safe and have low toxicity with almost no smell. When phosphoric acid however is added to the iodophors the disinfectant becomes a little more irritant and corrosive.

Key factors about iodine disinfectants (iodophors) include:

- Usually they have a high detergent activity.
- They are ideal for foot baths.

- They are brown in colour when very active becoming straw coloured when losing their activity (used in foot baths for this reason).
- They are very quick in action.
- They are very effective against viruses and bacteria.
- They are moderately active in the presence of organic matter.
- They tend to be more expensive.

Quaternary ammonia compounds (QACs)

These compounds may be used for cleaning and sterilising water systems and equipment and they are very efficient, particularly if the organic matter has been removed. They are not usually suitable for the disinfection of premises on their own because of the large amounts of organic materials present that immediately neutralise them. QACs are not compatible with soaps and they should not be mixed with other detergents. Some are used as antiseptics. They are more active against Gram positive bacterial organisms.

Key factors about (QACs):

- They usually have little or no effect against fungi and bacterial spores.
- Inactive in the presence of organic matter.
- Inactivated by soaps and disinfectants.
- No activity against viruses.
- Suitable for cleaning water systems and smooth surfaces.

Aldehydes

These substances such as formaldehyde, are very toxic but are good disinfectants in aerosol form. There are now alternative products available that are equally effective and much safer to use.

Peroxygen compounds

These are broad spectrum disinfectants that are highly active against most micro-organisms. They are based on combinations of peroxyacetic acids or other derivatives, hydrogen peroxide, organic acids and anionic detergents. They are powerful oxygenating agents.

Precautions to be taken when using disinfectants

Always:

- Follow the manufacturers' instructions carefully.
- Wear gloves and eye protectors when handling the concentrate.
- Wash concentrate off the skin immediately.
- Ensure the dilution is correct for the purpose being used.
- If there is contact with eyes wash immediately with copious amounts of water and seek medical help.
- Where foot baths are used ensure that these are cleaned and replenished regularly.
- Store in original container, tightly enclosed.
- Keep away from children.

Fumigation as the ultimate in disinfection

Guidelines for fumigation – using potassium permanganate and formalin

Formaldehyde is a noxious and highly toxic gas and exposure for only a short period can cause respiratory distress and with severe exposure ultimately coma. (Its use is forbidden in some countries).

Whenever fumigation is undertaken:

- There should be two people available.
- Place the bee hive and all bee equipment in a small space/room.
- One person should be in the room and one at the door to provide assistance if required.
- Wear protective goggles and a dampened face mask – these should be available to both persons.
- Use rubber gloves.
- Wet the hive before fumigating and seal any openings in the building.
- Always add formaldehyde to the potassium permanganate. Have the exact quantities to be added ready in separate containers and add slowly.
- Use high-sided metal containers for mixing the compounds. Once the compounds are mixed vacate the house immediately.
- Ensure no livestock will be exposed to fumes that might escape.
- If the person in the room gets into difficulty, the person at the door should open all doors and immediately pull the person out.
- Distribute the metal containers evenly through the room.
- Add 1000ml formaldehyde (40%) to 400g of potassium permanganate per 1000m^3 of air space.
- Leave the building/room shut for 12 hours.
- Open up and ventilate eight hours before use.
- Place a notice on the door warning people 'Fumigation in process'.

Oxygenating disinfectants sprayed as a fine mist can be used as an alternative to fumigation with formalin. They are much safer and easier to handle.

Washing hands

Disinfection of hands should always take place after handling bees and bee products, particularly where there has been faeces or dead bee contact. This is not least for personal hygiene reasons. The importance of washing hands after handling diseased bees can reduce the risk of spread of pathogens.

What should you use on the apiary

This would obviously depend on availability but the following should be considered:

- Foot baths – use an iodine based one.
- General disinfection of hives – use a phenol or organic acid based one.
- Water – use a QAC or chlorine based one.

- Concrete surfaces – use a phenol or organic acid based one.
- Virus infections – use iodophors or peroxygen complexes (formalin fumigation is also effective).
- Bacterial problems – use iodophors or peroxygen complexes.
- Hands – use QAC compounds or soaps.
- Trucks – use a government approved disinfectant that is highly active against the major notifiable and transmissible diseases in your country.
- Aerosols – use formalin, chlorine, iodine or oxidising agents, preferably the latter.

Heating

Heating the hive materials can be useful to control Nosema and other fungal infections. The hive materials need to be heated to over 50°C for 24 hours. This will kill some viruses especially the RNA viruses. DNA viruses may require more radical treatments.

Flaming

Because most hives are made of wood flaming the wooden surface is a useful tool when cleaning hive materials. However, it will only sterilise the surface of the wood, it will not penetrate into the tissues. Also take care not to burn too much and set the hive on fire. Ensure that you remove all plastic runners on frames to avoid melting the plastic with the blowtorch.

Freezing

Many of the pathogens of bees are other insects – wax moths for example. The hive materials can be placed in a freezer (–15°C) and left for 7 days to kill all insects and their eggs. Note this will not destroy bacterial or viral pathogens.

Possible sources of pathogen entry into your apiary and what you should consider in order to prevent it.

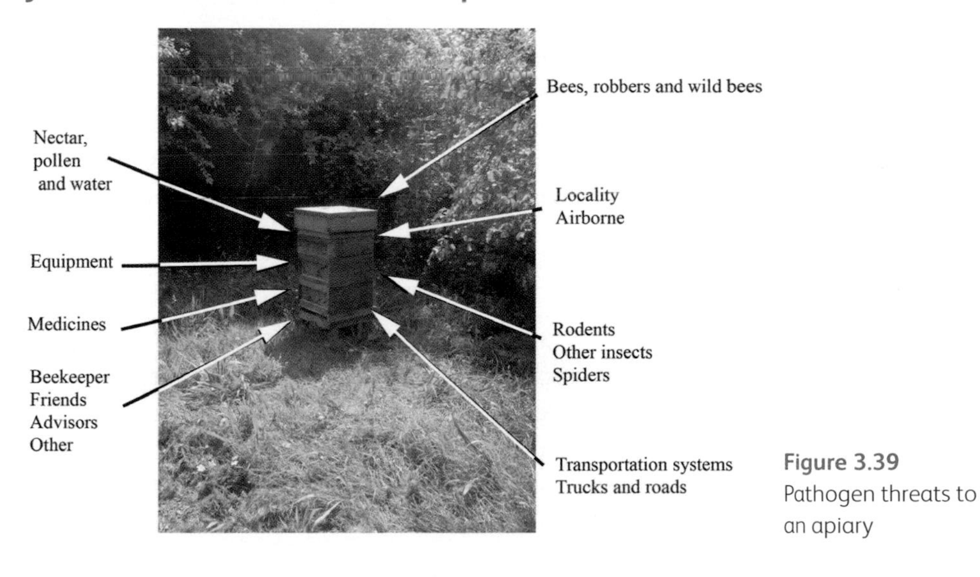

Figure 3.39 Pathogen threats to an apiary

Before the new apiary is populated with new bees:

- Is your apiary in a bee dense area?
- Is your apiary going to be neighbourhood friendly?
- What is the position of your bee unit relative to other 'infected' bee hives around it?
- How far away is the nearest infected large commercial apiary?
- Do you have an uninterrupted view of it?
- Is the land around you flat or hilly, bare or wooded, on the coast or inland?
- Can you rely on a future supply of queens of the same health status?
- What precautions can you take against contamination?
- Is there easy access to water?

When the apiary is being populated and afterwards – biosecurity considerations:

- Age of hive and other hives in location. Have new hives been introduced recently?
- Does the apiary practise migration pollination?
- Has a new queen been purchased and from where?
- Has a swarmed hive been collected and introduced recently?
- Location of hive in relation to other hives – Google Earth can be useful in this manner.
- Location of hive in relation to public path.
- Location of hive in relation to access by livestock.
- Note general appearance.
- Water source.
- Available food sources – note some food like oil seed rape (canola) may need frequent collection of honey.
- Security measures to stop pests and people – electric fencing.
- If electric fencing used are the wires clear of vegetation?
- Is the hive placed on a secure hive stand?
- Is hive stand clear of vegetation?
- Is hive stand prepared to stop ants and other insects from entering the hive?
- Note source of bee equipment – note source of wax for frame foundation.
- Have other beekeepers been recently – especially with their own equipment?
- Hygiene of bee examination equipment. Sterilise hive tools and frame scrapes after cleaning by soaking in 0.5% sodium hypochlorite for 20 minutes. Do not allow other beekeepers to bring their own equipment.
- Wear disposable gloves.
- Provide boots for all visitors.

Biosecurity introduction of new stock – the new queen and attendants:

- Keep good records and a calendar of events.
- Enquire about the health of the area where the new queen is coming from.
- Enquire if the source provides specific health information regarding their queen bees.
- When bees arrive keep them well separate from your hives and any bee equipment.

- Examine bees in detail – looking for any parasitic conditions (Varroasis (*Varroa destructor*) in particular) and for any deformities especially in wing structure.
- Sacrifice some of the workers and postmortem. Dissect bees and look for Acarina (*Acarapis woodi*), and test for European foulbrood (*Melissococcus plutonius)* and American foulbrood (*Paenibacillus larvae*) by lateral flow. Macerate bees and examine for Nosema (*Nosema apis* and *Nosema ceranae*).
- Calculate the crude protein content of the worker bees as an indication of general health.
- Set up new hive as far as possible away from your current established apiaries.
- After one month, examine hive in detail. Review health of new hive. If no problems are noted, relocate hive to your apiary.

Swarms

General rule: do not capture and use swarms of unknown health. They can prove to be a serious risk to your own apiary.

However, this can be harsh and beekeepers have a duty to protect the general public from the harassment caused by swarms. From a neighbourly view, capture the swarm, and then isolate and quarantine the bees. Monitor their health using separate equipment including clothing. Wear gloves when handling equipment and the bees. Ideally have a set of equipment designed for swarm control. Wash your hands, clothing, boots and equipment.

If the swarm hive has significant pathogens which you cannot eliminate, humanely destroy the bees. Note that the brood pathogens are left behind by the swarming bees.

Figure 3.40 Swarm on a tree

Figure 3.41 An unusual position for a new hive – inside the barbeque! Note the position of the capped brood

Reducing the pathogen load in a hive

There are a number of simple methods used to reduce the pathogen load in a hive. Each is described in more detail in Chapter 12.

Brief descriptions of three popular techniques are provided here:

Drone trapping

Drones are placed in the cooler parts of the hive. If a super frame is used in the brood box the worker bees will add freeform cells underneath the super frame and fill it with drones. These drones will attract *Varroa* mites. Once the cells are capped (at least 11 days post egg laid) the whole drone section can be neatly removed with a sharp knife and the drones and trapped developing *Varroa* destroyed by incineration.

Shook swarm technique

When the hive swarms it leaves behind the developing brood and thus most of the brood pathogens. A deliberate method utilising a new hive and new comb can be made to artificially move the queen and shake the nurse bees into a 'new' hive. The old brood can then be destroyed by incineration or freezing if you wish to save the wooden frames. It is generally advisable to destroy the brood and wax by incineration.

Genetic selection for good cleansing behaviour

Bees with a high hygienic behaviour appear to be able to protect themselves against a range of pathogens including *Varroa* and Chalkbrood.

Utilising a hygienic behaviour examination can be a significant step to controlling Colony Collapse Disorder and should be strenuously considered by all professional queen bee breeders.

It is possible to examine a hive/queen for hygienic behaviour with a 24 hour test (see Chapter 12).

Workshop

Many beekeepers have an active workshop. Ensure this not a source of pathogens. Many workshops can be extremely poorly organised and dirty.

Figure 3.42 Workshop

Figure 3.43 Ensure unused equipment is stored clean

Figure 3.44 Place all used equipment into a freezer for at least a day to kill arthropod pests

Killing bees

Individual/small groups of bees should be placed in the refrigerator for 20 minutes before euthanasia.

Unfortunately from time to time it will be required to kill the hive so follow this procedure:

1 Obtain 4.5 litres of water and add 250 mls of washing up-liquid.
2 Allow the bees to enter the hive
3 Seal the entrance
4 Apply the strong washing-up liquid by a hand held mister or pressure washer
5 Destroy the frames
6 Sterilise the brood box
7 Freeze the brood box for at least 7 days.

In emergency situations, it may be necessary to destroy the hive and all its surroundings either by burning or burying.

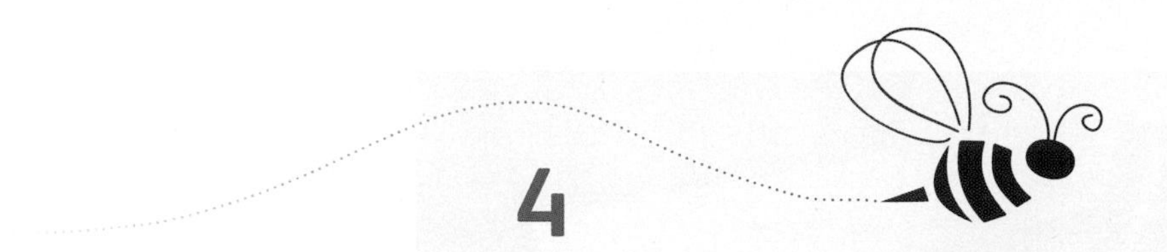

4

Treating sick and compromised hives

You are advised to consult your veterinarian and bee inspector when assessing information given in this chapter.

Some medicines are banned in various parts of the world. Ensure that your treatment plans are legal in your area.

In an ideal world the hive would never get sick. But bees and hives are living environments and are also seen as opportunities for other pathogens. Considering there can be 80,000 bees in an *Apis mellifera* hive in the middle of summer, it is amazing that so few bee diseases are recognised.

Bee treatments should only be undertaken after a proper diagnosis is made or if preventative, careful considerations of the undertaking has been made. Discuss any health issue with the local bee inspector or your veterinarian.

Being a member of a local bee club and attending workshops and meetings is a great way to learn about bees and bee diseases, but beekeepers also have a duty to protect the excellent positive properties of bees and their products for human beings.

The treatment of sick and compromised bees in the modern apiary can be complex, with a wide range of medicines available for a variety of conditions. This chapter looks at some of the complexities and interactions involved in the use of medicines so that treatments can be carried out efficiently and without risk.

Specific treatments for diseases are discussed later in their relevant chapters.

Legal requirements

Remember to keep medicine records of all treatments administered to the hive for five years. This applies even if you get rid of the bees. If you move the bees to another person, copy all their previous five year medical history and give these records to the new owners. But also keep a copy for yourself.

Based on EU directives.

Medicines must be used safely and correctly in food producing animals (and this includes honey bees) to ensure there are no residues in their products. Most countries across the world have strict controls over both the methods of prescribing and the uses of medicines. Consult with your veterinary advisors to understand the current situation in your own country.

The European Union rules adapted for the UK will be used as an example for how medicines are characterised.

In the UK there are currently four categories of medicines:

1. AVM-GSL Authorised veterinary medicine – general sales list.
   This may be sold by anyone (formerly GSL).
2. NFA-VPS Non-food animal medicine – veterinarian, pharmacist, Suitably Qualified Person (SQP).
   A medicine for companion animals which must be supplied by a veterinarian, pharmacist or Suitably Qualified Person (formerly PML companion animal products and a few P products).
3. POM-VPS Prescription-only medicine – veterinarian, pharmacist, Suitably Qualified Person.
   A medicine for food-producing animals to be supplied only on veterinary prescription, which must be prescribed by a veterinarian, pharmacist or SQP (either orally or in writing) and which must be supplied by one of those groups of people in accordance with the prescription (formerly PML livestock products, MFSX products and a few P products).
4. POM-V Prescription only medicine – veterinarian. A medicine to be supplied only on veterinary prescription, which must be prescribed (either orally or in writing) by a veterinarian to animals under his care following a clinical assessment, and which may be supplied by a veterinarian or pharmacist in accordance with the prescription (formerly POM products and a few P products).

When a medicine is supplied to the apiary the following information should be available:

- A description of the medicine
- The date of manufacture
- The date of dispensing
- The date of expiry
- The client's name and address
- The species to be treated
- The date of withdrawal
- The dose rate and instructions for use
- Name and address of the supplier
- Manufacturer's batch number
- The name and address of the veterinarian prescribing

A typical bottle label would appear as shown in Table 4.01.

Table 4.1 An example of a medicine label

FOR ANIMAL TREATMENT ONLY		
Terramycin powder Dosage: Tetracycline is administered at 200mg per colony administered every 4th day on 3 occasions. Warnings: The withdrawal period is 6 weeks. Do not used on hives producing honey for human consumption for 6 weeks.	Mr Jones, Apiary Date 4092014 Hive 1 Brood disorder Administer: Dr John Carr Carrsconsulting. com UK	Store at room temperature not exceeding 25 °C. Protect from light. Do not dilute. Keep out of the reach of children. Once a box has been broken open the contents should be used within 4 weeks. Return unused material. Wash hands after use. Avoid contact with eyes. For use: see leaflet: PL: 1596/4168A Pharmacy for bees company Ltd. Lot: 5511-03 Exp: Dec 982015

How medicines are prescribed

Most medicines have two names, one which describes the chemical which is the active principle, often referred to as the *generic name*, and the second, the manufacturer's own *trade name*. For example, oxytetracycline hydrochloride (OTC) is the generic name for a broad spectrum antibiotic. Terramycin is the trade name given to it by its pharmaceutical company. Other trade names for the same medicine include, tetramin, duphacycline, engemycin and so on. These vary from country to country. Throughout this book the generic names are used with references to some commonly used trade names.

You may be familiar with the trade names of medicines used locally but also try to remember the generic names, because this will help you to understand how they function and how to identify them irrespective of a trade name.

Understanding dosage levels

All medicines have a recommended therapeutic range per hive. This range is used by the veterinarian so that they can decide whether a higher or lower dose level is required.

Table 4.2 Dosage terminology

1000ηg (nanograms)	= 1μg (microgram) also mcg
1000μg	= 1mg (milligram)
1000mg	= 1g (gram)
1000g	= 1kg (kilogram)
1000kg	= 1T (tonne)

mg/kg = g/tonne µg/kg	= ppm (parts per million) = ppb (parts per billion)
mg/kg × 0.0001	= %
ppm × 0.0001	= %
1000µl (microlitres)	= 1cc (cubic centimetre) or 1ml (millilitre)
1000ml	= 1 litre

For conversions to imperial and other measurements see Chapter 12.

Controlling and storing medicines

Table 4.3 Storage of medicines

Medicines that should be stored in the refrigerator 2–8°C	Medicines that should be stored in a dark, cool place 18–25°C*
Any bottles that have been opened and are in use	Vitamins and minerals
Any other medicines where the label indicates this temperature requirement	Antibiotics
Vaccines (when they become available)	Disinfectants
	Other chemicals

*A dark cool place is not on a shelf at room temperature. In many parts of the world medicines would be too hot in the summer months. Pay particular attention to the temperature requirements of all medicine products.

To achieve the maximum response to medicines and prevent any abuses, discipline should be maintained in their control, administration and storage. Consider all medicines to be dangerous. Many become potentially toxic if the recommended levels of treatment are exceeded or if they are given in the wrong way.

Some medicines have a very narrow range between treatment levels and poisoning. Many medicines have no recommendation for the use in bees. In particular, in bumblebee species there will be little or no research undertaken to determine dose rate, efficacy or safety.

Light and heat destroy medicines and freezing also has an adverse effect, particularly on vaccines. Tetracyclines, for example, may break down in light and produce toxic compounds, especially as products age and become out of date.

Checklist for medicines

- Provide a locked room or cupboard for all your medicines.
- Make sure that all medicines and dosing equipment are kept well away from children and people not on the staff. Note medicines should be kept away from other animals.
- Provide a refrigerator for vaccines and other medicines as required. Use a maximum/minimum thermometer and record temperatures daily.
- Allow only certain designated people to have direct access to the main medicine store.
- Document all medicines in and out of the medicine store.
- Insist on empty bottles being returned before a second bottle is taken out.

- Agree with your veterinarian the minimum amounts that are required for a given period of time and follow the advice on usage.
- Make sure that all bottles are labelled for the correct use, that withdrawal periods are displayed and personnel are aware of them.
- Record the date the medicine bottle is first opened.
- Keep a daily record of all medicines used on the apiary.
- Make sure you have safety data sheets to hand in case of accidents.
- Ask your veterinarian to check your storage and usage of medicines regularly to ensure that the recommendations are being carried out.
- Check regularly that medicines are in date.
- Dispose of empty bottles and dosing equipment safely. Discuss this with your veterinary surgeon.

A FORMAT FOR RECORDING TREATMENTS

Bee Treatments

Date commenced	Hive Identification	Condition/ disease	Medicine used/ bottle number	Dose per day (ml) Number of doses	Withdrawal period (days)	Date of clearance	Administered by
				1			
				2			
				3			
				4			
				5			
				6			
				7			

Disposing of medicines

This must be carried out with care to prevent environmental contamination and accidental human or animal contamination.

There are two ways of safe disposal:

1. Empty bottles. These should be placed into a plastic bag and disposed of within the local authority guidelines or rules. Return to the veterinary practise from which you obtained the medicines.
2. Dosing equipment must be disposed of within the local authority guidelines or rules. Return to the veterinary practise you obtained the medicines from. It may be possible to incinerate but check with your veterinary practise first.

Figure 4.1 Poor disposal of medication strips in a bee hive

Withdrawal times

All chemicals used in the bee hive must be used with extreme caution and must not pose a risk to human health. Do not medicate during a honey flow and ensure all honey, pollen and royal jelly is only collected at least 28 days after the medication has all been used up.

Compounds used to treat bees

Miticide treatments

Mites are a major parasite of both honey bees and bumblebees. They weaken a hive and if there are additional problems the hive can then succumb to serious secondary viral infections. Mites therefore need to be controlled.

In this chapter chemical control will be reviewed. Biological control is described in Chapter 12. The problem with treating mites is that the difference in physiology between an insect (a bee) and a spider (a mite) is extremely subtle and it is likely that any treatment intent on killing or eliminating the mite will do significant damage to the bee as well.

As with all medicines resistance occurs. It is useful to check if your hive's *Varroa* are demonstrating resistance.

Miticide resistance test

Establish, in the first case, that the hive has a problem. Place a miticide strip in a colony. If after 24 hours there is a mite drop rate in the hundreds, the miticide's active ingredient will be having a sufficient effect to enable its use for that season. Of course, you need to be sure that the mite drop is not being removed by the bees or birds before you count the mites.

Amitraz

This is an insecticide which works on the insect neurology and leads to over excitation of the neurological system, paralysis and then death. Amitraz is particularly effective as an acaricide (miticide). Do not use during honey flow. A commercial example of Amitraz is Apivar®. Humans react to Amitraz and the toxic effects are similar to the effects in insects.

Fenpyroximate

This is an insecticide in the family of pyrazole acaricides. Treat each hive with 225g of fenpyroximate. Do not treat during the honey flow. Note these products are also toxic to bees. A commercial example of Fenpyroximate is Hivastan® .

Formic acid

Formic acid is an organic colourless liquid acid. It is highly pungent with a penetrating odour.

Formic acid can be used on the bottom boards. This can be useful to control tracheal mite. Administer 30ml of formic acid every week for three to five applications in the spring and then again in June. Air temperature needs to be above 22°C but below 30°C. Do not apply within 30 days of honey flow.

Grease patties

Grease patties are generally placed in the hive in autumn and winter. Grease patties interfere with the ability of the mite to recognise the larva and pupa.

To make a grease patty use 1 part (by volume) of solid vegetable shorting (fat) and 2 parts (by volume) white sugar. Store the product in the freezer.

Grease patties may be used around the year. If the grease patty is combined with other products, then withdrawal times may need to be applied.

Icing sugar

Icing sugar is finely ground sugar. It must be used fresh and very dry.

Icing sugar is sprinkled over the hive and the sugar settles on the bees. In their efforts to remove the sugar, the *Varroa* mites are also removed. Note the mites will fall off the bee and if there is a solid board the mites will walk back up into the brood area. If there is a false bottom the mites will fall out of the hive. This does not mean they will not walk back into the hive, but it will take longer and if you are monitoring the hive debris the fallen mites can be removed and destroyed.

The use of icing sugar may be very useful in the control of *Varroa* in a new queen and her entourage.

Icing sugar administration can be used during honey flow.

Menthol

This is a natural alcohol organic compound. It is a waxy crystalline substance which is clear or white in colour and is a solid at room temperature.

Menthol fumes are toxic to the mites. Fume all equipment with bees. Note that it requires warmth to make the menthol crystals fume as they are solid at room temperature. Place the crystals on the crown board/top board.

The menthol needs to be removed at least 30 days before honey flow.

Organophosphates

The major organophosphate used in apiaries is the compound Coumaphos. Organophosphates are non-volatile, fat-soluble insecticides. A commercial example of 3.2% Coumaphos is Perizin®.

Trickle 10–50ml of the solution over the frames. The bees will consume the emulsion and spread it by social feeding. The organophosphate enters the haemolymph and then is taken up by the mites when they drink the haemolymph.

This must be administered to the hive at the end of the season when there are no sales of honey and few brood around.

A commercial example of 10% Coumaphos is Checkmite+®. The Checkmite+® strip hangs down between frames in the brood area so bees can walk on it and pick up minute amounts of the active ingredient. Do not put the strip on the bottom board or tops of frames. Checkmite+® can be useful for small hive beetles as well as mite control.

There is some concern that the product may create some neurological problems in honey bees.

Oxalic acid

Oxalic acid is an organic compound. It is a colourless crystalline solid which dissolves in water.

Mix 6% of oxalic acid with a 30% sugar solution and apply it by the trickling and spraying method. Note that the products have to be used fresh and breakdown products can be the toxic compound hydroxymethylfurfuraldehyde. This product is interesting as it works by damaging the proboscis of the *Varroa* mite.

It is possible to use oxalic acid crystals. By heating the crystals we get evaporation and the bees are treated by fumigation.

This is not authorised in some parts of the world – for example, the UK. Oxalic acid can kill brood and therefore has to be used in broodless colonies, in the wintertime.

Oxalic acid is toxic to people. Oxalic acid must only be handled wearing gloves and goggles. The gloves need to be acid resistant. Wear plastic protective equipment when handling the product.

If oxalic acid is ingested this is very toxic. Symptoms include burning sensations, coughing, wheezing, laryngitis and other respiratory distress. Long term, there can be kidney damage. If the oxalic acid is ingested, seek medical help immediately.

Pyrethroids

Pyrethrin are two mild natural insecticides produced by *Chrysanthemum cinerariifolium*. The pyrethrin compounds are neurotoxic. They act as a natural insect repellent for the chrysanthemum. Note these are not specific miticides.

There are a number of synthetic pyrethroids being made as they are not persistent in the environment.

Pyrethroids are non-volatile, fat-soluble compounds and thus tend to concentrate in the wax products. They are virtually absent in honey.

Human reactions may happen rarely but if you have any of these symptoms cease handling the product and seek medicine help immediately. The reactions may include:

- Irritation of skin and eyes.
- Irritability to sound or touch, abnormal facial sensation, sensation of prickling, tingling or creeping on skin, numbness.
- Headache, dizziness, nausea, vomiting, diarrhoea, excessive salivation, fatigue.
- In severe cases: fluid in the lungs and muscle twitching may develop. Seizures may occur and are more common with more toxic cyano-pyrethroids.

Start treatment when the *Varroa* drop count reveals a heavy infestation. The strips are normally applied for up to six weeks. Note that these chemicals can build up in the hive and eventually will affect the health of the bees. Two years of application has been demonstrated to affect queen bee fertility. These products may be more useful in the autumn as part of the *Varroa* seasonal control programme.

Flumethrin

This pyrethroid insecticide is widely used in treating domestic animals against insects. A commercial example of 90% flumethrin is the Bayvoral® strip.

Fluvalinate

A commercial example with 10% Tau-fluvalinate is Apistan®.

Sucrose octanoate

Sucrose octanoate is a sugar ester.

Sucrose octanoate causes death in the insect/mite by desiccation.

Make a 0.625% solution of sucrose octanoate. Spray 15–20ml of liquid on each side of the frame. Repeat applications at intervals of seven to ten days for up to three times per infestation. This takes some considerable time to administer.

Do not apply if it is windy (i.e. above 10km/hour). Do not inhale the spray and wear gloves and goggles when applying. Once the application is complete, wash your hands well with soap and water.

Thymol

Thymol is a natural monoterpene phenol extracted from the thyme plant (*Thymus vulgaris*).

Thymol may be used as a miticide. Once the temperature is above 15°C, thymol products can be used. Use two applications, ten to 14 days apart. Thymol may get into honey and therefore the treatment cannot be used during honey flow. Therefore, spring treatment is ideal. Place the thymol on the top of the brood box in the middle of frames. A commercial example containing 25% thymol is Apiguard®.

Thymol combination products – thymol, eucalyptus oil, menthol and camphor

Once the temperature is above 15°C, thymol combination products can be used. Use two applications ten to 14 days apart. Thymol combination products may get into honey and therefore the treatment cannot be used during honey flow. Therefore, spring treatment is ideal. Place the thymol combination product on the top of the brood box in the middle of the frames. As a commercial example, Api Life Var® contains 74% thymol oil, 16.4% eucalyptus oil, 3.8% menthol and 3.8% camphor in a tablet.

Table 4.4 Review of possible antimite management techniques over the year

Technique	Winter	Spring	Summer	Autumn
Open Mesh Floor sticky paper	■	■	■	■
Drone brood removal		■	■	
Drone comb trapping		■	■	
Thymol products				■
Fluvalinate/flumathrin		■		■
Coumaphos		■		
Oxalic acid	■			
Icing Sugar Dusting			■	

Insecticides

Many other insects can cause harm to the beehive. But as bees are insects as well, medication used to eliminate other insects generally would cause harm to the bees.

Coumaphos

Small hive beetles are insects like bees and chemical control can be extremely difficult. Coumaphos (Checkmite+®) may be a useful compound.

Over-wintered brood supers are an ideal environment for insect parasites to colonise the hive equipment. All hive equipment needs to be thoroughly cleaned and disinfected before storage. Once cleaned, the stored equipment can be sterilised by a variety of compounds.

Ethanoic acid (acetic acid)

Place combs into a sealed box and fumigate with vapours from 80% ethanoic acid (acetic acid). Cover any metal on the boxes with Vaseline® to protect it from the acid. Fumigate for at least ten days. Air well for ten days before use.

Freezing

Hive equipment may be placed in a freezer for one to seven days to kill off any bee insect pests. This does nothing to control virus, bacterial or fungal pathogens.

Paradichlorobenzene

This may be used on hive equipment. After treatment, stack supers outside for at least a week before use. Do not use 'moth balls' with paradichlorobenzene (PDB) as this is carcinogenic and can leave residues in wax and honey in working hives.

Control with chemicals that do not get into the honey. There must be no presence of PDB in honey sold in the EU.

Antifungal treatments

Essential oils: Thymol

Thymol products may have an antifungal action and may be helpful against chalkbrood while they probably have little effect against Nosema.

Thymol is a natural monoterpene phenol extracted from the thyme plant (*Thymus vulgaris*).

Thymol may be used as a miticide. Once the temperature is above 15°C, thymol products can be used. Use two applications ten to 14 days apart. Thymol may get into honey and therefore the treatment cannot be used during honey flow. Therefore, spring treatment is ideal. Place the thymol on the top of the brood box, in the middle of frames. A commercial example, containing 25% thymol, is Apiguard®.

Essential oils: Satureja montana

This is produced from winter savory (*Satureja montana*). The essential oil produced is a phenolic compound. The oil (similar to thymol) can be used to control fungal pathogens. Satureja oil has been recommended for control of Chalkbrood.

The oil is used in the spring. Place the satureja oil on the top of the brood box in the middle of frames. Note Chalkbrood also self-heals as the weather warms up.

Essential oils: Origanum vulgare

This is another essential oil, similar to thymol and saturega, from the oregano plant.

It is likely that many essential oils from plants have beneficial effects on hives and more research is required in this field. It should be remembered that bees have capitalised on these oils for millions of years in their use of propolis to seal and help maintain a healthy environment in the hive.

Fumigillin

Fumigillin is a antimicrobial agent isolated from *Aspergillus fumigatus*. Fumigillin is effective against Nosema, but it has no effect on the spores of Nosema and must be ingested. Each hive requires 170mg of fumigillin. It is mixed into a syrup and is placed where it can be consumed by the bees. It is applied in the autumn after the honey has been collected. A commercial example of Fumigillin is Fumidil B® but it should be noted that this product has been banned in the EU.

Antibacterial medicines and their uses

Antibacterial medicines are either produced from the fermentation of moulds (antibiotics) or they are synthesised chemically. Antibiotics are generally ineffective against viruses and fungi. They have minimal effect against the protozoans or metazoans. They are therefore of limited value in bee treatments and must be used with extreme caution.

They act in one of two ways, by either killing bacteria, in which case they are called *bactericidal*, or by inhibiting bacterial multiplication, in which case they are called *bacteriostatic.*

Bactericidal antibiotics generally act quicker than bacteriostatic ones. Bacteria however, often multiply after a primary virus infection and antibiotics are used to control these secondary infections.

Antibiotics act in one of three ways:

1 They destroy the bacterial cell wall (e.g. penicillins, cephalosporins).
2 They interfere with the protein metabolism inside the cell (e.g. oxytetracycline, chlortetracycline, streptomycin).
3 They interfere with the protein synthesis of the cell nucleus (e.g. sulphonamides).

Only three antibiotics are widely used in the treatment of pathogens in bees. This is largely because there are few bacterial infections recognised. However, two of these conditions are reportable in most parts of the world – American Foulbrood and European Foulbrood and therefore destruction of the hive is likely to be the treatment of choice rather than medication. It is the role of the local bee inspector to decide on the required treatments.

Tetracyclines

These antibiotics are produced from streptomyces fungi and are widely used in bee medicine, especially in North America.

Tetracyclines include oxytetracycline (OTC) and chlortetracycline (CTC). They have a wide range of activity against Gram positive and Gram negative bacteria. They are bacteriostatic at low levels but may become bactericidal at high doses.

If treatment is allowed by the local bee inspector, mix the antibiotic with powdered sugar. Sprinkle the mixture on the edge of the brood-nest avoiding spilling directly into the brood. The nurse bees will consume the medication and feed it to the larvae.

The withdrawal period for tetracycline when used with honey bees may vary between countries. As a guide, allow a minimum of 42 days. Note the withdrawal period starts after the tetracycline has been consumed, not from the day you provided the last dose. If provided in a patty this can take ten days to consume.

Tetracycline is administered at 200mg per hive administered every fourth day on three occasions. The withdrawal period is six weeks. Do not use on hives producing honey for human consumption for six weeks. It is possible to make a patty of 800mg as a single treatment dose.

Macrolides – tylosin phosphate

This group includes:

Erythromycin
Tiamulin
Tilmicosin
Tulathromycin
Tylosin
Lincomycin
Valnemulin

They are mainly active against Gram positive bacteria and are very good against spiroplasmas.

Tylosin phosphate is used to control some bee pathogens. Tylosin phosphate is added at 200mg added to 20g sugar to a hive. Treat once a week for three weeks. Complete treatment at least four weeks prior to honey flow.

There are a number of other antibiotics available for the treatment of animals, including bees, especially under the supervision of a veterinary surgeon.

Other types of antibacterial medicine

Aminoglycosides

These antibiotics contain sugars and include:

Apramycin
Framycetin
Gentamicin
Neomycin
Spectinomycin
Streptomycin

They are very active against Gram negative bacteria. They are bactericidal and are poorly absorbed from the intestinal tract. The use of streptomycin is banned in some countries.

Cephalosporins

Most of these medicines are poorly absorbed from the intestine and may have limited use in bees. They would include:

Cephalexin
Ceftiofur

Ceftiofur has a wide range of activity and is an excellent medicine for the treatment of respiratory bacterial disease.

Penicillins

All the penicillins are bactericidal. Penicillins have no effect against spiroplasmas.

Quinolones

These are very active against both Gram positive and Gram negative organisms.

Sulphonamides

There are approximately 30 different ones available. Sulphonamides are often combined with synthetic substances called trimethoprim and baquiloprim. They are then termed potentiated sulphonamides and have a wider spectrum of activity. Sulphonamides are bacteriostatic but have a wide range of activity against both Gram positive and Gram negative organisms.

Antibacterial sensitivity tests

In most farm animals the use of antibacterial sensitivity tests are part of the routine investigation to determine which antibiotic is the best to use. In bee medicine this is extremely uncommon as veterinarians are rarely involved in treating bee conditions. However, the practise should be more widely used and should be encouraged. Laboratories would also have to become more familiar with handling bee samples.

Antibiotic sensitivity testing allows the veterinarian to decide which medicine is most likely to provide a satisfactory outcome. Some antibacterial medicines are more active against Gram positive than Gram negative bacteria and others are the reverse. This gives a guide to the choice of medicine to be used. Medicine sensitivity tests carried out in laboratories give a better guide against specific infections, but this is not perfect. Some bacteria which are sensitive in laboratory tests are not sensitive in diseased animals – they may be multiplying in sites where the medicine cannot reach them, or the antibiotic is not reaching them in high enough concentrations.

In its simplest and commonest form the test is carried out by growing the bacteria in a growth medium and then suspending it in a saline solution. A thin film of this is spread over the surface of a culture plate and left to dry. Discs of cardboard impregnated with different antibiotics are placed on the surface and the plate is then incubated. The medicine diffuses out into the growth medium radially. If the organism is killed by the antibiotic there is a clear zone of no growth around the disc. If the organism is resistant to the medicine it grows right up to the disc. Such tests usually take 24 to 48 hours. Some bacteria, however, take weeks to grow.

Application of medication products in the beehive

Always read all the instructions provided with the medication. If you have any issues, discuss application methods with your veterinarian and bee inspector. Note that you should always dispose of all medication products appropriately.

Figure 4.2 Application of a medication powder to the hive

Figure 4.3 Grease patty placed on the top of the brood frames

Figure 4.4 and 4.5 Showering and spraying medication product into a hive

Figure 4.6 Medication in a block preparation

Figure 4.7 Pouring a medication product

Figure 4.8 and 4.9 Various medication strips applied directly to the frame

Figure 4.10 Various medication strips applied to the hive lower board

Alternatives to treatment

Cleansing behaviour

Bees with a high hygienic behaviour appear to be able to protect themselves against a range of pathogens, including *Varroa* and Chalkbrood. Utilising a hygienic behaviour examination can be a significant step to controlling Colony Collapse Disorder. It is possible to examine a hive for hygienic behaviour with a 24 hour test.

1. Identify a frame with a solid area of capped worked pupae (pink eye stage).
2. Shake the nursery bees from the frame.
3. Place the frame horizontally on a solid surface.
4. Using a 6cm circular ring, kill all the pupae using a sharp pin to penetrate the cappings and the pupae. Alternatively, the pupae may be killed by freezing with liquid nitrogen or it may be possible to remove the circle/square of pupae and kill them by freezing for 48 hours in a freezer and then returning the piece of frame.
5. Replace the frame into the hive.
6. Re-examine the frame 24 hours later.
7. If 95% or more of the pupae have been removed and the cells cleaned the queen may be considered 'hygienic'.

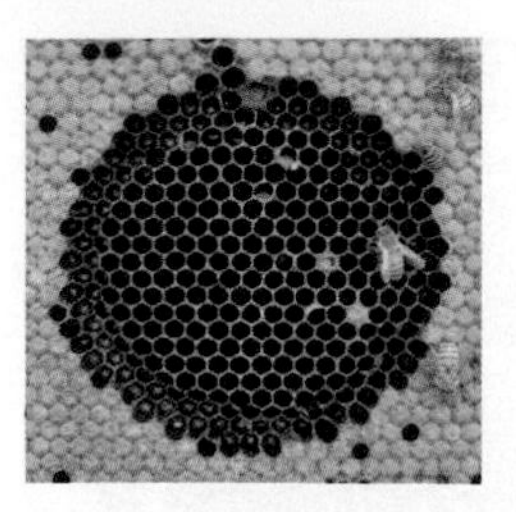

Figure 4.11 Hygienic bees

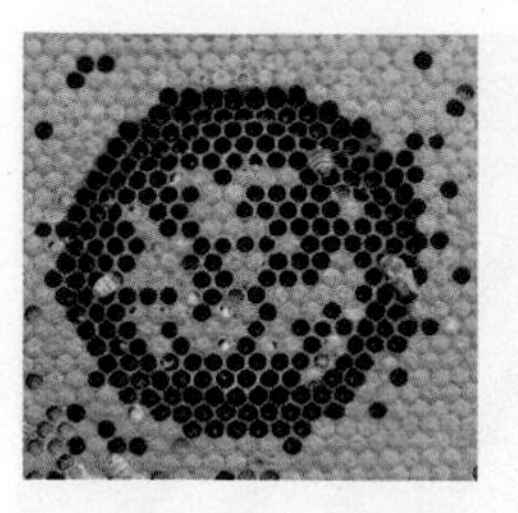

Figure 4.12 Unhygienic bees

8 If the queen is considered hygienic she and her drones should be utilised as part of the future breeding programme.

Table 4.5 Treatment options summary

Medication	Trade name	Indications	Dose and routines	Frequency	Withdrawal
Amitraz	Apivar	Mite control	Strip hung between brood	Treat for 10 weeks	Not during honey flow
Thymol	Api Life Var Apigard Thymovar	Mites and fungi	25% oil solution. Use at about 15°C	10 to 14 days apart	Not during honey flow in the spring
Fenpyroximate	Hivastan	Mite control	225g in a Patty	Not during honey flow	May be toxic to bee
Formic acid	Mite-away	Mite control	Air above 22°C	Slow-release pad	
Fumagillin	Fumadil B	Nosema		In syrup	Not in EU
Menthol	Api Life Var	Mite control		Menthol crystals	Not when honey supers in place
Coumaphos	Checkmite+	Mite control	1.4g coumaphos	Strip hung between brood for ten weeks	Note may reduce queen lifespan. Not during honey flow
Oxalic acid		Mite control	6% oxalic acid with 30% sugar solution	Single treatment only in sugar	Toxic and only in winter
Pyrethroid	Bayvarol Apistan	Mite control	0.7g on the strip	Strip hung between brood for 8 weeks	Not during honey flow

Medication	Trade name	Indications	Dose and routines	Frequency	Withdrawal
Tetracycline		Antibiotic	200mg per colony	Every fourth day for 3 treatments	6 weeks. Not during honey flow Residues may persist for a year in honey
Tetracycline		Antibiotic	800mg per colony patty	Once	6 weeks. Not during honey flow Note residue
Tylosin Phosphate	Tylan	Antibiotic	200mg added to 20g sugar	Once a week for 3 weeks	4 weeks. Not during honey flow

Final note

If you have any concerns regarding treating your bees, consult with your local veterinary surgery and bee inspectors. Do not treat blind, as many of these products will kill or at least wound your bees. In addition, medicines can produce residues which are harmful to human health. Honey and other bee products are excellent health products and must not contain any harmful residues.

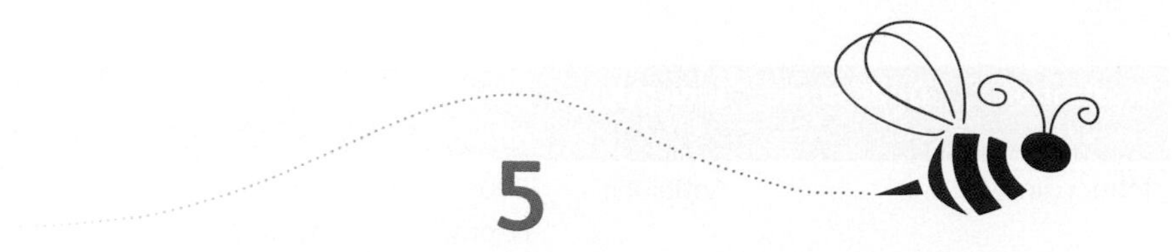

5

Managing the health of the brood

The queen can lay around 2,000 eggs per day during the peak of summer when food is freely available. In an average productive hive, in the middle of summer, there could be 9,000 uncapped and 20,000 capped brood. Uncapped brood will contain eggs for three days and then for five days pearly-white larvae progressively fill the cell.

There are three cell types in *Apis mellifera*

- The smaller 5mm diameter worker cell with a flat cap.
- The larger 7mm diameter drone cell with a domed cap.
- Queen cells and play cells, which are large peanut shaped cells easily differentiated from the two other cells.

To get a worker larva through the first five days each larva has to be fed 1,500 times a day! With 9,000 mouths to feed, this is 13.5 million feeds per day. It takes hundreds of visits to a flower to collect sufficient nectar to take a larva from L1 to L5. Note there is a difference of eight days between the egg being laid and the L5 being capped. A lot can happen in the countryside in a week! Sudden changes in food resources can be a major cause of stress on a colony, especially in the unpredictable and often wet spring months.

After nine days post-lay in female cells and in 11 days post-lay in drone cells, the workers place a wax cap over the cell and the larva continues its development without receiving any more nutrients from the attendant workers. Capped cells can be easily differentiated from cells with capped honey cells, as the capping colour is yellow because there is an air gap between the larva and the capping. The honey cells are normally wider (7mm) and the capped honey cells are white. Cells filled with pollen have the colour of the pollen which gives a guide to the flowers being visited by the foraging worker bees.

Normal brood

It is vital to understand the normal appearance of brood to be able to recognise when the brood is abnormal. The important feature is the pearly-white colour of the healthy larva and the position in the cell and how the larva progressively fills the cell. Review Chapter 1 for more information on this.

Queen cells

Three types of cells can be recognised which may be associated with the future queen of the hive:

1 Queen play cell – the worker bees will practise developing queen cells with a structure called a 'play cell' which is an enlarged cell outside the main frame structure but without an egg or developing larva inside.
2 Swarming queen cell at the bottom of the brood frame.
3 Superseding queen cell – a cell which is hanging from the middle of the brood is often an emergency superseding cell.

Figure 5.1 Play cell

Figure 5.2 Swarming queen cell at the bottom of the brood frame

Figure 5.3 Detail of the swarming queen cell

Figure 5.4 Superseding cell in the middle of the brood frame

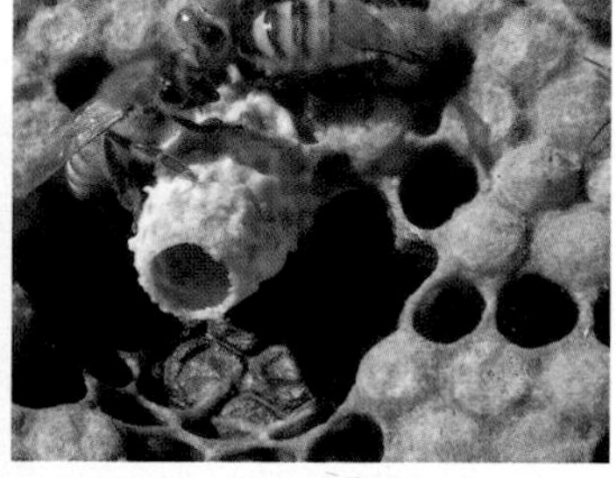

Figure 5.5 Detail of a superseding cell

Worker cells

These are the bulk of the brood frame. Worker cells are about 5mm across with a flat/slightly domed capping over the cell once the larva successfully reaches L5.

The worker brood is kept at 36°C. The humidity of the brood is kept at 40% RH. The cappings are even and brown in colour because there is an air gap underneath the capping. Honey is also capped but here there is no air gap and the capping appears white in colour.

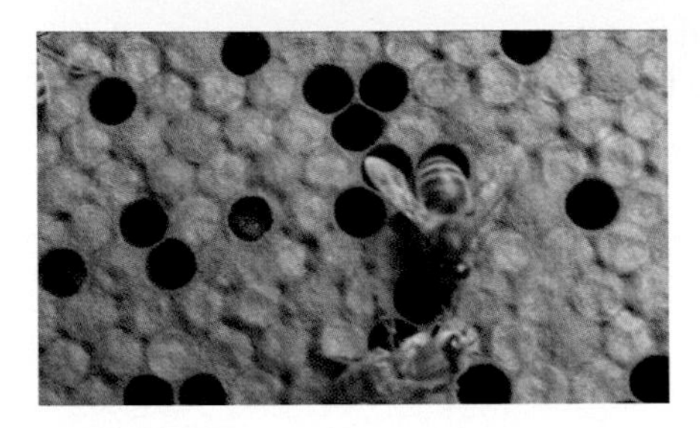

Figure 5.6 Worker cappings flat, even and brown in colour

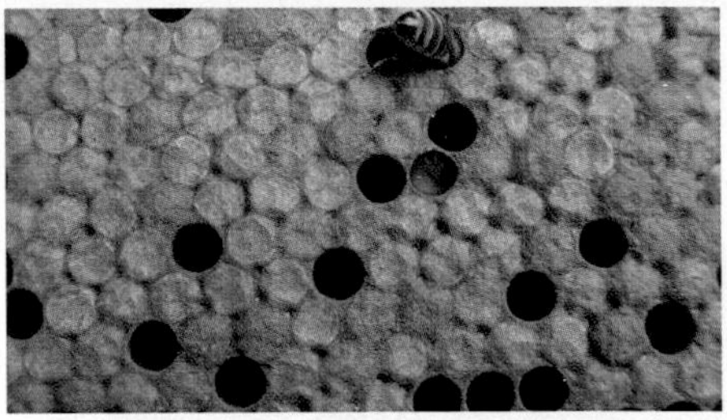

Figure 5.7 Detail of the worker cell capped.

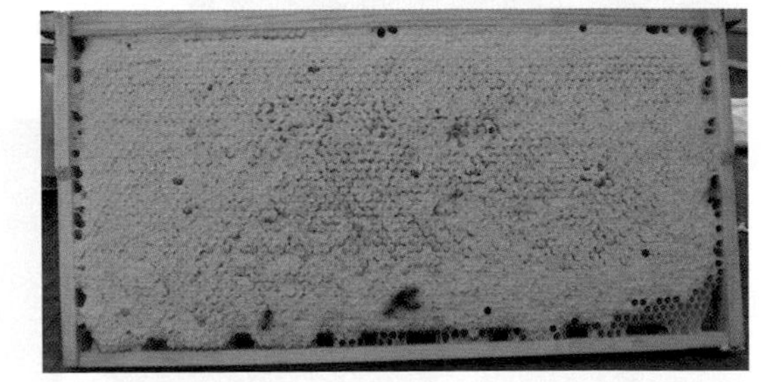

Figure 5.8 Honey cells are capped but the lack of an air gap makes the capping appear the original white

Drone cells

Drone cells can be easily recognised by being bigger at 7mm across and have a clear domed appearance, bulging out of the cell. The drone cells are situated at the edges of the brood frame where it is kept cooler, at 35°C.

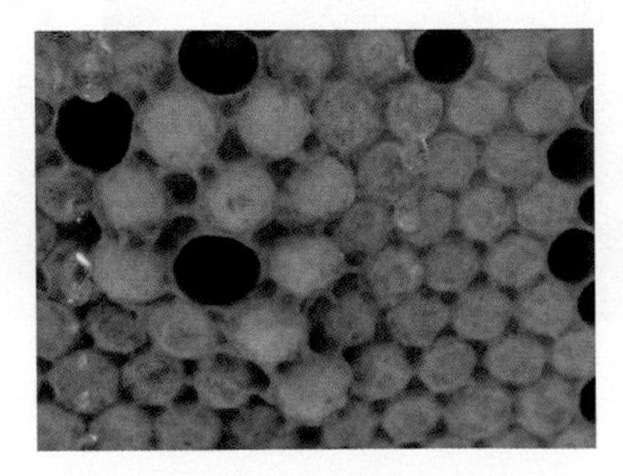

Figure 5.9 The larger cells of the drone. When capped the cap has a characteristic domed appearance

Bumblebee brood

Once the nesting site is selected, an area is cleared by the queen to form a small chamber. Pollen is collected and moulded into a mass upon which the first wax cells are made. The queen then lays her first batch of eggs.

The eggs hatch after 4–6 days and the larvae feed on the pollen mass and additional nectar and pollen as supplied by the queen. During this time the queen may forage for short periods of time. The wax cells are progressively expanded to accommodate the growing larvae. After 10–20 days (depending on the temperature) the final instar is complete. The larvae defecate, spin a cocoon and pupate.

The queen then recycles any wax and pollen to make another brood nest in which she lays her second batch of eggs.

As the season develops some species of bumblebee will start laying down pollen stores to assist larval development. The colony increases in size from a few dozen workers to around 200 workers. As the colony increases, eventually male (from unfertilised eggs) and future virgin queen bees are produced.

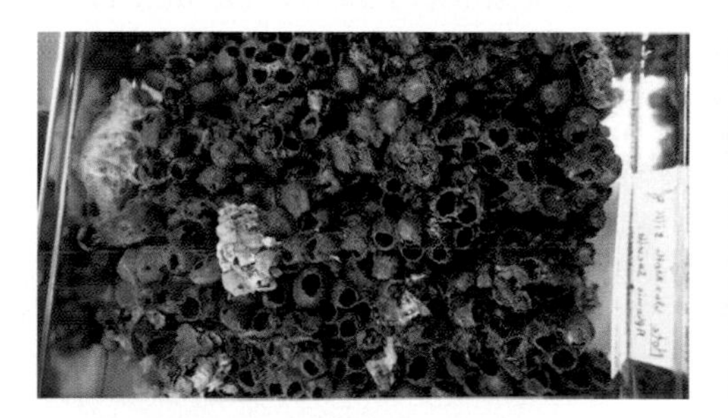

Figure 5.10 *Bombus terrestris* brood

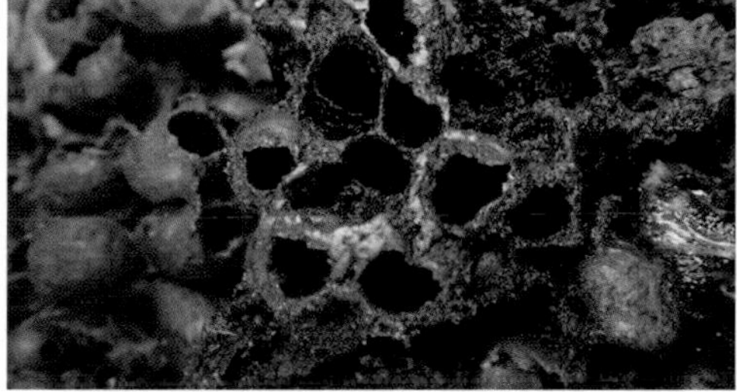

Figure 5.11 Detail of *B. terrestris* brood cells

Disorders of the brood

American foulbrood in *Apis mellifera* hives

American foulbrood (AFB) is caused by the bacteria *Paenibacillus larvae.*

This is a slender gram +ve rod with slightly rounded ends with a tendency to grow in chains. The bacteria produce spores making it very resistant. The bacteria does not grow on nutrient agar, it requires Difco brain-heart infusion. There are four types of AFB recognised:

ERIC I worldwide
II Europe
III Chile
IV uncommon

Another bacteria has also been recognised *Paenibacillus pulvifaciens*, but this may just be a variant of P. larvae.

Clinical signs

American foulbrood can cause problems at any time of the year. On opening the hive an unusual smell may be noticed, which is like a decayed glue pot – hence 'foulbrood'. The workers may also be unusually aggressive towards the hive examination. Examination of the brood frames will reveal that the capped brood distribution will be spotty and not even. The worker bees are trying to remove the dead and dying pupae. The bacteria infect the 1–2 day old larva, but the clinical signs are seen in the pupa which die at 12–16 days of age, thus after capping. The bacteria cannot affect larvae older than 3 days of age (6 days after lay). While some larva may become infected through previously infected cells, the majority of larvae are infected through the honey they are being fed within the first couple of days of life. It may only require 10 spores to infect a larva and on dying they produce billions. Close examination of the remaining capped brood may reveal some of the cappings are discoloured and convex (sunk inwards). Some may be punctured by workers who recognise a problem and have started to remove the dead pupae. Examination of the remaining pupa reveals that they are dull and often light brown to black instead of pearly white. In some cells the pupa has decayed to become only a dried black scale, which may be stuck in the cell and difficult to remove. One characteristic is when a toothpick is put into the affected pupa the dead pupa sticks to it and the resultant 'mucus material' appears ropey. Interestingly the tongue (glossa) of the dead pupa may remain as a fine thread.

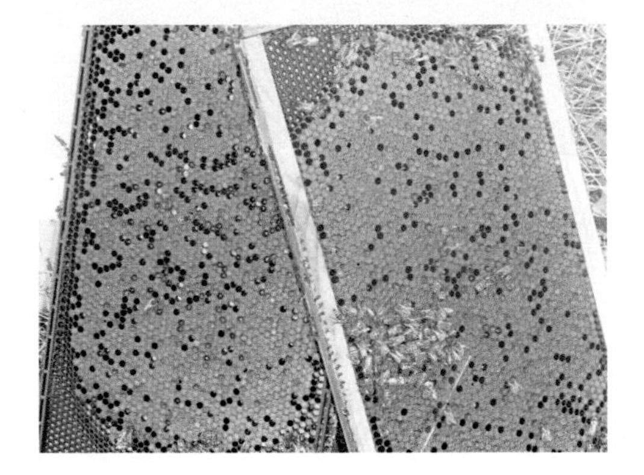

Figure 5.12 Spotty brood – pepper pot

Diagnosis

The "Rope test" using a toothpick on the dead pupa is indicative, but not diagnostic as most brown decomposing pupa may do this. Definitive diagnosis is through culture of the organism. A rapid hive test is available using lateral flow devices looking for antibodies in the bees (see Chapter 12). This will be used by the local bee inspector. American Foulbrood diagnosis is generally done in Government laboratories.

There are dogs being trained to detect the smell of American Foulbrood.

A PCR test is available. AFB can be stained with 0.2% carbol fuchsin stain where they will appear to be ellipsoidal and thick rimmed bacteria.

The dead dried scales of an ABF infected larva will fluoresce in ultraviolet 360nm. Note some moulds and pollens may also fluoresce.

Similar diseases

European foulbrood – note larvae die before being capped.

Note there are other Paenibacillus species which may secondarily affect European foulbrood infected cases. These include *Paenibacillus albei* and *P. apiarius.*

Treatment

Inform the local authorities. In many countries this is a notifiable disease and in several countries the only control measure will be to destroy the hive and colonies by fire. If treatment is allowed by the local bee inspector, antibiotics may be useful, including tetracycline hydrochloride, tylosin phosphate or sulphathiazole. When applying antibiotic medications, mix the antibiotic with powdered sugar. Sprinkle the mixture on the edge of the brood-nest, avoiding spilling directly into the brood. The nurse bees will consume the medication and feed it to the larvae.

Tetracycline is administered at 200mg per colony administered every fourth day on 3 occasions. The withdrawal period is six weeks. Do not use on hives producing honey for human consumption for six weeks. It is possible to make a patty of 800mg as a single treatment dose.

Tylosin phosphate is administered at 200mg added to 20g sugar. Treat once a week for three weeks. Complete treatment at least four weeks prior to honey flow.

Management, control and prevention

Employ strict biosecurity as the bacteria produce spores, which are extremely resistant and can survive for 40 years. *Paenibacillus larvae* can be found in infected wax, wood, hive tools, honey, pollen and on honey bees. Therefore, do not share equipment and avoid second hand equipment. The bacteria causes no problem in the adults who remain infected for many weeks spreading the bacteria.

In nature the bacteria is spread between colonies by robbing and drifting. The bacteria gain access to the larvae through infected honey. Once the hive and colonies have become infected the only real method of destruction is to burn equipment which has become infected and sterilise all metal equipment.

Zoonosis (infectious to man)

None

American foulbrood in *Apis cerana* hives

American foulbrood (*Paenibacillus larvae*) can be found in *Apis cerana* hives. At times the infestation can be high without any obvious clinical effect. In *A. cerana* it appears that the nurse bees detect the infection and remove the infected larvae before they can pupate thus removing the bacterium before it can sporuate.

Bald brood

Examination of the brood frame reveals that the cappings to the worker cells are abnormal. There is a round hole in the cappings. The damage is often in a straight line. The head of the developing pupa is visible under the damaged cap, however, they appear normal and are alive. The pupae will still continue pupation and emerge normally. The workers tend to ignore the exposed pupae.

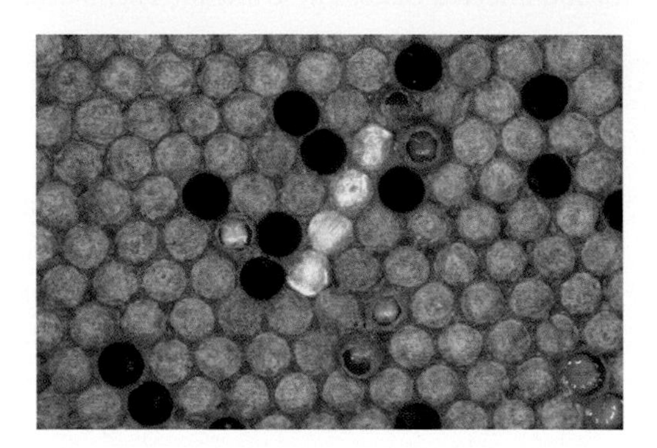

Figure 5.13 Developing pupae visible inside their cells after their cappings have been removed

Courtesy The Animal and Plant Health Agency (APHA), Crown Copyright

Diagnosis

The most likely cause is the presence of larvae of the Greater Wax Moth (*Galleria mellonella*) chewing through the cappings of pupating larvae.

A more unusual cause is a genetic defect where cappings are absent or deformed.

Management control and prevention

Remove wax moths from hives. Kill the Greater Wax Moth larvae. If you drop them on the floor they will crawl to any another hive within 50 metres! Freezing the frames will destroy any eggs and larvae.

Where there is suspicion of a genetic cause it will be necessary to introduce a new queen.

Black queen cell virus (BQCV)

Black queen cell virus (BQCV) can affect many species of Apidiae for example Adrena spp; Bemix spp; *Bombus* spp; *Megachile rotundata; Nomia melanderi* as well as *Apis mellifera.* In *A. florea* and *A. dorsata* the virus can be isolated without clinical signs. In *A. cerana* hives it can be associated with serious losses. The black queen cell virus is a Dicistroviridae, which is a RNA virus.

Clinical signs

The condition is more often seen in the summer, when future queens are being laid. The condition affects the queen larva which turns black and dies. The queen larva may resemble sacbrood.

If the queen larva survives, it is likely that the future queen will be of poor quality.

The virus may also affect larvae of workers. This may reduce the viability of the future workers. The condition may be associated with concurrent clinical Nosema infestation of the hive.

Diagnosis

Appearance of the larvae. The Black queen cell virus appears to be particularly attracted to the alimentary tract.

Management, control and prevention

The primary method of control is to manage Nosema. Regular brood comb changes reduces the risk. Shook swarm techniques may be used if the virus is diagnosed in the hive (see Chapter 12).

Chalkbrood

Chalkbrood is caused by a fungus *Ascosphaera apis* in the honey bee and by *Ascosphaera aggregata* in the bumblebee. The condition is also called 'Chalk', describing the appearance of the dead larva.

Megachile rotundata is very important to the pollination of the USA alfalfa seed industry. The high mortality associated with *Ascosphaera aggregata* results in some 3×10^8 bumblebees having to be imported into the USA from Canada – half the US requirement.

Clinical signs

Chalkbrood commonly occurs in the wet spring as the hive starts to prepare for the summer. When you approach the hive you may see chalk mummies deposited on the landing board or on the ground surrounding the hive. Occasionally you may see a worker bee leaving the hive with a chalk mummy in its mandibles. Examination of the hive debris will also reveal a number of chalk mummies.

The larva is infected at day 1 to 5 and will become mummified within 2–3 days after infection. The larva dies after capping and will be seen to be a hard white or black pointed oval. Examination of the brood frames reveals scattered individual upright chalkbrood mainly at the bottom of their cells. The chalkbrood may rattle if the frame is gently shaken and will not be stuck to the cell walls. Examination of the sealed brood will reveal almost no torn cappings. Heavily affected hives may raise less drone brood, which could affect fertility of the colonies. The condition is seen more in older hives than in younger hives. In bumblebees the mortality may be 15–50% of the developing larvae.

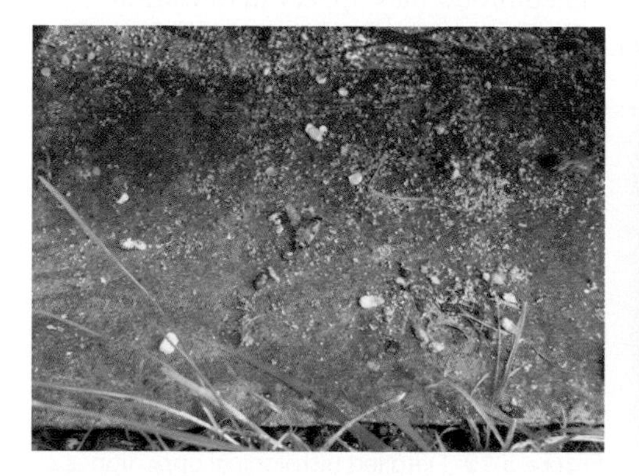

Figure 5.14 Chalk mummies seen in the hive debris

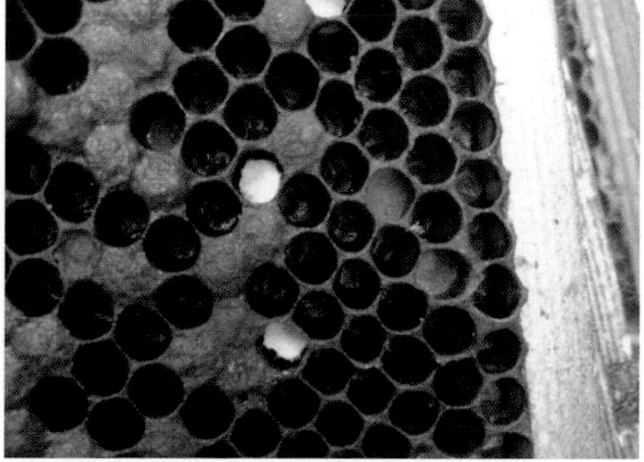

Figure 5.15 Chalk mummies inside cells

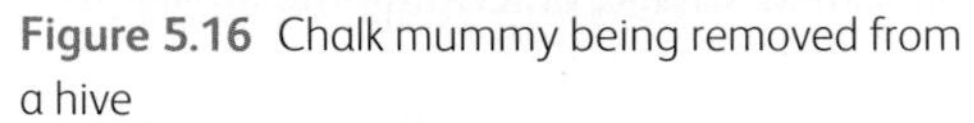

Figure 5.16 Chalk mummy being removed from a hive

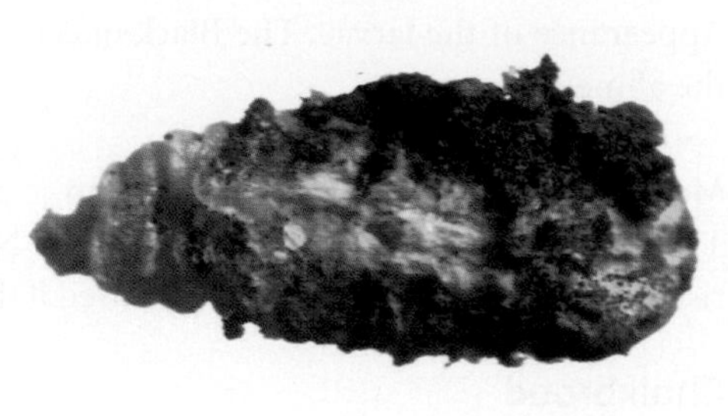

Figure 5.17 Gross detail of a larva which died of chalkbrood with evidence of black sporulation areas

Chalkbrood is often seen in association with sacbrood.

Diagnosis

The larva ingests the spore. In the ventriculus, germination is stimulated by the presence of CO_2. The mycelium develops in the body cavity and the hyphae break out posteriorly, thus leaving the head intact. The larva becomes covered by fluffy cotton-like mycelia. Type + and – strains are required for sporulation. If the mummy is just white it is only infected with one type. If the chalkbrood mummy is also covered in black spots this indicates that the larva was infected with both sexual types and that the fungus is sporulating. The fruiting bodies (50–150μm) develop on aerial hyphae outside the dead larvae. The individual spore balls are 9–19μm in diameter.

It is possible to culture the fungus but this is rarely necessary.

Histological examination of the larva can be assisted with Lactophenol-Acid Fuchsin or lactophenol cotton blue to assist visualisation of the fungal hyphae and fruiting bodies.

Similar diseases

Sacbrood may be occurring at the same time. Chilled brood may appear in dull white dead larvae but will not be hard. Stonebrood are very similar but the mummified larva is very hard.

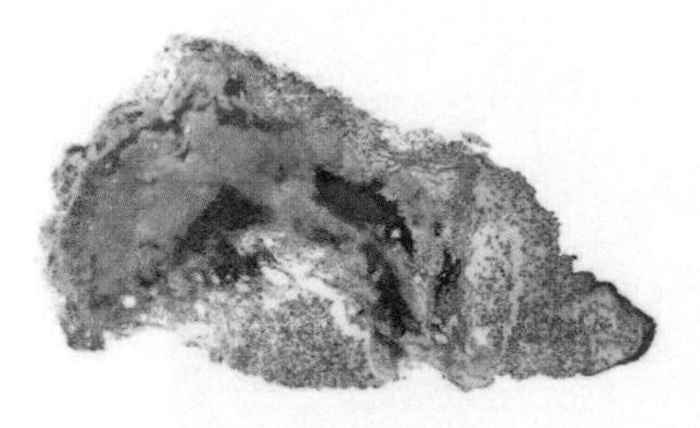

Figure 5.18 Transverse section through a chalkbrood mummified larva stained with H&E

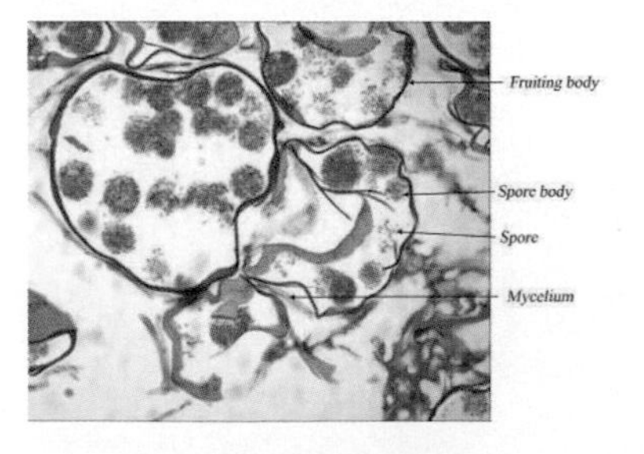

Figure 5.19 Detailed histological appearance of a sporulating *Ascosphaera apis*

Treatment

The condition generally self-resolves once the weather warms. However, resolution can be assisted by increasing the ventilation though the hive. Clean frames when not in use. Adding *Satureja montana*, an essential oil at 0.01% to the hive may have some benefit.

Management, control and prevention

Do not move infected frames. In bumblebees do not reuse the previous year's nesting materials. Ensure the hive is well ventilated, avoid condensation and damp hives. This can be difficult in wet springs. Do not examine the hive unnecessarily in cold weather. Having a slatted top board helps to reduce chilling of the brood. Keep the number of workers to brood ratio optimal. If chalkbrood is problematic it may be necessary to change the queen to a more hygienic queen. As a test for hygienic behaviour, eliminate queens with hives showing chalkbrood. Interestingly, this may also improve the honey bee's *Varroa* resistance.

The fungi *Ascosphaera apis* are in the honey and are then fed to the larvae. The spores of these fungi are extremely resistant and can persist for 15 years in mummified larvae. The spores may survive four years in the environment. Sterilise all hive materials before disposal and burn all hive debris, do not just spread it on the ground around the hive. The fungi are spread by drifting and robbing.

The spores may contaminate the bodies of the bumblebee, which then infects the pollen, which is unwittingly fed to the larvae, which die. Imported bumblebees should be quarantined and their larvae allowed to mature before being introduced to large scale production.

Chilled brood

This is associated with a sudden reduction in temperature, which the worker honey bees did not anticipate or were unable to cope with. The susceptible young brood die. This is recognised by the worker bees and rapidly removed. As you approach the hive, the dead larvae may be seen on the landing ramp or on the floor outside the hive. Note, however, birds and ants are well aware of chilled brood and will rapidly snap up the tasty morsels. So an hour after sunrise there may be no clinical signs remaining.

Clinical signs

The larva that is chilled will die and turn black within one day. Multiple ages of brood will die at same time. The dead larva is normally removed within 24 hours of death.

If the brood is chilled, it may not die. But there can be abnormalities of development. These abnormalities may include deformed wings but subtle changes in brain development may also

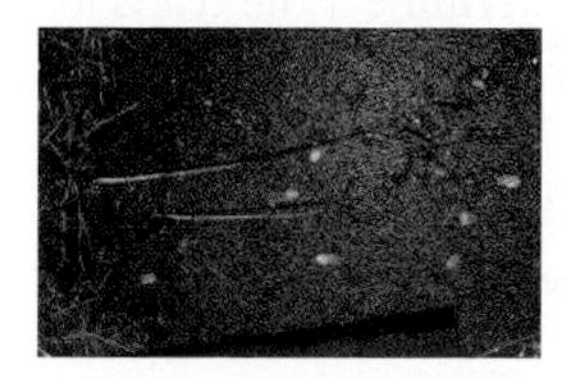

Figure 5.20 Chilled brood on the floor outside the hive early in the morning

affect learning and memory – vital skills in foraging. These changes may have a significant impact on colony collapse.

Diagnosis

The larvae and pupae are dead with no other clinical signs. The problem only lasts during the cold season.

Similar diseases

In Chalkbrood and Stonebrood the larvae are hard and mummified. In Sacbrood the larva is liquefied.

Treatment

None.

Management, control and prevention

Avoid chilling of the hive. Use blankets and insulation where necessary. Do not examine the hive unnecessarily in cold weather. Having the top board slatted reduces the chilling effect of hive examinations.

Figure 5.21 Slatted top board

Figure 5.22 Insulation under the top board

Drone laying queen

When you examine the brood frame you will notice large dome cappings over a lot of the smaller worker cells. These are drone pupae trying to develop within worker cells. The queen has laid unfertilised eggs in worker cells. This is generally because she has run out of sperm or there has been a fault in her sperm delivery system from the spermatheca. Check the legs of the queen as well as she needs to measure the cell with her legs to determine if the cell is for drones or workers. The adult drones that develop in the smaller worker cells will also appear smaller than normal.

Because of the lack of workers being born, the remaining cells and brood may be neglected and dying. There may be an abnormally large number of drone bees compared to normal workers.

Figure 5.23 Large domed cappings in small worker cells – arrow 1. Large number of drone bees compared to workers—arrow 2

Management, control and prevention

Replace the queen. Ensure that the new queen is healthy and is well mated. If the queen is mated by artificial insemination, ensure that over 20 drone semen collections were placed into the virgin queen to fill her spermatheca. Check the legs of the queen.

European foulbrood

European foulbrood is associated with the bacteria *Melissococcus plutonius*. This is a Gram positive cocci (round) bacteria which can occur in chains. This can look very similar to streptococci bacteria in mammals. Occasionally the bacteria can be pleomorphic (have a variety of shapes). The bacteria can be difficult to grow, it may not appear for two to four days and requires anaerobic (oxygen free) atmosphere to grow.

Melissococcus plutonius is present in most countries but is not found in New Zealand. In many parts of the world this is a notifiable condition and the local bee inspector needs to be informed as soon as you have any suspicions.

To separate this from American foulbrood, remember E for Early: before capping.

Note that European foulbrood may be found in *Apis cerana* and *A. laboriosa* hives.

Clinical signs

When you open up the hive you may notice an unusual smell or odour. The smell is sour hence the term 'foulbrood'. When you examine a frame of brood you will notice the brood pattern is spotty and not even. This has been described as a crossword puzzle appearance. Closer examination of the larvae will reveal that they are not the expected pearly-white but are dull, yellow and sometimes black. They may appear granular. The larvae are often curled in unusual shapes, often in a C shape, and do not fill the cell. Some of the younger larvae may be seen dead at the bottom of the cell. *M plutonius* bacteria compete for food so larva dies of starvation. They are not stuck to the walls of the cell. Examination of individual larvae may reveal that their intestines can be seen through their skin (hypodermis). Normal larva has a white intestine line whereas with EFB the intestines are yellow. The larvae which have survived to capping may appear normal as do the workers, drone and queen. The larvae are infected though the honey and the bacteria is found in the ventriculus of L2 larvae (five days after laying). The bacteria then infect the rest of the intestines.

The condition is more often seen in the springtime but can be seen at any time if the hive is infected and stressed. Larval starvation may be a trigger factor.

Diagnosis

The clinical signs and appearance of the larvae are the first indications. Field diagnosis is achieved through a lateral flow device to look for antibodies (see Chapter 12). The presence of the bacteria is then confirmed by PCR and culture. The odour of the hive may be influenced by other bacteria notably *Achromobacter eurjdice* and *Enterococcus faecalis.*

Similar diseases

American foulbrood. Queen half-moon syndrome.

Treatment

In many countries this is a notifiable condition and must be reported to the bee inspector who will coordinate control. In most cases the pathogen is controlled by destroying the hive and colonies by burning.

If the bee inspector allows treatment, tetracycline hydrochloride and tylosin phosphate may be used. Provide in spring and autumn in sugar.

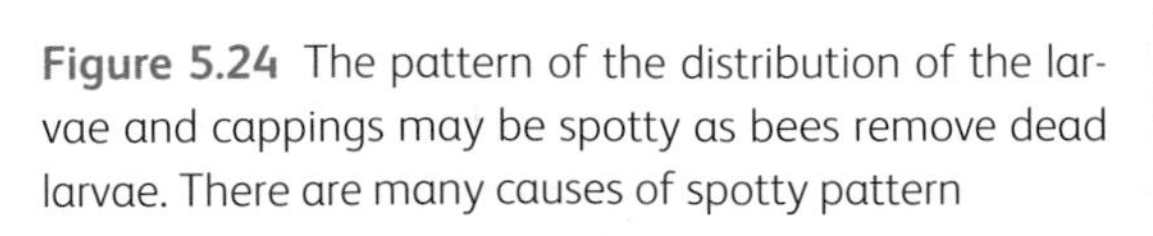

Figure 5.24 The pattern of the distribution of the larvae and cappings may be spotty as bees remove dead larvae. There are many causes of spotty pattern

Courtesy The Animal and Plant Health Agency (APHA), Crown Copyright

Tetracycline is administered at 200mg per colony every fourth day for three occasions. The withdrawal period is six weeks. Do not used on hives producing honey for human consumption. It is possible to make a patty of 800mg for single dose.

Tylosin is added at 200mg added to 20g sugar. Treat once a week for three weeks. Complete treatment at least four weeks prior to honey flow.

Management, control and prevention

European Foulbrood predominates when the colony is under stress. Therefore:

- Maintain a strong colony with lots of food. Do not be greedy when taking the honey bees' stores in the autumn.
- Annual re-queening is advised, but when one has a good queen it is difficult to do this. After 3 years, many queens need to be replaced.
- Keep a good log of the hive's activities.
- Ensure that the total crude protein content of bees is above 40%. This is a measure of the protein level in the body. A starving bee will have less protein. When bees drop to less than 30% total crude protein they become more susceptible to EFB.
- Control of the bacteria can be achieved using the shook swarm technique (see Chapter 12).

The spores are resistant for three years. Larvae are infected through honey. Transmission between hives is by drifting and robbing. Maintain the highest level of biosecurity. Do not keep hives too close to each other, less than three metres.

Euvarroa

Euvarroa are a species of mite which is very similar to *Varroa*. The *Euvarroa* species are smaller than *Varroa sp*. The species of *Euvarroa* found on *Apis florea* include *E. sinhai, E.wongsirii* and *E. adreniformis*. There are very few abnormal clinical signs.

Fly pests of bees

Many species of Apidae can become parasitised by various flies which ultimately kill the host. Bumblebees can be particularly prone to *Sarcophagidae: Brachicoma devia* (UK) and *Brachicoma sarcophagina* (USA).

The female *Brachicoma devia* lays her larvae next to bumblebee larvae. When the bumblebee larvae spin their cocoons and pupate, three or four *B. devia* larvae enter and suck out the contents of the bumblebee pupae. The developed larvae then leave the killed bumblebee pupae, pupate themselves and leave the nest.

Brachicoma devia also infest other Apidae and wasp larvae.

Foundation wire

In the natural hollow tree hive, the comb is made incorporating natural obstacles. The workers are not stupid and will only support cells which are likely to return a healthy bee. The layout of the wires supporting our (not their) foundation comb may interfere with the honey bee's willingness to use the cells above the wires. Note that workers will even move eggs, which the queen had laid, out of an unsuitable cell. This leads to a line of empty cells. Note this is different from bald brood where upright pupae can be seen in the cells.

Galleriasis

Wax moth (*Galleria mellonella* and *Aphomia sociella*) infestation may be sufficiently high that the developing wax moth larvae create so many silk threads, it can make it difficult for emerging bees to escape the cell. The emerging workers are thus found dead, trapped with the wax moth silk threads.

Laying workers

When you examine the brood frames you notice a large number of domed drone cappings over worker cells. Abnormally small drone pupae will be seen within the worker cells. There will be a progressive shortage of worker bees and thus the unsealed brood may be neglected and dying. One characteristic, which may help to differentiate from Drone laying queen syndrome, is that multiple eggs may be present in cells, adhering to sides of the cell. These are laid by the workers.

Figure 5.25 Galleriasis of a bumblebee nest

The cause of the problem is that there is no queen in the hive and she died before the workers could make an adequate queen cell.

Diagnosis

Place a frame with eggs and young larvae (from a healthy hive) in the middle of the hive. If there is no queen, the workers should start making queen cells within days.

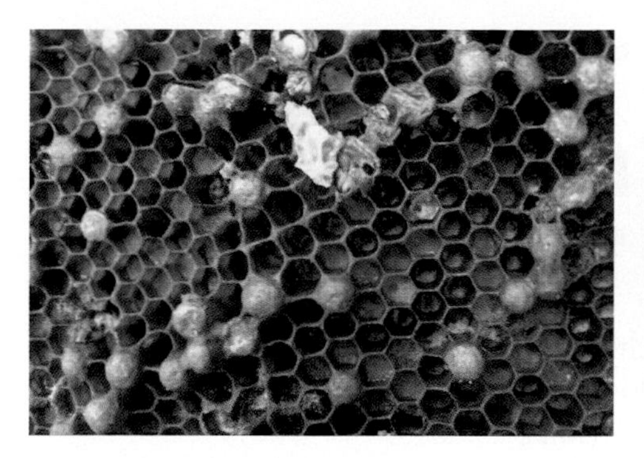
Figure 5.26 Drones in worker cells

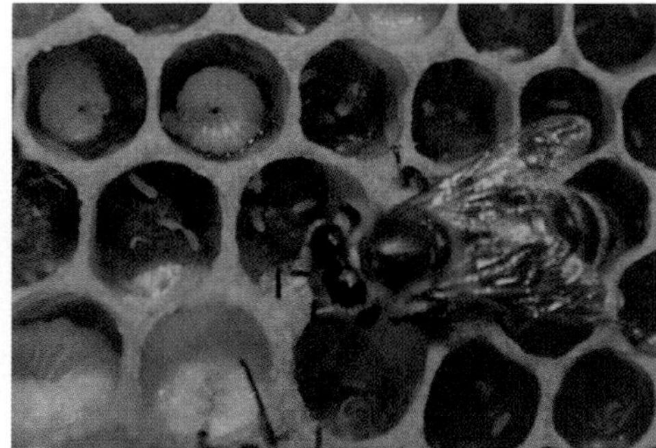
Figure 5.27 Multiple eggs laid in the cells

Treatment

If the hive is otherwise healthy, move the colony 20m from its original position. Remove all the brood frames and shake bees off each comb onto the ground. Allow the bees to find their way into the colony. This will select only the healthy worker bees and reduce the number of workers laying eggs.

If the hive appears unhealthy, it is better to kill the hive. Ensure you sterilise all equipment used on and in the hive before reusing on another hive. Ensure that all the frames and boxes are properly sterilised. They may require to be placed in a freezer to kill off any bee pests.

Management, control and prevention

The best prevention is to monitor the hive closely in the late spring to ensure that the queen and hive has not swarmed, leaving a weakened hive. Check the hive regularly for eggs, capped and uncapped larvae. If there are eggs, the queen was present three days ago!

Queen half-moon syndrome

Examination of the brood frame reveals that the L3 and L4 larvae are twisted into a C shape (half-moon). The larva colour has changed from the pearly-white to a yellow or light brown and then dark brown with their tracheal lines present. Examination of other cells reveals the many cells contain multiple eggs, often in chains.

Drone larvae and pupae may be seen bulging from smaller worker cells. These may have been laid by workers. After emergence the problem queen has received poor nutrition and may be superseded. Replace the queen and add healthy workers to replenish the food stocks.

Purple brood

When the forage bees collect and use the pollen and nectar from *Cyrilla racemiflora* (Southern Leatherwood USA) the larvae and pupae can become purple in colour.

Sacbrood

When you examine the brood frames you may notice unusual scattered cells. There is no abnormal smell. When you remove the larva from the cell you notice that it looks like a sac

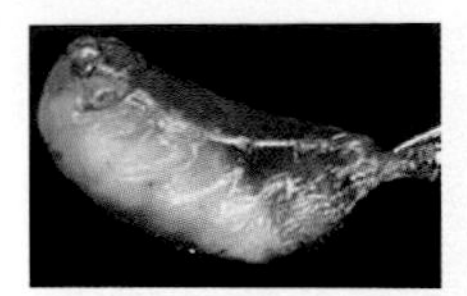

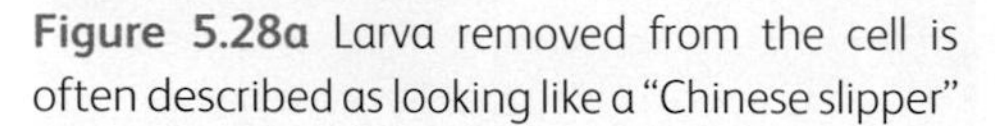

Figure 5.28a Larva removed from the cell is often described as looking like a "Chinese slipper"

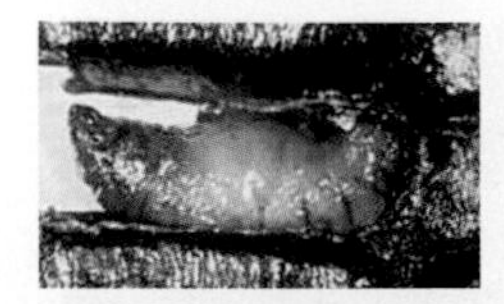

Figure 5.28b Sacbrood L5 larva *in situ* in a cell

of liquid. Sacbrood is caused by a virus, which is unusual as it has a Latin name – *Morator aetatulas*. This is a picornovirus-like virus. This virus is an RNA virus and enveloped. Note that this virus can affect *Apis mellifera* and *Apis cerana.*

Apis cerana can suffer from epidemics of a related virus called the Thai sacbrood virus.

Clinical signs

Scattered around the brood frame will be larvae at stage L5 upright in the cell and the cell may be capped. The larvae colour changes from pearly-white to grey and finally black. When larvae are removed from the cell they often appear to be in a sac filled with water. Closer examination of the larvae may reveal that the head is not properly developed and may be darker than the rest of the body. The larva is not attached to the cell wall and is loose in the cell and easily removed. Pupation fails as the larva is unable to dissolve the sac-like skin.

Diagnosis

The larva is white and spongy – chalk-like. Diagnosis is made through the appearance of the larva and the retarded head development. Check the total protein content of worker bees.

Similar diseases

With Chalkbrood and Stonebrood the larva is white and firm. Chilled brood's larva is dead but otherwise normal.

Treatment

There are no specific treatments available. Recovery occurs spontaneously once nectar flows are strong. Low protein availability in the spring may be a contributing factor.

Management, control and prevention

Being enveloped, the *Morator aetatula* virus is destroyed by disinfectants and soaps. Good hygiene of hive tools is therefore essential in control. This should also include clothing, gloves and boots. Maintaining good healthy hives is essential in control. It is important to note that the concentration of hives in an area may reduce food source availability and may lead to a general weakening of some hives. Large numbers of hives also increase the pathogen load in an area by robbing and drifting. Hive health can be maintained by re-queening the hive regularly. Have regular and routine brood comb changes, for example 3 frames per year per brood box in a modified Langstroth hive, to maintain a healthy number of new frames. Note the width of the cell reduces with each larva being developed in the cell and thus creates a correspondingly smaller worker.

Having natural form comb in top box design hives may help control sacbrood as the bees have to keep making clean natural comb.

Spotty or scatted brood

This can occur with a number of conditions. It is often seen in early spring and with a new queen or with a supersedure of a failing queen.

- Young queen
- Chalkbrood
- Sacbrood
- American foulbrood
- European foulbrood
- Supersedure – failing queen

Stonebrood

Stonebrood is caused by a common fungus from the Aspergillus family. A variety of Aspergillus spp. may be identified. The colour of the larvae or pupae may assist in the diagnosis of the condition. In the case of bumblebees, many other fungi may also be commonly found.

Clinical signs

When you open the hive there will be no abnormal smell. Examination of the brood frames reveals a few larvae or pupae which are dead and black or some other colour.

Table 5.1 Colours of larvae and possible causes

Colour of Stonebrood larva	Possible causal agent
Yellow/green	*A. flavus, A. ochraceus, A. nidulans or A.glaucus.*
Blue green/grey	*A. fumigatus*
Black	*A. niger*

Closer examination of the larvae or pupae reveals they have a collar-like ring near their head. The larvae die and become black and extremely hard. Eventually the fungus erupts and forms a false skin, which is white. In bumblebees, the overwintered queens may be infested by environmental fungi and die. The larva die in 2–4 days post-infection.

Diagnosis

Infection is through the skin or gut so the larval and pupal stages can be affected, unlike chalkbrood where only the larva becomes infected.

The fungus can be cultured. With care a wet mount of larvae can be prepared and examination of the neck of the larvae may reveal a ring around the neck showing mycelia penetrating throughout the larvae. Note that this fungus is zoonotic (it can affect people).

Similar diseases

Chalkbrood (more sponge-like) and Chilled brood.

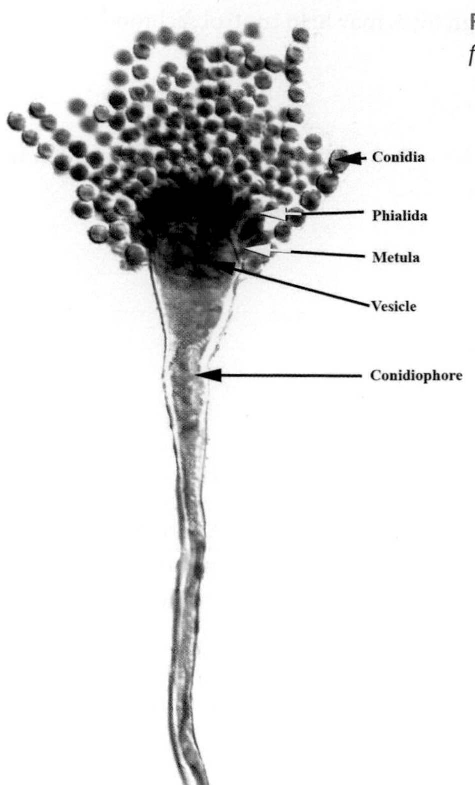

Figure 5.29 The major features of an *Aspergillus flavus* sporulating body

Treatment

Improve hive ventilation, hygiene and general hive health. Control *Varroa* and reduce hive stress factors. The spores are in the environment so maintain hive health.

Management, control and prevention

The larva ingests spores from the honey being fed by the worker bees but can also be infected through their skin. The problem occurs predominantly in stressed bees. It is important to note that the spores are in the environment anyway and only waiting for the hive to become suitably stressed.

Zoonosis

Aspergillus can infect man causing severe respiratory disease. Take care handling any stonebrood larvae. It is best to destroy the stonebrood larvae by incineration. Only handle with gloves and wash your hands afterwards. Avoid contact with the face before carefully washing your hands.

Tropilaelaps

Tropilaelaps are found commonly in parts of Asia.

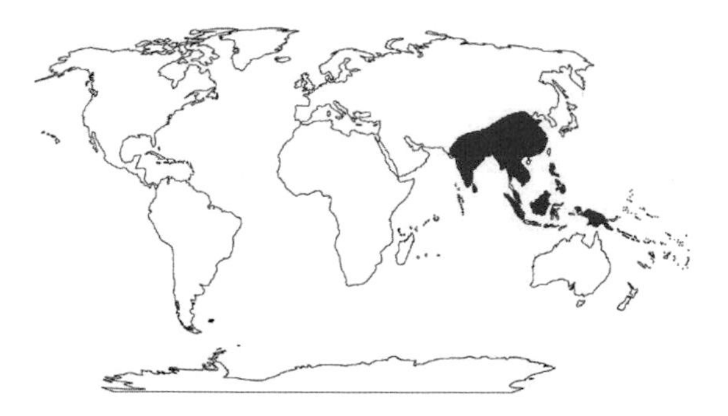

Figure 5.30 Geographical spread of Tropilaelaps spp.

Tropilaelap mites are 1mm long and 0.6mm wide. Of the two major species, *T. koenigerum* is slightly smaller than *T. clareae*.

Tropilaelaps is a parasite of the *Apis* species, the natural host is probably *Apis dorsata* (Giant Honey Bee).

There are four recognised species of Tropilaelaps:

Tropilaelaps clareae infesting *A. mellifera* in the Philippines.
Tropilaelaps mercedesae infesting *A. dorsata, A. laboriosa* and *A. mellifera.*
Tropilaelaps koenigerum infesting *A. dorsata* and not *A. mellifera.*
Tropilaelaps thaili infesting *A. laboriosa.*

The mite can live on a number of *Apis* species – *A. cerana* and *A. florea* – but has not been seen to replicate. *Tropilaelaps mercedesae* is probably the greatest threat as it has a wide species association.

Figure 5.31 Adult female Tropilaelaps dorsal surface

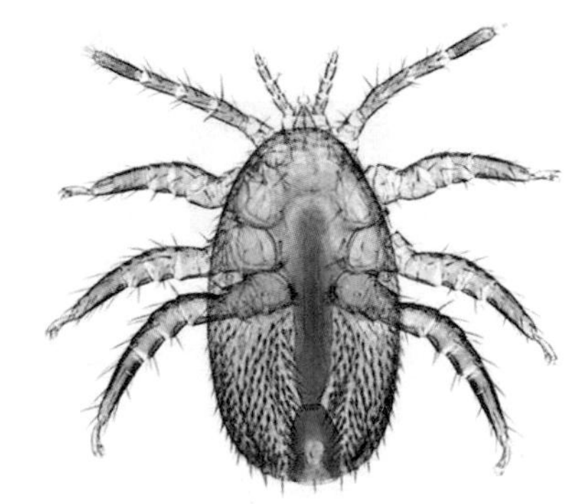

Figure 5.32 Tropilaelaps ventral surface

Figure 5.33 Tropilaelaps on pupa

Clinical signs

The adult female mite infests an L5 larva as its cell is about to be capped.

The larva pupates and the mite feeds off the pupal haemolymph. The mite lays eggs. The first egg which hatches is a male and the subsequent mites are female. The male mates with his sisters. The life cycle is much faster than in *Varroa* by a rate of 25:1. There is no preference for worker or drone infestation, unlike in *Varroa.*

The adult female mite cannot feed off adult bees and will die within three days without brood to feed upon. Therefore, the phoretic risk is much lower than with *Varroa.*

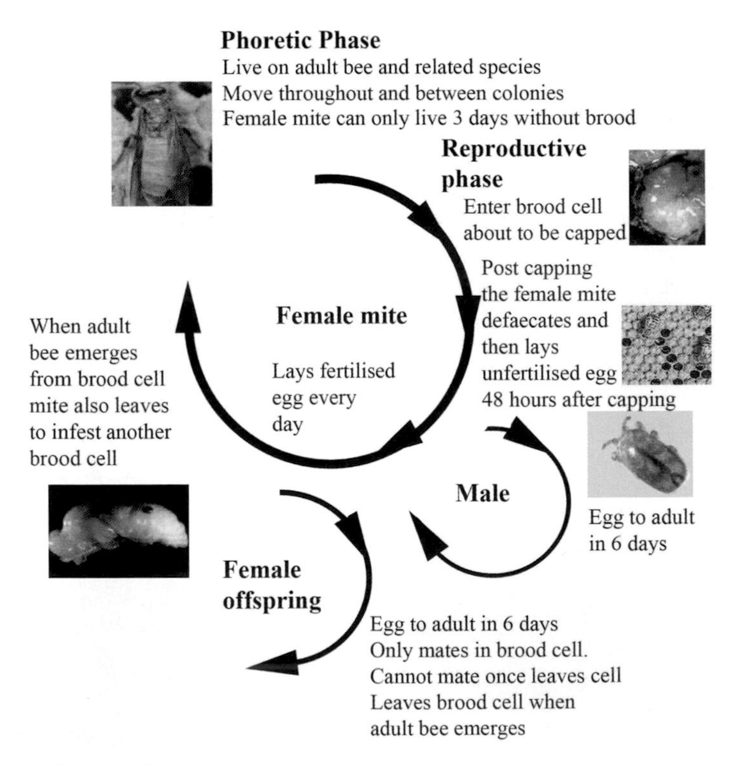

Figure 5.34 The life cycle of Tropilaelaps

From this a generation time can be calculated:

Table 5.2

Time	Event	Time	Event	Days since egg laid (day 16 queen would emerge)	
2 days	Male laid	8 days	Male matures	W17	D19
3 days	Female laid	9 days	Females mature	W18	D20
4 days	Female laid	10 days	Females mature	W19	D21
5 days	Female laid	11 days	Females mature	W20	D22
6 days	Female laid	12 days	Females mature	WE	D23
7 days	Female laid	13 days	Females mature	D24	
8 days	Female laid	14 days	Females mature	DE	
9 days	Female laid	Female not mature in time to be mated			

W# = Worker and day post-lay. D# = Drone and day post-lay. WE= Worker Emerges. DE= Drone Emerges

On average, the female mite produces three new females (as well as herself) for each larval worker infested and six female mites for each larval drone infested.

There can be many different female mites infesting the same larva at the same time – 14 have been recorded. It is even possible for new females to produce a mated female offspring before drones mature.

The African honey bee with an emergence two days earlier than the European honey bee would produce only one or two mature female mites, not four!

Secondary infections

Mites have the ability to affect their host with a number of different secondary infections, mainly viruses. In Asia, hives infested with Tropilaelaps produce offspring with more deformed wings. High infestations may result in the death of entire colonies. However, for the infestation to get out of control the hive has to be stressed.

Diagnosis

Mites seen in the hive. Note the difference between Tropilaelaps and *Varroa*:

- Tropilaelaps is smaller and elongated. The horseshoe shape on the rear is very characteristic.
- As Tropilaelaps cannot live on the adult, there are fewer adult females in the hive and so visualisation by looking at the worker bees is unlikely.
- The nymph stages of Tropilaelaps are brilliant white. Mites found inside the hive are both male and female, although male mites are more often found in capped cells.
- The adult mites move quickly and may be difficult to catch. This can be assisted using a wet fine bristle brush. A small mouth aspirator may be useful. They move much faster than *Varroa* mites.

Examine capped brood for the presence of the mite using a capping fork.

Look at the hive debris. The bodies of Tropilaelaps may be recognised. In debris from *Apis dorsata,* many of the Tropilaelap mites will show signs of trauma from the worker bee mandibles.

Similar diseases

Braula coeca may resemble Tropilaelaps. However, *Braula* is an insect with six legs whereas Tropilaelaps, as a mite (arachnid – spider), has eight. *Varroa* is another species of mite which infests Apidae but is a different shape.

Treatment

Mites can be controlled without brood – cold climates may limit spread of mites.

Chemical

Tau-Fluvalinate (Apistan) and Flumethrin (Bayvarol) are both synthetic pyrethroids. The strips are normally applied for up to six weeks. Note these chemicals can build up in the honey and eventually will affect the health of the bees. Two years of application has been demonstrated to affect queen bee fertility.

Apigard – Thymol

Once the temperature is above 15°C, thymol products can be used. Use two applications ten to 15 days apart.

Api Life Var – Thymol, eucalyptus oil, menthol and camphor

This is used twice, 14 days apart.

Organophosphates – Coumaphos (Checkmite+)

The checkmite+ strip hangs down between frames in the brood area so bees can walk on it and pick up minute amounts of the active ingredient. Do not put the strip on the bottom board or tops of frames.

Apivar (Amitraz – Formamidine)

This may be considered with veterinary control.

Mechanical

Eliminating brood for more than five days will eliminate mites.
Queen trapping – see Chapter 12.
Artificial swarm technique – see Chapter 12.
Shook swarm technique – see Chapter 12.

Others

Reducing drifting.

Management, control and prevention

Employ biosecurity in countries which are negative. While low numbers are present on the adult, they could still act as a carrier (fomite/phoretic). Note that Asian bees may be inadvertently transported in cargo. But also note that the mite only lives for three days without brood, making cargo less of a biosecurity risk.

If the mite is detected early in a new colonisation, the hive could/would be destroyed. It would be possible to control it by eliminating brood for more than five days.

Tubular brood

The Lesser Wax Moth – *Achroia grisella* – will burrow in straight lines at the base of the cells producing tunnels throughout the hive.

Varroosis

In most countries *Varroa* is currently the major health problem for domesticated *Apis mellifera*. It is still absent in most of Oceania, but it is a problem in most of the rest of the honey producing areas of the world.

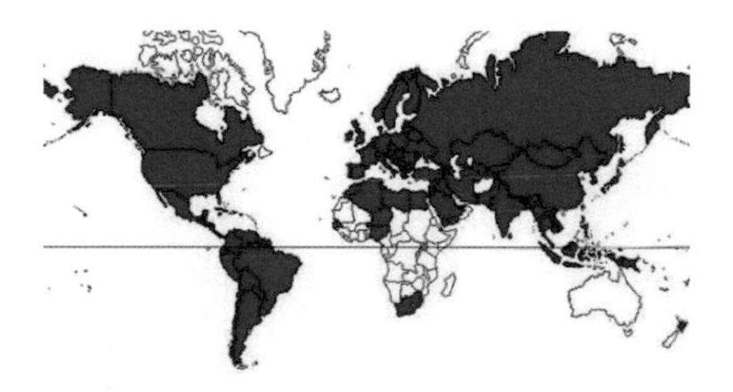

Figure 5.35 Global distribution of *Varroa*. Positively confirmed countries/areas indicated in red

The problem has arisen because *Varroa destructor* recently moved from its natural host *Apis cerana* into domesticated *Apis mellifera* hives.

The *Varroa* mites can also infest other native bee species – bumblebees (*Bombus pennsylvanicus*) for example. However, while they do not multiply on other bees, they can act as a transportation system (phoretic) which is important as it complicates the control of *Varroa*.

It should be remembered that mites and bees have been living together for many millions of years. Bumblebees and other *Apis* species have many mites which have various levels of predation.

There are many species of *Varroa* affecting *Apis*: *Varroa destructor, V. jacobsoni, V. underwoodi* and *V. rinderi*. *Varroa rinderi* appears to be a specific parasite of *Apis koschevnikovi*. Note that there are also other closely related *Euvarroa* species which have been found on *Apis florea* – *Euvarroa sinhai, E.wongsirii* and *E. adreniformis*.

Varroa infestation of a hive can result in almost no clinical signs or, in a combination of other issues which contribute to result in severe stress on the hive. *Varroa* mites may also infect *Apis*

mellifera with a range of secondary viruses and may play a significant role in Colony Collapse Disorder.

Clinical signs

Most hives are infested with *Varroa*, with no or few clinical signs. If the mite remains at a low level, the hive can be extremely productive and successful. Absence of the mite is extremely difficult. Once an area is infested, beekeepers can only reduce the infestation, they cannot eliminate the mite. The mites move between hives by robbing; the drifting of honey bees and bumblebees can act as a means of transport called the phoretic phase of the mite's life cycle.

The *Varroa* mites are easily recognised in the hive debris. The female mite is quite large at 1.5mm in width. In the summertime, it may even be possible with careful examination to see the mite on the worker honey bees. Note do not chill the hive in trying to find the mite as you will only make the situation worse.

As a clinical problem the mites are classically seen in stressed hives in late autumn.

While *Varroa* can contribute to the collapse of a hive, it can take 3–4 years for the mite numbers to build up in a hive to ultimately result in the collapse of the hive.

The mite is oval. All the brown mites you see are adult females. Down the microscope it will be clear the mite has eight legs. The mite is a member of the spider family not an insect. In some cases the carapace (dorsal surface) becomes stretched because of oocytes inside the female.

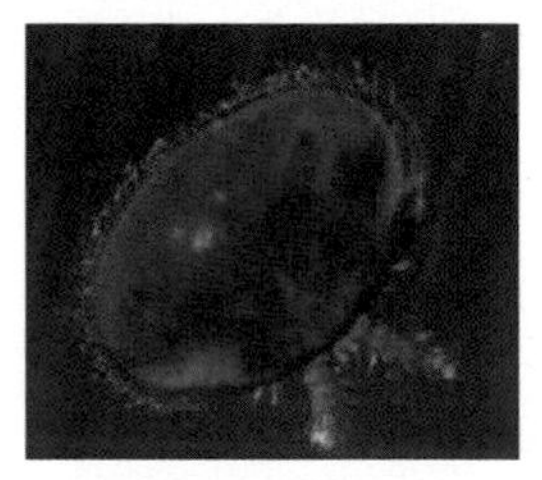

Figure 5.36 The dorsal surface of the mite

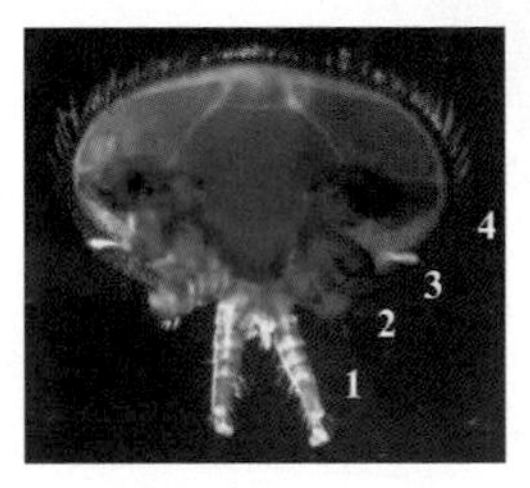

Figure 5.37 Ventral surface of the mite

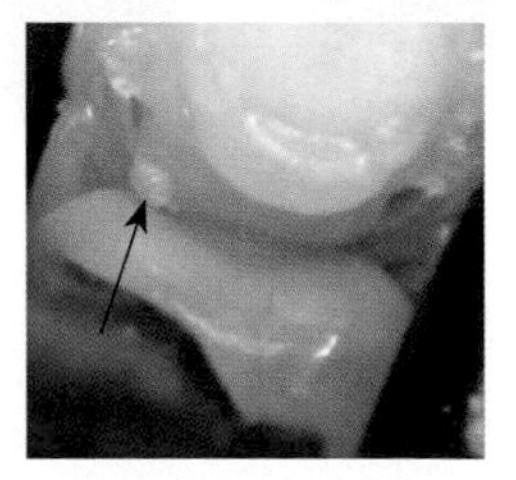

Figure 5.38 Young mites may be pale/white.

Secondary infections

While feeding on the haemolymph of the pupa, the *Varroa* mite may assist in the transmission of a number of viruses. In particular, deformed wing virus is easily seen when the brood is being examined.

There are a number of other viruses which are also associated with *Varroa* infestation.

Viruses which have been associated with transmission with *Varroa* are:

Acute bee paralysis virus
Black queen cell virus
Deformed wing virus
Israeli acute paralysis virus
Kashmir bee virus

Figure 5.39 Can you spot the worker with the deformed wing?

Slow bee paralysis virus
Varroa destructor macula-like virus
Varroa destructor virus –1

These all may contribute to collapse of the colony and may be classified as Colony Collapse Disorder (CCD), albeit CCD might be associated with a specific pathogen as it is recognised that *Varroa* is not essential in the pathogenesis.

There is evidence that *Varroa* create an immunodeficiency crisis within the haemolymph of the bee.

Life cycle

The adult female mite infests the L5 larva as the larva is about to be capped. The adult female defecates in the cell as it is being capped. The female mite lays her first egg, which is unfertilised, about 60 hours after capping. This first egg laid is a male. Subsequent eggs, laid every 25–36 hours, are fertilised and are female. The newly-born male mates with his sisters.

The adult female mite can live for months when no brood is available to allow multiplication. With brood, the life expectancy of the adult female mite is 27 days.

On average, the female mite produces a new female (as well as herself) for each worker larva infested and two female mites for each drone pupa infested.

Evolution of varroosis and lessons from Apis cerana

In *Apis cerana* the *varroa* mite is restricted to drone cells and therefore has limited impact on the survivability of the infested colony. *Apis cerana* have various adaptations to manage *Varroa* infestations. The capping of drone brood is hardened and so weakened infested drones are unable to open the cocoon caps and the *Varroa* die in the capped cell. The workers also seem able to detect the infected pupae and will uncap and remove them. *Apis cerana* also display a range of more intense grooming behaviours and are able to remove or damage the mites by biting them. In addition they perform a "dance" to call up help from other workers to remove the mites. The combination of these factors enables *Apis cerana* to remove 98% of mites from their colonies.

Note the African *Apis mellifera* races have a shorter development period for the female queen and worker although the drone still requires 24 days to develop. However, this shorter development cycle may make a difference to reduce the explosion of *Varroa* within a hive during peak honey flow periods. Examination of hives infested with *Varroa* in Africa have found few mature *varroa* associated with emerging worker brood.

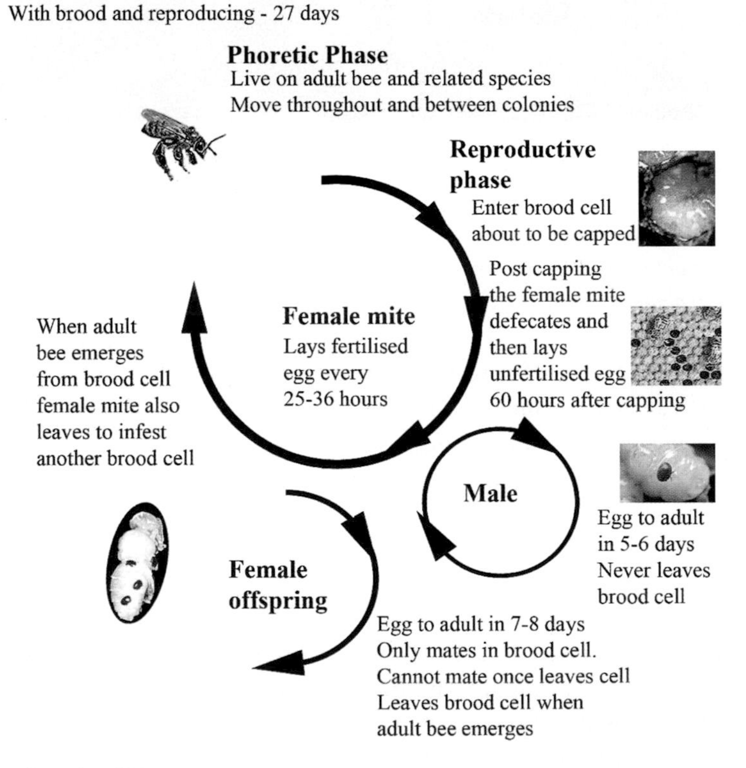

Figure 5.40 Life cycle of *Varroa*

From this a generation time can be calculated:

Table 5.3

Time	Event	Time	Event	Days since egg laid (16 days post-lay queen emerges)	
3 days	Male laid	8 days	Male matures	W19	D19
4.5 days	Female laid	12.5 days	Females mature	WE	D20.5
6 days	Female laid	13 days	Females mature	D23	
7.5 days	Female laid	14.5 days	Females mature	DE	
9 days	Female laid	Female not mature in time to be mated			

Worker and day post-lay. D# = Drone and day post-lay. WE= Worker Emerges. DE= Drone Emerges

This explains why the queen larvae are rarely infested with *Varroa* mites, but the longer pupation of the drone allows for large number of mites to reach maturity and fertility. On the African honey bee, with an emergence two days earlier than the European honey bee, no mature female *Varroa* would be produced.

Transmission

Horizontal

Through drifting and robbing, particularly when there is a concentration of hives in an area. They can live on other bees and bumblebees can act as fomite carriers. They can also be carried through swopping combs between hives and on hive tools.

Vertical

Through the brood. New queens may be infested.

Diagnosis

Diagnosis is relatively easy as the mites are easily visualised, especially in the hive debris. A hand lens will assist and it can be confirmed using a dissection microscope.

Figure 5.41 *Varroa* easily seen on uncapped drone brood

Figure 5.42 *Varroa* absent on uncapped emerging worker brood from African honey bees

Visualisation of the mites

1 Drone pupae
 Drones are preferentially affected as they have a longer pupation and more mated females can be produced. Examine capped drone brood using a capping tool and look for the clear presence of mites on the drone pupa. This will not reduce the viability of the hive.
2 Windscreen washer fluid
 Take 500ml of car windscreen washer fluid (alcohol). Shake around 500 bees into the solution (one frame). Shake the fluid and bees. The bees are killed in the fluid. Count the mites which appear. The mites will generally float to the top. If less than ten, then hive infestation is acceptable. If more than 40, the hive needs treatment.
3 Ether roll
 Collect 200–300 bees into a jar and anaesthetise with ether (ether may be obtained from a car store as it is used in air to start engines). Use a 1–2 second blast. Roll the jar for ten seconds and the mites will dislodge and adhere to the side of the jar. The remaining bees can be spread on a white paper to observe additional mites, if required.
4 Icing sugar
 Collect 200–300 bees in a jar. Replace the lid with another lid cut with a #8 sieve in the top. Introduce 10g of icing sugar through the sieve. Roll the jar around. Allow the jar to sit for a couple of minutes. Pour the sugar and mites out of the jar through the #8 sieve onto a clean sheet of paper.

Monitoring the hive for the number of mites

Monitor mites with a screened bottom board with a sticky surface – spray on cooking oil for example. Review the interpretation of the number of mites per examination. During the winter one new mite per day represents around 50 mites in the colony.

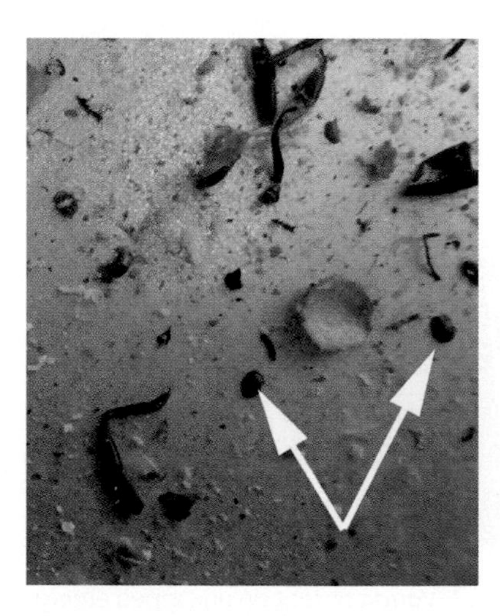

Figure 5.43 Hive debris examination – *Varroa* arrowed

Figure 5.44 Mite seen on the worker bee

Routine monitoring for Varroa *infestation*

Monitor for *Varroa* four times a year – the 'June gap' (UK) after the spring flow is a vital time. If more than ten mites fall to the hive debris a day in June, treat the hive.

At post-mortem, various histological changes may be seen including reduction in the hypopharyngeal glands of workers. There may be a reduction in the size of the fat bodies. In careful examination, the bite wound may be seen on the pupa. This can be assisted by staining the white pupa with Typan blue, where the bite wound may reveal itself as a blue dot.

Similar diseases

Braula coeca may resemble *Varroa*. However, *Braula* is an insect with six legs whereas *Varroa*, as a mite, has eight. Tropilaelaps is another species of mite which infests Apidae but is a different shape.

Treatment

None of the treatments achieves elimination, only control.

It is vital to read and follow all manufacturer's advice regarding withdrawal times and bee product sales. Some of these techniques may not be legal in your area – check with your veterinarian or bee inspector first.

Figure 5.45 Medicated patty in Autumn *Varroa* control

Figure 5.46 Mite control using insecticide strips

Chemical treatments

*Tau-Fluvalinate (*Apistan*) or Flumethrin (Bayvarol)*

These compounds are both synthetic pyrethroids. Start treatment when the sticky boards reveal a heavy infestation. The strips are normally applied for up to six weeks. Note these chemicals can build up in the honey and eventually will affect the health of the bees. Two years of application has been demonstrated to affect queen bee fertility. These products may be more useful in the autumn as part of the *Varroa* seasonal control programme.

Apigard – Thymol.

Once the temperature is above 15°C thymol products can be used. Use two applications ten to 15 days apart.

Api Life Var – Thymol, eucalyptus oil, menthol and camphor.

This is used for six weeks.

Organophosphates – Coumaphos (Checkmite+).
The Checkmite+ strip hangs down between frames in the brood area so bees can walk on it and pick up minute amounts of the active ingredient. Do not put the strip on the bottom board or on tops of frames.

Apivar (Amitraz – Formamidine)
This may be considered with veterinary control.

Organic treatments

Oxalic acid
This is not authorised in some parts of the world – for example the UK. Oxalic acid can kill brood and therefore has to be used in broodless colonies – in the wintertime. Mix 6% of oxalic acid with 30% sugar solution and apply by the trickling method. Note that the products have to be used fresh and breakdown products can be toxic to the bees – hydroxymethylfurfuraldehyde. This product works by damaging the proboscis of the *Varroa* mite.

Formic acid
Formic acid can be used on the bottom boards. Administer 30 mls every week for 3 to 5 applications in the spring. Then apply again in June. The air temperature needs to be above 22°C but below 30°C for the treatments to be effective.

Coriander (Coriandrum sativum)
Varroa do not like the smell of coriander pollen or nectar. This may help bees relieve themselves of the mites.

Biological

A variety of control measures are being designed under the title of Integrated Pest Management (IPM) to reduce the reliance of beekeepers on chemical control measures, which may create potential resistance problems and chemical residues in honey.

Resistant lines of bees – hygienic bees
Use more resistant lines of bees – some subspecies have slightly shorter pupation periods, which reduce the *Varroa* population, and thus its impact. Test and select for hygienic behaviour in queen bees so the bee naturally cleans up brood which is infested. Not selecting queens from chalkbrood colonies may also help. This is described in detail in Chapter 12. Selecting hygienic bees will prove to be extremely useful for a whole range of bee pathogens.

Drone brood sacrifice – drone trapping
Hang a super frame in the brood box at the end of the brood area. The worker bees will prepare free-form drone cells underneath. This is because this would be the cooler part of the brood area. Drones are preferentially laid in this area.

Monitor this area and when the drone cells are capped the entire drone brood area can be removed by a sharp knife. Burn the wax and drone pupae to kill the adult female mites.

Figure 5.47 Drone brood sacrifice

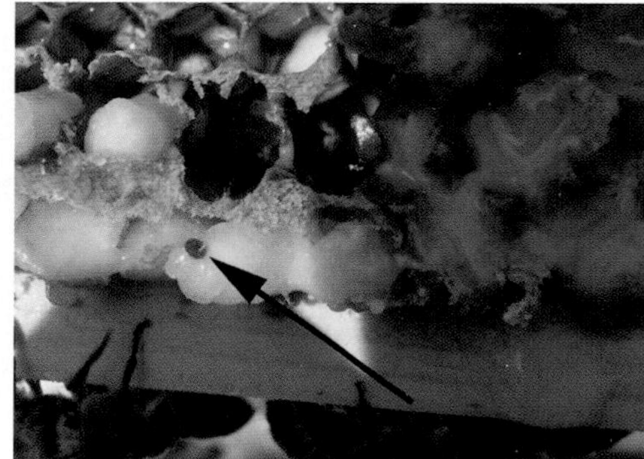

Figure 5.48 Mites on sacrificed drone

Queen trapping

Control the queen's movement within a closed frame queen excluder. This then controls where the eggs are laid and thus controls the age of pupae. This seals *Varroa* mites in capped cells and they are then removed from the brood. This should only be used in the spring or early summer (see chapter 12).

Competitive biological control

For competitive biological control using a predator mite – *Hypoapis* mites have been tried with some success. Predator fungi – such as Beauveria or Metarhizium fungi are also under trials.

Mechanical

Artificial swarm technique

Because this can be used to control a number of pathogens it is described in more detail in Chapter 12.

Shook swarm technique

Because this can be used to control a number of pathogens it is described in more detail in Chapter 12.

Powdered sugar dusting (using new dry icing sugar)

This encourages better grooming of the bees. Have sticky paper on the screening board to capture any falling mites. The mites start appearing within 15 minutes of application. Note use a new packet of icing sugar as humidity can affect the efficiency. Particle size needs to be 5μm to attach to the mite.

Screening bottom boards – sticky boards.

With a moderate to heavy mite infestation the sticky board will have 150 to 500 mites stuck to it.

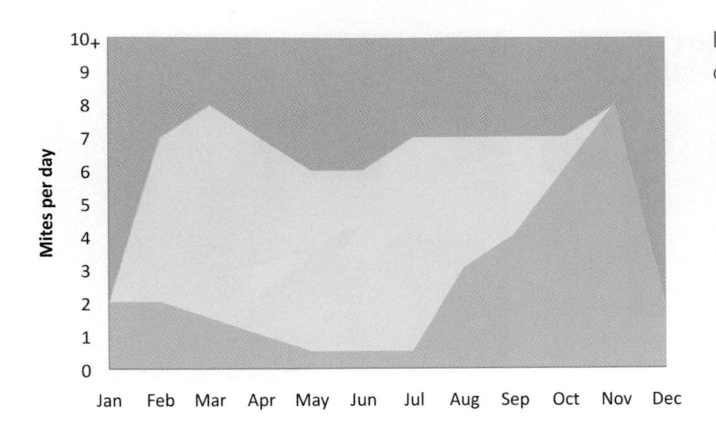

Figure 5.49 Guide to mite count

Reducing drifting

Note that drones deliberately will drift into other hives! Consider the hive's locality. Mark clearly the hive entrance to assist bee recognition of their own hive. As honey bees age they make more mistakes in hive recognition.

Management, control and prevention

Mite drop count

Using the daily mite drop technique, the hive *Varroa* risk can be estimated. When in the green area, the hive *Varroa* can be just monitored every 2 weeks. Once in the yellow area, monitor the hive weekly for changes in the *Varroa* count. If the count is in the pink area, treat the hive for *Varroa.*

Table 5.4 Review of possible *Varroa* management techniques over the year

Technique	Winter	Spring	Summer	Autumn
Open mesh floor, sticky paper				
Drone brood removal				
Drone comb trapping				
Apiguard – thymol products				
Apistan/Bayvoral – tau-fluvalinate/flumathrin				
Checkmite+ – Coumaphos				
Oxalic acid				
Icing sugar dusting				

Wasp parasites

Mutilla europaea is a parasitic wasp that enters a honey bee hive or a bumblebee nest. This wasp is very ant-like and about 14 mm long with an orange–red thorax with black hairs. The black abdomen has three (female) and two (male) silver white bands. The wasp larvae consume the developing honey bee or bumblebee larva. Once they emerge, the adults feed on the honey stores in the hive. The adults may make a hiss-like chirp which may be noticed when the hive is opened.

Other wasps may parasite on brood, notably *Melittobia spp* in bumblebees. *Philanthus bicinctus* is a predator of bumblebees in North America.

6

Managing the health of the adult bee

The life cycle of the honey bee

Once the bee emerges from the cell it quickly dries itself and then starts on its adult life cycle.

The African *Apis mellifera* queen may emerge at 14–15 days post-lay. The queen will live for up to 5 years, although her reproductive performance will progressively fail as her spermatheca empties of viable semen. The majority of queens will be replaced after 2–3 years. In marked queens, the age of the queen can be estimated by the colour used to mark the queen.

Marking	Year ending
White (grey)	1 or 6
Yellow	2 or 7
Red	3 or 8
Green	4 or 9
Blue	5 or 0

Figure 6.1 Queen marked white – from 2011

Figure 6.2 Queen marked red from 2013. Note some of the marker is on her left compound eye – this must be avoided

The African *Apis mellifera* drone emerges at 22–24 days as with the European drone.

The drone will live for a few weeks to up to 4 months within the hive. As it develops all the sperm are moved into the vas deferens and the testes shrink. The drone contributes nothing to the maintenance of the hive. When matured (after 10–12 days) he will fly out of the hive and seek out a courting arena. If he is unsuccessful in mating, he will return to hive (or an adjacent hive) in the evening (well before dark), eat and rest before flying to a courting arena the next day. In the autumn the remaining drones are thrown out of the hive and will be found dead around the hive in the morning. Early birds will often quickly eat up this bounty of big and fat bees.

Note different drone species fly at different times of the day.

Table 6.1 Time of day of mating (various species of *Apis* in Borneo and *A. mellifera* in Europe)

Species	Start	End
A. nuluensis	1045	1315
A. andreniformis	1200	1400
A. cerana	1415	1530
A. koschevnikovi	1700	Dusk
A. dorsata	1900	Dark
A. mellifera (Europe)	1400	1700

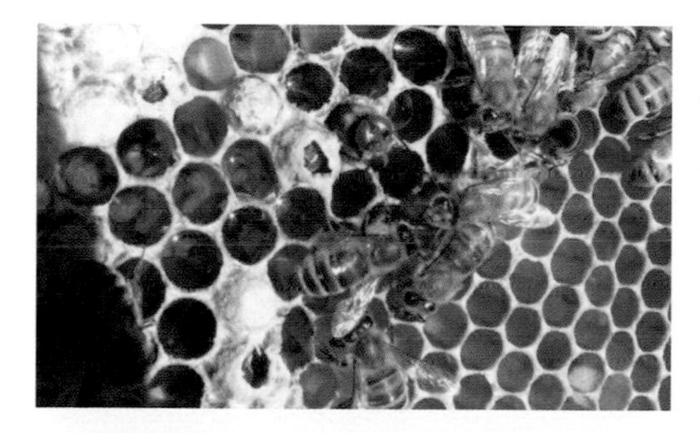

Figure 6.3 Emerging drone

Figure 6.4 Drone about to go on a mating flight

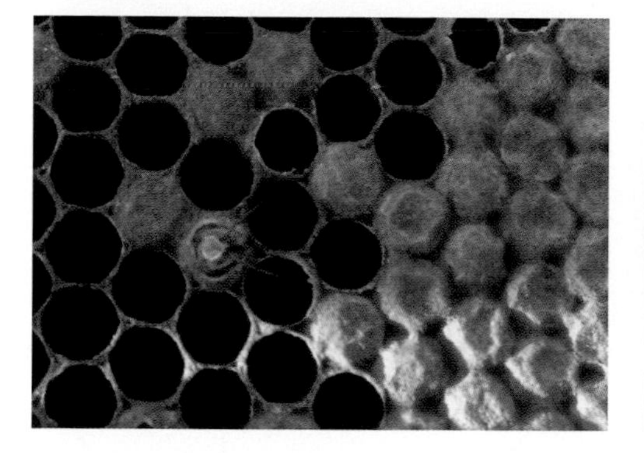

Figure 6.5 Emerging worker

Figure 6.6 Newly emerged worker wings wet and cloudy – note lower bee with detached wings

African *Apis mellifera* workers emerge at 19 days. This can be significant in the control of *Varroa*.

The worker bee then has a maturation cycle to perform during her short life cycle of only 35 days during the summer. Thus from being laid as an egg to death, the life cycle for a working honey bee is 56 days in the summer.

Working honey bees in the winter live longer as there is less 'work' to be done and the primary responsibility is survival of the queen until spring.

Table 6.2 The worker honey bee life cycle midsummer.

Days after emergence and age since lay		
1	22	Cell cleaning
2	23	
3	24	Queen care – nursing (feeding young) and wax production
4	25	
5	26	
6	27	
7	28	
8	29	
9	30	
10	31	
11	32	Wax work
12	33	Nectar processing
13	34	Guarding
14	35	Undertaking
15	36	
16	37	
17	38	
18	39	
19	40	
20	41	
21	42	Foraging:
22	43	Water
23	44	Nectar
24	45	Pollen
25	46	Propolis collection
26	47	Colony defence (soldiering)
27	48	

Days after emergence and age since lay		
28	49	Note in the wintertime the workers live longer as there is less foraging.
29	50	
30	51	
31	52	Also bees are able to revert to nursing etc when demands are high in the hive
32	53	
33	54	
34	55	But note the hypopharyngeal glands will normally atrophy in older workers.
35	56	

In the winter (or the wet season) the hive's primary function is keeping the queen alive. On warm days worker bees may fly out of the hive on cleansing flights to defecate. At the end of winter the hive has around 5–6,000 workers and a queen. In the summer the hive can expand to 80,000 workers, drones and a queen (plus developing queens). Over winter the workers may live 4 to 6 months depending on the duration of winter.

Figure 6.7 Summer hive

Figure 6.8 Winter hive

Flying patterns of adult bees

Cleansing flight

Cleansing flights occur when honey bees fly to defecate. They may be seen during a warm day during the winter months. In the other months, cleansing flights will be missed among the normal flying of forage workers and drones. The number of workers cleansing will generally be low.

In the spring, worker bees will remove dead and dying larvae – especially chalkbrood – by physically removing the dead larvae and flying outside the hive before disposing of the bodies.

Figure 6.9 Worker bee on a cleansing flight in winter. Note the mouse excluder

Figure 6.10 Removal of a chalkbrood larvae by a worker bee – arrowed

Learning flights

On a warm day bees may be seen flying in patterns close to the hive. It may even be mistaken for a hive about to swarm. On close examination these are workers learning to become forages. They are orientating themselves to the hive and their surroundings. These workers are about six weeks old (three weeks post-emergence). It is important to note these bees when sampling bees. These are not old forage bees. Therefore, they may give a false impression of the health of the hive.

Foraging flights

At around six weeks of age (post-lay) the worker bee graduates to become a foraging worker. She becomes more independent and will have to find food, water and potentially a new home for the hive. Some of this is done following directions given by other workers, sometimes totally independently. The bees communicate the location and direction of the new resource by dancing within the hive (see Chapter 2).

Bees on forage flights move in a very deliberate manner. Once they have left the landing board the bees will fly a short distance from the hive and then move in a straight line very fast.

Figure 6.11 Orientation workers flying outside the hive

When the foraging worker returns home it may be carrying pollen, honey, water, soot and/or dirt from a new potential home or propolis.

The life cycle of the bumblebee

There are numerous species of bumblebees (*Bombus*) each with its own particular life cycle. But the common biology follows a general style which is described below.

Queen bumblebee life cycle

The life cycle of the bumblebee is generally annual. Mated young queens overwinter in the ground, perhaps in an abandoned small mammal nest. Depending on ground temperature (and thus the time of the year) the mated queen comes out of hibernation and emerges to forage. *Bombus terrestria* is one of the earliest species to emerge at around 6°C – thus around mid-March in northern Europe. By contrast, *B. hortorum* emerge when the ground temperature reaches 9°C – around May in northern Europe.

The queens forage on available nectar and pollen. During this time the young queen's ovaries will develop rapidly. The queens forage during the day and roost at night under moss or other vegetation. Eventually the queen will start seeking out an area to establish her nest. Fighting over optimal sites is common and dead queens may be found around ideal sites.

Once the nesting site is selected, an area is cleared by the queen to form a small chamber. Pollen is collected and moulded into a mass upon which the first wax cells are made. The queen then lays her first batch of eggs.

The eggs hatch after 4–6 days and the larvae feed on the pollen mass and additional nectar and pollen as supplied by the queen. During this time the queen may forage for short periods of time. The wax cells are progressively expanded to accommodate the growing larvae. After 10–20 days (depending on the temperature) the final instar is complete. The larvae defecate, spin a cocoon and pupate. The queen then recycles any wax and pollen to make another brood nest in which she lays her second batch of eggs.

As the season develops, some species of bumblebee will start laying down pollen stores to assist larval development. The colony increases in size from a few dozen workers to around 200 workers. As the colony increases, eventually drone (male) (from unfertilised eggs) and virgin queen bees are produced.

The young virgin queens develop and leave the nest. However, they generally return at night. In the nest, they do not contribute to the work of the nest but will eat pollen and honey. During this time they build up their fat body into a large structure in their abdomen.

During the day the virgin queens will fly to mating areas and be mated. In some species the mating will leave a scar on the string apparatus. For example in *B. platorum* there are two marks and in *B. hortorum* there are four marks

Once the young queen has built up sufficient fat stores and filled her honey stomach with honey she leaves her birth nest for good and looks for a place to overwinter and hibernate.

The worker bumblebee life cycle

Adult workers emerge after two weeks as pupae. The adult workers dry within hours and their wings harden within a day. The new workers start foraging and assisting the queen. The queen may still assist in the foraging.

As the number of workers increases with each brood, eventually the queen stops leaving the nest. Unusually within insects the adult size of bumblebees can vary depending on the nutrition provided to the larvae. Thus the first workers, only fed by the queen are smaller than subsequent generations.

The drone bumblebee life cycle

The males become adult and leave the nest and do not return. They occupy a territory which they scent mark. These scent marked areas attract virgin queens and mating takes place. The males die before the onset of winter.

Ageing bees

Queen honey bees

The easiest way to age a bee is to colour mark her thorax different colours to indicate different years. As the bee ages the wings become more ragged. The small hairs (microtrichia) on the wings and hairs on the abdomen become worn off and so the queen becomes progressively hairless. The process of egg laying also progressively droops the abdomen and the end of her abdomen can show wear by friction from the edge of cells.

Worker honey bees

It is possible to age bees by looking at their wings. As the bee ages the wings become more ragged and the small hairs (microtrichia) on the wings become worn off. Using histology the worker bee's life cycle stage can be estimated looking at the development of the wax glands.

The hypopharyngeal gland development may also provide an indication of the stage of life cycle occupied by the worker bee.

Bumblebees

An old queen can be easily differentiated from a young bee. The old queen will have ragged wings with loss of the hairs on the wings (microtrichia). Dissection will reveal the active ovaries and a small fat body, which changes colour from white to yellow, and then eventually to brown.

Disorders of the adult bee

Acute bee paralysis virus (ABPV or APV)

Acute bee paralysis virus (ABPV or APV) is caused by a Dicistroviridae. This is an RNA virus. Acute bee paralysis virus affects both honey bees (*Apis mellifera*) and bumblebees (*Bombus terrestris*).

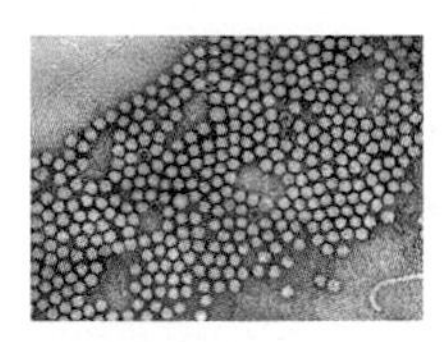

Figure 6.12 Acute bee paralysis virus in the electronmicroscope.

Clinical signs

The condition is more likely to be seen in the summer. Infected bees have reduced hair cover and appear black and with bloated abdomens. Adult bees may present with trembling. The position of the wing is abnormal, asymmetric or perpendicular to their bodies. Infected bees may be seen crawling on the ground and up grass stems just outside the landing zone. Death of the bees is rapid in 3–5 days. The virus also kills pupae which will limit the development of *Varroa*. In common with a number of bee viruses, ABPV is believed to play a role in Colony Collapse Disorder (CCD).

Diagnosis

ABPV is caused by a Dicistroviridae which is an RNA virus. The virus is related to Israeli acute paralysis virus (IAPV). Antisera will also cross-react with Kashmir bee virus.

Diagnosis can be made through PCR. This is a technique that looks for an organism's genetic code (DNA or RNA). A positive PCR result indicates that the organism (virus for example) is present but not that the organism is viable.

The virus appears to have a tendency to infect the brain, hypopharyngeal gland, semen and fat body specifically (trophism).

Similar disorders

Poisoning and other paralysis condition of bees
Israeli acute paralysis virus (IAPV)
Chronic bee paralysis virus (CBPV)
Slow bee paralysis virus (SBPV)

Treatment

There are no specific control measures. Control the number of *Varroa* mites in a hive.

Management and Control

Control *Varroa* infestation. Ensure the hive has plenty of food and is not placed under extreme environmental stress.

Do not reuse combs from infected colonies. Sterilise the inside of the super by scorching with a blow torch.

If the condition is extreme destroy the colony and burn all the bees and combs. Sterilise all equipment and clothing.

Air sac mites of bumblebee – *Locustacarus buchneri*

Locustacarus (Bombacarus) buchneri is an internal mite parasite of adult bumblebees. This lives in the large tracheal air sac located in the first abdomen segment of bumblebees. The gravid female becomes very round and sac like before laying her eggs. As the numbers of mites increase, especially towards the end of the season the bumblebee may become very weak. *Locustacarus buchneri* may contribute to the eventual demise of the bumblebee.

Alpha-1, 2.1 and 2.2 bacteria

These are a bacterial group related to bacteria found in ants. They are often found in the intestinal tract of a few honey bees in a hive. They are not associated with any known pathogenic process.

Amoebiasis – *Malpighamoeba mellificae*

Amoebiasis is associated with an amoeba parasite *Malpighamoeba mellificae* – a protist amoeba – which in the honey bee infests the Malpighian tubules of the adult bee.

Clinical signs

The condition is more often recognised in the spring. There may be an association with Nosema infestation and Bee virus X, although this may be seasonal. If the hive is watched closely the bees have a mild yellow diarrhoea. The bees may be practising more cleansing flights than normal. These signs are also seen with Nosema. The pathogen is transmitted between bees by ingestion of infective cysts.

Diagnosis

Primary trophozoites develop in the ventriculus epithelial cells. Secondary trophozoites and mature cysts are found in the Malpighian tubules. The mature cysts can be seen on histological examination or on direct examination of smears of the Malpighian tubules. The cysts are 5–8μm in diameter.

Malpighamoeba mellificae may be revealed by gently crushing the Malpighian tubules with a cover slip and looking down a light microscope.

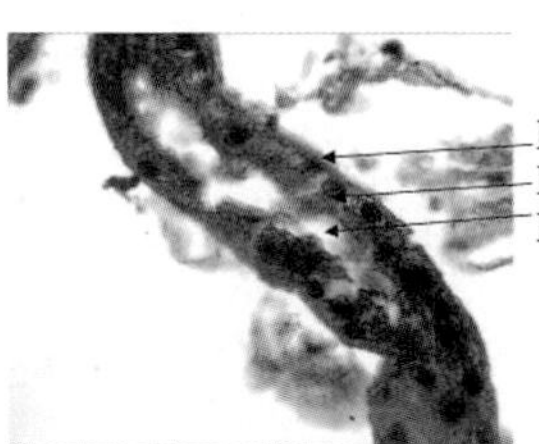

Figure 6.13 Normal Malpighian tubules

Treatment

None specific.

Control

Good hive management and reduction in environmental stress. Control of clinical Nosema. Glacial acetic acid may be helpful to sterilize comb and equipment.

Ants

Ants are a natural enemy of the bee. It is important to ensure that ants cannot gain entrance to the hive. The weaver ant (*Oecophylla smaragdina*) is a particular enemy. The weaver ants will attack any lone bee, including the Giant honey bee (*Apis dorsata*) and if more than a couple of weaver ants get a hold on the bee, the bee will be torn apart. If the weaver ant finds a hive, the guard bees must remove them physically from the locality to stop other ants following the pheromone trail. *Apis andreniformis* create 2–5cm wide sticky repellent resin bars before and after the hive on the branch.

Apicystis bombi

Apicystis bombi is a microsporidia, a unicellular parasitic fungi, similar to Nosema. *Apicystis bombi* may be isolated from Bumblebees. These microsporidia are larger than Nosema spp. at 11–14 x 3–5μm. *Apicystis bombi* may be found in the fat bodies and may be associated with a very white but indistinct fat body.

Aphid lethal paralysis virus

This is a common intestinal virus of Aphids. This virus may pass to honey bees in the late summer, when honey bees feed on honeydew, the secretion of the aphid. No clinical symptoms are related to this virus. Aphid lethal paralysis virus is a dicistovirus. The virus is related to the Berkeley bee virus.

Arkansas bee virus

The Arkansas bee virus is a picornovirus-like RNA virus infecting *Apis mellifera*. This virus has not been associated with any specific clinical problem.

Bee louse – *Braula coeca*

The bee louse is called *Braula coeca*. These can be found on any honey bee but they prefer the queen bee. The clinical signs are extremely mild. The adult honey bees appear not to be

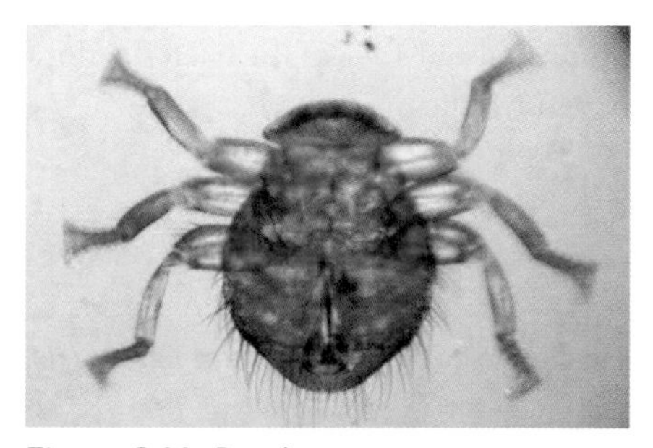

Figure 6.14 *Braula coeca*

Figure 6.15 The *Varroa* mite (eight legs) and the *Braula* fly (six legs) note the similarity in size

clinically affected. The honey bee and *Braula* appear to live in harmony. The *Braula* steals food from the mouth of the bee. The eggs are laid on the honey capped cells. The larvae burrow into the cappings, giving them a cracked appearance and hence can damage the appearance of the capped honey. The *Braula* is blind.

The bee louse is actually incorrectly named as it is actually a flightless fly and therefore another insect with six legs not eight. *Braula* may be confused with *Varroa* as they are similar in size – but *Varroa*, as a mite, has eight legs and identification is easy down the microscope.

Methods to control *Varroa* will generally control *Braula*. In addition, place combs in the freezer and freeze combs for seven hours as this will kill any *Braula*, including eggs. Attempts to control *Varroa* have virtually eliminated *Braula coeca* from many parts of the world.

Braula coeca is not present in Australia and is, therefore, a notifiable condition.

In *Apis laboriosa* there are two closely related species *Megabraula onerosa* and *Megabraula antecessor*. Both appear to live in harmony with *A. laboriosa*.

Bee virus X

This virus is associated with *Malpighamoeba mellificae* (Amoebiasis) in *Apis mellifera*. It has been reported that in the later winter it may shorten the bee's life expectancy. This is as yet an unclassified spherical RNA virus.

Bee virus Y

Bee virus Y is associated with Nosema in *Apis mellifera*. No specific signs are associated with the virus. However, peak isolation occurs in the early summer.

This is as yet an unclassified spherical RNA virus.

Berkeley bee virus

This is a picornovirus like RNA virus. No specific signs are associated with the virus.

Bifidobacterium

Bacteria of the Bifidobacterium genus are normally microaerophilic (requiring a low oxygen atmosphere), but the species found in bees may also be aerobic (requiring oxygen). The bacteria are Gram positive rods and they sporulate. Bees have specific species associated with them. *Bifidobacterium asteroids*, *B. coryneforme* and *B. incidum* are commonly found in *Apis mellifera*.

Bifidobacterium are found in most workers and compose about 15% of the bacterial flora of the bee's intestine. They are not associated with any pathogenic process.

Big Sioux River virus

Big Sioux River virus is a common intestinal virus of Aphids. This may pass to honey bees in the late summer when bees feed on honeydew. No clinical symptoms are related to this virus. Big Sioux River virus is a RNA dicistovirus. The virus is related to *Rhopalosiphum padi* virus.

Chronic bee paralysis virus (CPBV)

Chronic bee paralysis virus (CPBV) is found in *Apis mellifera*. The virus is associated with behavioural changes with a reduction in hair colouration. This is as yet an unclassified ovoid RNA virus. This condition of the bee is not new as it was described by Aristotle 2000 years ago! The paralysis form is also called 'mal des forêts' or sickness of the forest.

Clinical signs

There are two clinical syndromes.

Type 1

The adult bees present with an abnormal trembling. The bees may have a paralysis that limits flight. Infected bees may be crawling on the ground and up grass stems. Examination of the adult bee may also reveal a bloated abdomen.

Type II

Infected bees have reduced hair cover and appear black and shiny – 'little black syndrome'. The hairs are removed by other bees in the hive. Death occurs in about seven days.

Diagnosis

Pathological examination of the abdomen may reveal bloat associated with distension of the honey stomach with liquid. PCR examination of the bee confirms the diagnosis.

Similar conditions

Poisoning and other paralysis conditions of bees
Acute bee paralysis virus (APBV)
Israeli acute paralysis virus (IAPV)
Slow bee paralysis virus (SBPV)

Treatment

There is no specific treatment.

Management and control

Do not reuse combs from infected colonies. Sterilise the inside of the super by scorching it with a blow torch. Do not move hive tools between hives or apiaries. Destroy the colony and burn all bees and combs.

Chronic bee paralysis virus satellite

In cases of Chronic paralysis bee virus in *Apis mellifera*, a small virus particle is sometimes seen. Its clinical significance is not known.

Cloudy wing virus (CWV)

Examination of the bees (*Apis mellifera*) reveals that the wings are not transparent but cloudy and discoloured. This discolouration occurs in heavily infected bees with Cloudy wing virus.

The individuals are weakened and soon die. The virus is quite common, for example occurring in 15% of UK hives. The *Varroa* mite is a vector in the transmission of the virus. This is as yet an unclassified spherical RNA virus.

Management and control

Do not re-use combs from infected colonies. Sterilise the inside of the super by scorching with a blow torch. Do not share hive tools. Keep *Varroa* mite populations under control. In severe cases, destroy the colony and burn all bees and combs.

Colony Collapse Disorder (CCD)

Hives will fail for a great range of reasons and most are perfectly normal. About 15% of hives are expected to be abandoned (fail) overwinter.

Natural causes of colony collapse could include:

- The queen runs out of semen stored in the spermathateca. If she fails to produce fertilised eggs the hive will fail due to the lack of workers.
- The hive swarms too many times leaving a young queen with too few workers to survive a winter.

One of the problems with Colony Collapse Disorder is that there is no really clear definition of the problem. The beekeeper knows that the hive or hives are failing or are found to have failed in the spring. When the hive mortality exceeds 20% then beekeepers naturally suspect more than Nature's normal culling. Depending on your definition of colony collapse disorder, this problem may be the same as other conditions. Similar conditions have been termed autumn collapse, May disease, spring dwindle, disappearing disease and fall dwindle disease. Many of these conditions have been described for over 100 years, confounding the definition of CCD.

Clinical signs

The condition is recognised more in the spring. When hives are examined in the spring they are found to be empty of bees. Colony Collapse conditions are seen around the world and may well have different causes in different locations.

In the autumn, examination of the hive reveals that the queen, drones and brood are present. But there is a general lack of workers, especially foraging workers. The brood looks normal and there is no unusual hive odour and the brood cappings look normal. There are no mass deaths of workers outside the hive. The hive's stores appear to be in place with both pollen and honey available. However, with a lack of forage workers, the hive is progressively going to fail through starvation. The hive has not been robbed.

As yet there is no clear causal agent. The most common belief is that the condition is associated with a number of agents – environmental and pathogenic. There are reports that the hive's stores are also not attacked quickly by wax moths or small hive beetles but these reports are unsubstantiated. In addition, the colony appears to be reluctant to consume offered food, however, if the hive's stores are in place offering additional food is not restoring the bee's need. This might indicate that the hive is not starving, but where have all the foragers gone?

Definition

The US definition requires:

- Unusually large amount of brood for the adult bee population
- Few or no dead worker bees in or near the hive

Although these two conditions really go hand in hand the brood is near normal whereas there is a general lack of workers. Careful monitoring of the hive will reveal a logarithmic progressive reduction in forage flights in worker bees.
There have been numerous suggestions as to the causes of CCD:

Pathogens

In North America there have been associations between Nosema and Invertebrate Iridescent Virus (IIV). CCD is not reported in Australia and interestingly IIV and the *Varroa* mite is not reported from Australian bees.

Many studies point to *Varroa* as a major stressor on the hive, particularly in combination with a number of viruses which the *Varroa* mite is known to transmit. These viruses may act as an immunosuppressive on the honey bees.

Chemical

Modern farming methods use powerful chemicals to reduce pest infestation of crops. Many of these pests are arthropods and chemicals used to control arthropods will weaken bees often

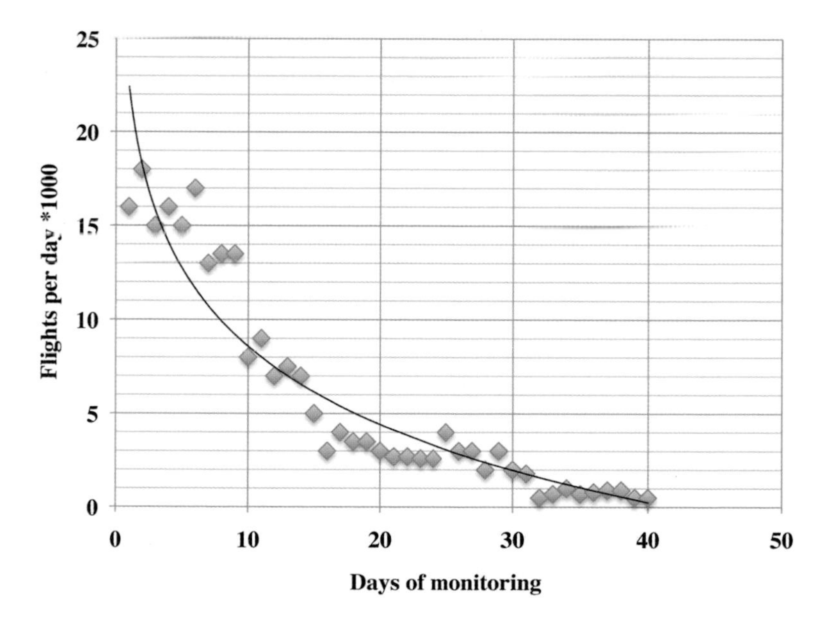

Figure 6.16 Plot of the number of foraging worker bees' flights from a failing hive with Colony Collapse Disorder (CCD)

causing mass deaths away from the hive. In the European Union neonicotinoides have been banned as part of the control measure for CCD.

Farming of crops

Poor nutrition can be associated with monoculture or feeding high fructose to supplement winter stores.

Figure 6.17 and 18 Review the environment of the hive. Monocrop culture can lead to swaths of land which is unusable for bees

Figure 6.19 Sudden massive food availability encourages the queens to lay

Figure 6.20 One month later – when the larvae have developed into workers – the food supply has gone. The hive now has 4 times as many mouths to feed and no easy food source

Genetically modified crops

The use of genetically modified crops has also be blamed; for example, using *Bacillus thuringiensis* (Bt) corn. The corn produces a protein which is poisonous to certain insects – the European corn borer, for example. However, depending on your preferred definition, CCD is described in countries using no genetically modified crops.

Genetics

There are relatively few breeders in the world and the rest of the bee population is mass produced from these few bee lines. Possible poor rearing conditions of queens weaken the queen bee's fat stores and she is unlikely to have a long life. The introduction of artificially inseminated queens, who may have few sperm in their spermatheca, results in early reproductive failure of the queen. Some behavioural traits may be genetically encoded. For example, hygienic behaviour may also assist colony survival, pathogen and stress control. Thus bees selected for specific hygienic behaviours may not succumb to CCD as easily as other hives. Africanised *Apis mellifera* are claimed to be more resistant to CCD. The Sub-Saharan African *A. mellifera* subspecies are certainly smaller and more aggressive.

Human causes

Greedily taking too much stores in the autumn and not monitoring the bee's nutritional requirements or by providing stores too late means that the bees do not have the right conditions to convert the supplemental feeding into preserved honey. It takes time for nectar to be converted into usable honey.

Environmental

Harsh long winters and cold wet springs both weaken a hive.

Climate change

Climate change has been blamed, but until we have a clear definition of the problem this is too vague. Note there are some reports of CCD-like problems for over a hundred years. But similar reductions in the numbers of related species are also being seen.

Others

There are multiple other proposed causes, from wind farms to electromagnetic issues. Again, until we have a clear definition of the clinical signs of Colony Collapse Disorder it will be impossible to investigate these claims.

Similar diseases

Note all 'pathogens' are found in normal healthy hives including *Varroa*. A specific hive and probably all hives will 'fail' and 'collapse' at some point. Many conditions both pathogenic and environmental will cause hive collapse. Do not rush into the diagnosis of CCD.

Previous reported collapse conditions: autumn collapse, disappearing disease, fall dwindle disease and May disease (which also may be associated with Spiroplasmosis) and spring dwindle (another name for Nosema) may all be versions of CCD.

Treatment

There is no known treatment for CCD.

Management and control

CCD can reoccur if an old hive is used. Do not combine 'sick' and healthy hives.

Maintain good biosecurity. Monitor the weight of the hive and the total protein concentration of the bees. Do not over examine the hive and cause chilling. Do not steal too much honey in the autumn. Note that bees require protein (pollen) as well as energy (sugar) when attempting to restore hive health.

Be aware of the local monocrop farming system and monitor the pollen coming into the hive. Is the pollen of different colours (multiple sources) or only one colour (one source)?

Crithidia

Crithidia are a form of trypanosome. These are single cell prosites with the ability to move by a tail at the rear of the cell and are variously called flagellates. They are very common in bees. They can be easily seen down the microscope looking at wet drops of ventriculus contents and watching for movement.

Bumblebees may have *Crithidia bombi* and *Crithidia expeki.* These typanosomes may be found in the ventriculus. Infection reduces new queen survival and colony reproduction. Workers are weakened.

It appears that *C. bombi* can affect the behaviour of the workers which forage more slowly and seem to be slower learning how to collect pollen and nectar and associating different flower patterns. This change of behaviour may be associated with the activation of the bumblebee's immune system. There is no specific treatment of Crithidia infestations.

Honey bees may be found to be infected with *Crithidia mellificae,* found in the gut lumen and gut epithelium, but they do not appear to cause any overt disease.

Dead bumblebees around a nest site

Fighting over optimal sites is common between newly emerged bumblebee queens. The loser dead queens may be found on the ground around ideal sites. Hibernated bumblebees may found with dead with holes in the abdomen following great tit (*Parus major*) predation.

Deformed wing virus (DWV)

Deformed wings can be seen frequently when you open the hive. The deformities obviously cause problems with the bee's ability to fly. However, note that the worker bee will still be able to function as a member of the hive, until the time when it is required to become a forager. Deformed wing virus (DWV) is an Ifavirus, an RNA virus. The virus can be transmitted by *Varroa.* The virus is able to replicate in the *Varroa* mite, and may play a role in CCD.

Deformed wing virus can be isolated from *A. florea* and *A. dorsata* without any clinical signs. Deformed wing virus can cause serious disease in *A. cerana* colonies.

Figure 6.21 Worker with a deformed wing

Figure 6.22 Detail of deformed wing

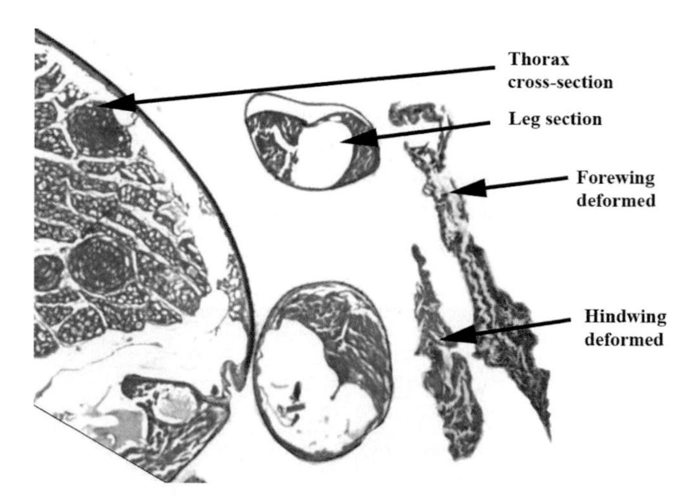

Figure 6.23 Histology of a deformed wing. Note the wing appears premature – not inflated

Clinical signs

Bees have deformed or poorly developed wings. The pupae are infected at the white eye stage. Bees will ultimately die from the virus. There may also be changes to the shape of the bee with a shorter abdomen, more rounded, discolouration and even some paralysis.

The clinical signs are more often seen in the autumn.

Diagnosis

Confirm the presence of the virus in the hive by PCR. The condition is more often seen in the autumn.

Similar conditions

Cloudy wing virus.

Chilling can result in deformities to the wings.

Note K-wings where the wings are normal in appearance but are not held together properly.

Treatment

There is no specific treatment for Deformed wing virus affected bees.

Management and control

Keep *Varroa* population under control. This is so common that the hive may have to live with the virus. Do not reuse combs from infected colonies and practise good hive biosecurity. Reduce the various stressors on the hive. Sterilise the inside of the super by scorching it with a blow torch.

If the problem becomes very aggressive within the hive, destroy the colony and burn all bees and combs.

Deformed wings

Deformed wings may also occur if the brood is chilled during development.

Drowning

It is always alarming when feeding bees over winter how many unfortunately drown in the feeding liquid. Cover the surface of the frame feeder with dry new straw to provide a platform for the bees.

Figure 6.24 Worker bees drowned in feeding syrup

Figure 6.25 Straw floating on the surface of a frame feeder to reduce drowning

Egypt bee virus

Serologically related to Deformed wing virus (DWV) it is asymptomatic (and does not cause any clinical deformity of the wing).

Filamentous virus – also called the F-virus or bee rickettsiosis

Filamentous virus is a baculovirus-like DNA virus.

Filamentous virus may be associated with Nosema infestation in honey bees (*Apis mellifera*). The virus multiplies in the fat bodies and ovarian tissues of workers. When severely affected the haemolymph becomes milky white associated with the number of virus particles! Peak infection is seen around late spring with a trough in the autumn.

Experimentally, the virus is considered non-pathogenic.

Fish

When a bee lands on water, it is always at risk from being eaten by fish such as brown trout (*Salmo trutta*). Bees can be found occasionally in the stomach contents of the fish.

Fly pests of bees

Many species of Apidae can become parasitised by various flies which ultimately kill the host bee.

Flies which result in the death of the bee

Senotainia tricuspis may be an important parasite of honey bees and bumblebees in Russia and the Ukraine and has been introduced into Australia.

Conopid flies – 'big-headed' flies

Parasitic flies are extremely common, for example *Conops quadrifasciatus*. There are numerous species, over 20 in the UK alone. The adult female waits on a flower and 'captures' a bumblebee when it comes to feed. The female fly moves to face the bumblebee.

The fly then flies off and grapples with the bumblebee while in flight. The female pierces the abdomen through the intersegmental membrane and lays an egg. The egg is long and narrow with a hook attachment at one end. The egg hatches. The larva attaches itself to an air sac. The fly larva develops within the abdomen of the bumblebee, eventually filling its abdomen. The bumblebee often dies in her nest. The larvae pupate and overwinter within the husk of the bumblebee's abdomen to start the cycle again. Interestingly, some of the conopid larvae may affect the bumblebee's behaviour encouraging them to dig in the ground, perhaps to benefit their pupation.

With *Thecophora* spp., the final instar larva will push its head through the petiole into the thorax and feed on the thoracic contents. Pupation then continues in the abdomen.

In the United Kingdom the common species are *Physocephala rufipes* and *Sicus ferrugineus*.

Figure 6.26 *Physocephala rufipes*, 20mm long

Figure 6.27 *Sicus ferrugineus*, 18mm long

Physocephala paralleliventris parasises *A. cerana, A koschevnikovi* and *A. dosata* in Borneo. *Apocephalus borealis* may parasitise sufficient bees to play a role in CCD in the USA.

Figure 6.28 *Apocephalus borealis* adult laying an egg in a honey bee

Figure 6.29 *Apocephalus borealis* emerging from the honey bee

Phorid flies

These tiny humpback flies can cause devastation to hives of stingless bees. There are several genera of Phorid flies which attend the colonies of Meliponini (stingless bees). These would include Pseudohypocera, Aphiochaeta, Melitophora and Melaloncha. The larvae are capable of exterminating a colony within days living in the pollen and honey pots.

Tachinid flies

Rondaniooestrus apivorus is a grey fly which deposits its larva in the flying bee.

Flies that co-inhabit the hive/nest

Volucella bombylans is a hoverfly. The adult *V. bombylans* mimics the bumblebee both in overall shape and behaviour. The fly will even raise a hind leg as a 'defence' posture and mimics the buzzing sound of bumblebees. The female lays her eggs in the bumblebee nest. The female *V. bombulans* may even be killed by the bumblebees while laying her eggs. The larvae feed on the debris from the nest and pupate and overwinter in the nest. They appear to cause no problems to the bumblebee larvae.

Frischella perrara

The bacteria *Frischella perrara* is a Gram positive bacteria which is rod shaped and may form filaments and chains. The bacteria can be isolated from honey bees. It has not been found in bumblebees. It produces translucent colonies about 1mm in size in three days in an anaerobic environment. The bacteria is present in most workers and comprises over 10% of the gut bacteria. This is not associated with any pathogenic process.

Gilliamella apicola

The bacteria *Gilliamella apicola* can be isolated from the intestine of honey bees and bumblebees. This is a Gram negative, rod-shaped, non-motile bacteria which may form chains. It is negative for catalase, nitrate and oxidase. The bacteria produce large white colonies about 2.5mm in diameter in two days in a microaerophilic environment. This is extremely common in the bee intestine and may account for 20% or more of the intestinal bacterial content. This is not associated with any pathogenic process.

Invertebrate iridescent virus (IIV)

These viruses are widespread in insects. Invertebrate iridescent virus (IIV) is an Iridovirus DNA of *Apis mellifera* and *Apis cerana*. The adult bees will huddle in small groups and are unable to fly. There has been good correlation with cases of CCD with IIV associated with Nosema in the USA. However, it should be noted that IIV is found in strong colonies also. Type IIV-6 is called the Chilo virus. This virus is absent in Australia.

Iridescent viruses form a crystalline mass when purified and will appear blue when illuminated with a bright light.

Israeli acute paralysis virus (IAPV)

Israeli acute paralysis virus may infect many species of Apidae. This is a Dicistroviridae (RNA). The virus causes death in about four days in infected individuals. The virus is related to Acute bee paralysis virus (ABPV).

The virus plays a role in CCD. Many believe the presence of this virus may be considered a marker for eventual CCD in the hive.

Many island nations may be able to eliminate bee viruses including Israeli acute paralysis with intense and effective biosecurity programmes. New Zealand, for example, may be negative to IAPV but still needs to monitor new introductions carefully. The advent of better laboratory diagnostics assists in these programmes.

K-wing

The bee presents with disjointed wings, with the hind wing and forewing not joined together by the hamuli. Note this can be normal and 'affected' bees can shake themselves and reattach the two wings. Note the wings are 'disjointed' before being folded back along the abdomen. This may not affect the ability to fly.

Figure 6.30 K-wing, note the hind wing is in front of the forewing

Figure 6.31 Two K-wing workers

KaKugo virus *(KV)*

KaKugo virus is an Iflavirus (RNA) found in *Apis mellifera.* The KaKugo virus is related to Deformed wing virus. Clinically, infection with this virus has been said to increase aggression in guard bees.

Kashmir bee virus (KBV)

The Kashmir bee virus is a Dicistroviridae (RNA). The virus can be found in a wide variety of honey bee and bumblebee species – for example, *Apis mellifera, Apis cerana* and *Bombus terrestris.*

Diagnosis can be complicated as the tests may cross-react with Acute paralysis bee virus (APBV). Kashmir bee virus (KBV) is also related to Israeli acute paralysis virus (IAPV).

The virus appears to weaken the bee causing death in about three days. It may be more problematic in the autumn. The *Varroa* mite acts as a vector. This virus is often found in brood and adults.

The virus has a tissue trophism for the alimentary tract, epidermis, trachea, haemocytes and oenocytes.

Management and control

Destroy the colony and burn all bees and combs. Do not reuse combs from infected colonies. Sterilise the inside of the super by scorching it with a blow torch. Keep the *Varroa* population under control. This may be the same virus as IAPV. Sequencing of the virus will determine the relationship.

Kleptoparasitic beetles

A kleptoparasitic beetle is a beetle which lives on the food resources of another animal. In the honey bee there are a number of kleptoparasitic beetles and they are normally of only minor significance to the health of the hive. At one time these beetles were call 'bee lice' or 'blister beetles' as they can produce a substance causing blistering in skin.

Lactobacillus

There are a number of lactobacillus spp. found in the intestines of bees. These are Gram positive bacteria which are pleomorphic, with rods and short coccobacilli. Colony morphology is variable. The bacteria are microaerophilic. Lactobacillus will occur naturally on flowers. Lactobacillus is an important part of the protective microbiota of the bee gut. They are not associated with any pathogenic process.

Lake Sinai virus 1 and Lake Sinai virus 2

Lake Sinai Virus 1 and 2 are found in *Apis mellifera.* The viruses are related to Chronic bee paralysis virus (CBPV). The incidence of Lake Sinai virus 1 peaks in summer, whereas Lake Sinai virus 2 peaks in winter. These viruses may be also related to Bee X and Bee Y viruses. There are no specific clinical signs associated with these viruses.

Macula-like virus

Macula-like virus is an RNA virus belonging to the family Tymviridae of *Apis mellifera.* As yet there are no pathogenic clinical signs attributed to the virus.

Mites of bumblebees – *Parasitellus fucorum*

Mites of honey bees are discussed separately – under Tropilaelaps and *Varroa* in Chapter 5 discussing brood disorders.

Bumblebees are frequently found with mites. Commonly large light brown mites are found over the body of the bumblebee, often at the back of the thorax. *Parasitellus fucorum* is one of the common species. However, despite the name, Parasitellus is not a parasite but a commensal living in the nest debris. Interestingly these mites may themselves be parasitized by other mites – notably *Scutacarus acarorum.* If the farmed bumblebee has evidence of Parasitellus it is possible to put the bumblebee in the fridge (5°C) for 10 minutes. Most of the mites will fall off and the remaining mites can be removed with a small forceps. The mites are destroyed by incineration and once warmed up, the bumblebee will return to normal.

Nosema

Nosema may be referred to as 'spring dwindle' and dysentery.

Nosema are microsporidia which is a unicellular parasitic fungus. Various species of Nosema infest different species of *Apis*. There are two important species affecting *Apis mellifera*: *Nosema apis* and *Nosema ceranae*. The spores are large oval bodies 4–6µm long and 2–4µm wide. *N. ceranae* is slightly smaller than *N. apis*.

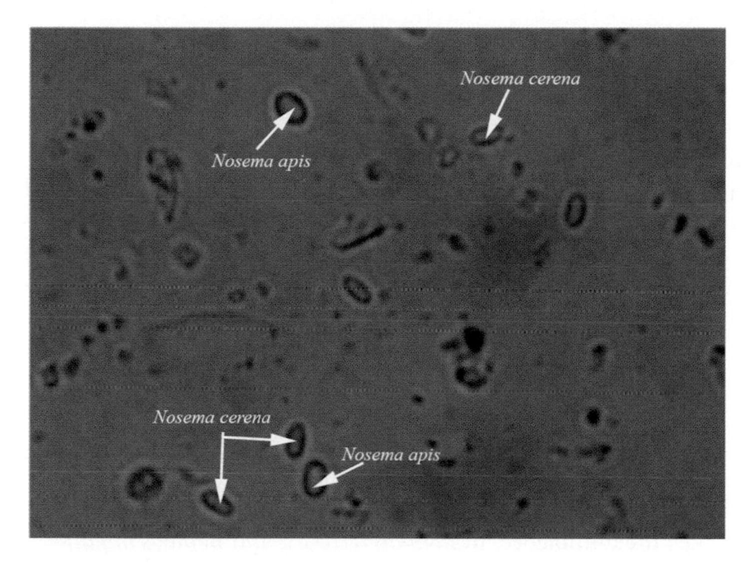

Figure 6.32 Mixed Nosema infection from the ventriculus contents of *Apis mellifera*. This mixed infection is serious

Nosema bombi affects bumblebees and the microsporidia is 4–5µm long and 3µm wide.

Another microsporidia, *Apicystis bombi,* may also be isolated from bumblebees. These are larger than *Nosema spp*. at 11–14 x 3–5µm.

Nosema spores are extremely long-lived and are not destroyed by freezing. Newly emerged bees are free from the infection; the larval and pupal stages are not affected. Thus, as the bee ages (forager bees for example) it is more likely to become infested. The bees become infected through ingestion of faecal materials. Infestation levels are then higher in the spring after being confined to their hive. It only takes about 100 spores for the hive to become infected. All of the bees can become infected – the queen, drone and workers.

The spores are resistant for five weeks in dead bees, a year in faeces and 2–4 months in wax and honey. The spores can be killed by heating or with acetic acid.

Clinical signs

Honey bees – Apis mellifera

The clinical signs of Nosema are generally seen in the winter and spring after periods of bad weather – the condition was called 'spring dwindling'. In honey bees Nosema affects the ventriculus resulting in intestinal problems. The damaged intestine results in dysentery which can be seen as yellow stripes on the outside of the hive or even inside the hive.

The clinical signs are 'dysentery' with clear signs of defecation in the hive, which is unusual. Worker bees will usually defecate outside the hive on 'cleansing flights'. The clinical signs mainly affect workers, but can affect drones and the queen. The worker may appear to have an increased width to its abdomen. If the queen is affected, egg production drops as the queen's ovaries degenerate. This is likely to trigger a response in the workers leading to supersedure of the queen.

The clinical signs of dysentery may be more associated with *N. apis* rather than *N. ceranae*. There can be a large number of dead bees in the apiary, but in general bees die away from the hive especially with *N. ceranae*. Weakened worker bees are unable to fly. They move instead by crawling with disjointed wings (K-wings).

The life expectancy in infested bees is only three weeks, which is significant for the queen. When there is a combined infestation, death can occur in only two weeks.

Secondary infections

Nosema also appears to play a role in the clinical expression of Black Queen Cell Virus.

The loss of workers, especially with *N. ceranae* where workers die away from the hive, indicates that some of the CCD diagnosed cases are related to Nosema together with a possible role of Invertebrate iridescent virus (IIV).

Bumblebees

Nosema infestation may weaken the bumblebee. In *Bombus terrestris* and *Bombus impatiens*, Nosema reduces colony growth. Nosema does not appear to infest *B. occidentalis*.

In bumblebees, Nosema affects all body tissues not just the ventriculus. Adults are less commonly infested.

Honey bees

Individual bee

Queen bees can be examined by placing them on a small petri dish and allowing her to walk freely. Within an hour the queen will have defecated – drops of clear, colourless liquid which can then be transferred by a capillary tube. This is something that could be considered prior to introduction of a purchased new queen.

Hive basis

Signs of dysentery evident

Place a glass slide near the hive entrance and collect faecal material from worker bees. Scrape off the deposit, mix with water and make a wet mount. Look for the characteristic rice grain appearance of the Nosema spores.

Hive examination

Collect a sample of forage bees from the landing pad. These are the older bees and are more likely to be infested. Note avoid young training bees which can be recognised as hovering around the landing pad.

- Collect 30 bees (about a matchbox full).
- Place the bees in the refrigerator/freezer for 20 minutes to anaesthetise the bees.
- Place 15–30ml of cold tap water with the bees in a crucible.
- Quickly kill the bees by macerating them with the 15–30ml water.
- Examine one drop of fluid on a slide/cover slip.

Results:

0–20 per X400 field of view	Mild infestation
20–100 per X400 field of view	Moderate
Over 100 spores per field of view	Heavy infestation

The number of spores can be calculated using a haemocytometer. Bees may have 10 million spores/ml (see Chapter 12 for more details).

Dead bee gross examination

Whole bee

Take a whole bee and macerate it in water. Examine the fluid to reveal the Nosema spores. Phase contrast microscopes may make the spores easier to see.

- Intact spores are white
- Germinated spores are black.

Intestines only

The intestinal tract and in particular the ventriculus needs to be examined.

- The digestive tract can be easily obtained from a bee.
- Remove the bee's head.
- Grasp as much of the stinger as possible with a pair of fine tweezers and then with a steady, gentle pull withdraw the entire digestive tract. Place the intestinal tract on a glass slide and locate the ventriculus.

In the healthy worker honey bee the ventriculus is the large straw brown organ in the middle of the intestine with clearly visible circular constrictions. In the queen and drone the normal ventriculus is normally white/colourless. In infested honey bees the ventriculus is white, soft and swollen and the constrictions are obscured.

Note: When you examine bees in detail other organisms may also be recognised – for example, Amoeba cysts (*Malpighamoeba mellificae*). Other protozoa that may be recognised include flagellates (trypanosomes) of the *Crithidia spp.*

Microscopic examination of the ventricular contents

Histological examination of the ventriculus may reveal numerous spores being released. See Chapter 1 for more detail on the normal anatomy of the ventriculus.

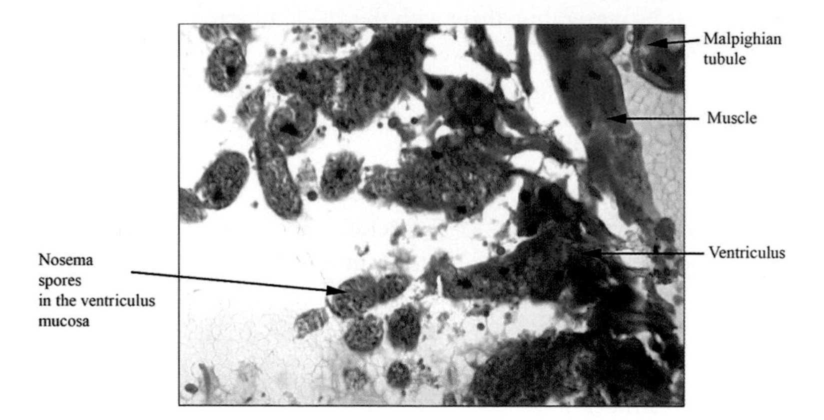

Figure 6.33 The ventriculus of the honey bee infested with Nosema and stained by H&E and examined down the microscope

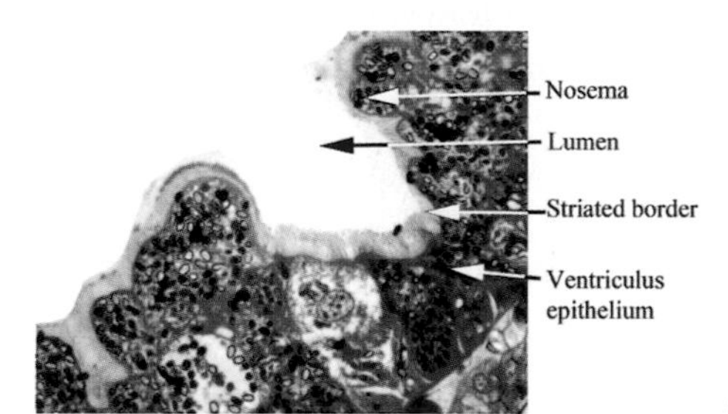

Figure 6.34 The ventriculus stained with toluidine blue clearly reveals Nosema infestation when examined down the microscope

Note with the normal stains (e.g. H&E) the spores may be difficult to see. In tissues stained with toluidine blue, the spores are black. Note the merozoites are unstained. When examining the whole bee it might be noticed that there is an under development of the hypopharyngeal glands. This is probably related to starvation associated with the action of Nosema on the function of the bee's intestines.

Bumblebees

The Nosema spores can be found infesting all tissues throughout the body except the ovaries and proctodeum.

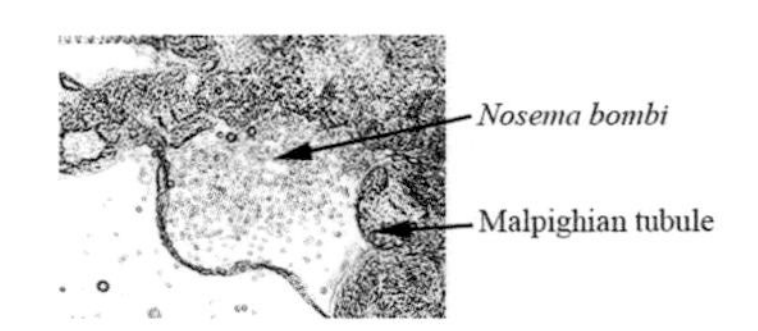

Figure 6.35 *Nosema bombi* revealed from the crushed malpighian tubules down the microscope

Apicystis bombi may be found in the fat and may be associated with a very white but indistinct fat body.

Similar conditions

CCD and other colony collapse problems. Dysentery may be seen after a long protracted cold and wet winter/spring.

Treatment

There are specific treatments available such as Fumidil B (Fumigillin) but note this has been banned in the EU. There is some evidence that thymol may be effective. Hives will clinically self-heal over the season.

Management and control

Be careful when manipulating the frames. If bees are trapped and squashed, they will spread Nosema.

Ensure all bee keeping equipment is clean. Disinfect honey comb and utensils with acetic acid. When frames are not in use it is possible to heat treat for more than 50°C for 24 hours to kill spores. Do not use other people's bee equipment on your own hives. Provide visitors with clothing and hive tools. Identify each hive with its own hive tool. Do not exchange combs between colonies.

Do not reuse old combs from colonies that have died out. Destroy all nesting material and infected colonies, especially with bumblebees.

Reduce stressors on the hive. Ensure that the bees are not starving. Ensure that the crude protein content of bees is above 40% (see Chapter 12). When bees drop to less than 30% they become more susceptible to Nosema.

Pseudomonas – Septicaemia

Pseudomonas is a rare secondary infection. The bacteria is called *Pseudomonas apisepsis*. Pseudomonas is a Gram negative rod. This may also be found under the bacteria *Bacillus apisepsis*. Pseudomonas apisepsis may actually be a virulent P. aeruginosa in the honeybee.

Clinical signs

Pseudomonas affects adult bees. Dead or dying bees present with a putrid odour. As you examine the bees, the bees fall apart. The haemolymph may be white and cloudy.

Bees die within 24 hours of being affected.

Diagnosis

The disease results in a destruction of the connective tissues of the thorax, legs, wings and antennae. Prepare a bacterial smear by removing the wing from the thorax.

Dip the wing base into a drop of water on a microscope slide. Stain the slide with Gram's stain (see Chapter 12).

Culture

This should be done by a veterinarian or a qualified lab. Culture the pseudomonas by streaking the base of the wing across blood agar. Spread out the streak and place in an incubator for 24 hours. Note Pseudomonas can affect people.

Treatment

Antibiotics may be useful. Isolate the organism and perform an antibiotic sensitivity test first. However, the bacteria are very common and treatment should be directed at controlling the stress level and general hygiene of the hive. Only medicate with antibiotics when prescribed by your veterinary surgeon.

Management and control

Review hive environment and biosecurity. Do not use other people's bee equipment on your own hives. Provide visitors with clothing and hive tools. Identify each hive with its own hive tool.

Zoonotic implications

Note that Pseudomonas can affect man.

Robber flies – asilid predator

A number of fly species will attack bumblebees. *Asilus crabroniformis* is a large fly that catches the bumblebee on the wing. The fly inserts its proboscis into the bumblebee and injects a paralysis agent. The robber fly then sucks out the haemolymph and the lifeless body is discarded.

Slow bee paralysis virus (SBPV)

Examination of the worker honey bees (*Apis mellifera*) reveals some bees which are paralysed in the front two pairs of legs. There is a late paralysis in bees, about 34 days post-lay, with death around day 36 days old as compared with the acute paralysis problems. Examination of the pupae reveals many dead within sealed cells. This is an as yet unclassified spherical RNA virus described as picornavirus-like. There is some association with *Varroa.*

Diagnosis

PCR examination.

Similar conditions

Acute paralysis virus (APBV)

Israeli acute paralysis virus (IAPV)
Chronic paralysis virus (CPBV)
Poisoning

Treatment

None specifically.

Management and control

Keep the *Varroa* mite population under control. Sterilise the inside of the super and brood box by scorching it with a blow torch. In severe infection destroy the colony and burn all bees and combs. Do not re-use combs from infected colonies.

Snodgrassella alvi

The bacteria *Snodgrassella alvi* can be isolated from the intestines of honey bees and bumblebees. This is a Gram negative, rod-shaped non-motile bacteria. It produces white round 1mm colonies in two days when grown in a microaerophilic environment.

This is an extremely common bacterium in the intestines of bees. The bacteria may contribute over 30% of the intestinal bacterial content. This is not associated with any pathogenic process.

Sphaerularia bombi

Sphaerularia bombi is a nematode that frequently infests bumblebees. About 50% of queens may be infested with the parasite.

The nematode releases eggs into the haemolymph. Inside the egg the larva goes through two instars. The third instar is hatched into the haemolymph. The larva is 1mm long and can be seen in the haemolymph in their thousands. The third stage progresses and migrates into the intestine and it is passed out in the faeces.

The third stage develops into adults and mates in the soil around the hibernation area. The mated females wait in the hibernation area to infest a queen bumblebee.

Clinical signs

During hibernation and early spring the condition may be recognised. Fifty percent of queens are often infested. The queen's behaviour and reproductive capacity is impacted. The nematode stops the development of the corpora allata and thus sterilises the queen. If there are any larvae and pupae they are not affected. The adult female nematode develops during the hibernation phase of the queen.

Diagnosis

In postmortem examination, the large (clearly visible) adult female nematode is found in the abdomen. The nematode (actually the egg sac) is generally seen as a white sausage structure 10–20mm long in the abdomen

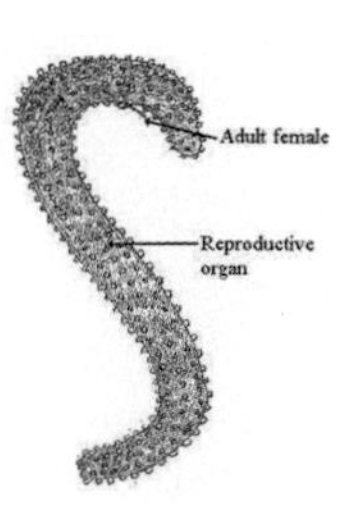

Figure 6.36 Drawing of *Sphaerularia bombi*

of the queen bumblebee. The main visible structure is actually the reproductive tract of the nematode. Note the extremely small adult female size attached to the egg sac.

The larvae and eggs of *Sphaerularia bombi* can be found in the haemolymph.

Treatment

There are no specific treatment options.

Management and control

Infested hibernation areas need to be sterilised.

Spiroplasma

Spiroplasma apis and *S. melliferum* have been isolated from *Apis mellifera* with minimal overt disease signs. Other cases may be associated with high mortality occurring within four days of infection. It is possible these are associated with 'May' disease.

Here in the spring a large number of sick adults cluster around the hive. They may have swollen abdomens. The bees may also have a quivering behaviour. The hive normally recovers spontaneously.

S. melliferum has also been commonly isolated from bumblebees.

Tracheal mite

Tracheal mite infestation of *Apis mellifera* has become a rare condition as the treatments offered to control *Varroa* infestation will also control tracheal mites. The tracheal mite *Acarapis woodi* is one of a number of similar mites, the others cause no pathological effects.

Acarapis woodi is 143–174µm long in the female and 125–136µm long in the male.

There are other species of *Acarapis* which live on the honey bee's skin – *A. externus*, *A. vagans* and *A. dorsalis*. *A. vagans* is found more commonly on the hind wing of drones.

There is a range of mites which live in and around the bumblebee. A mite similar to *Ascarpis woodi* is *Locustacarus buchneri* which infests the bumblebee's respiratory tract.

The mite moves between bees in the hive. Once inside the trachea, the female mite lays an egg. This first egg laid is an unfertilised male. Subsequent mites are female. The male mates with his sisters.

Life cycle

Males mature in 11–12 days. Females mature in 13–16 days. The life cycle lasts around 12–15 days.

The mites pierce the trachea and thin walled tracheal air sacs and live on the haemolymph.

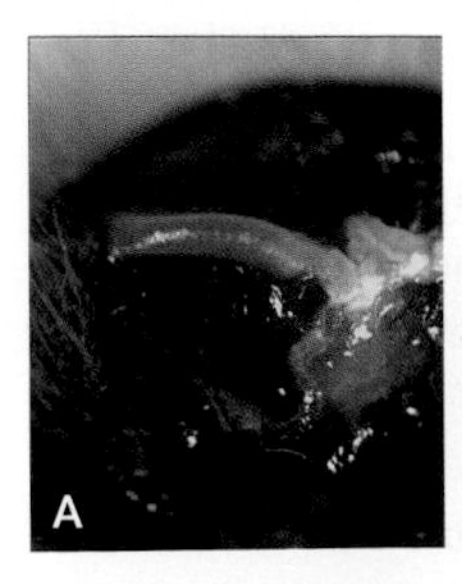

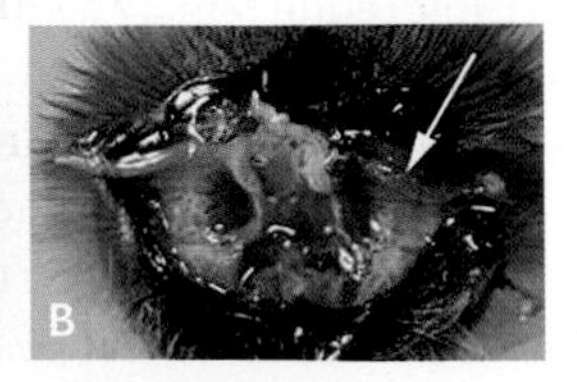

Figure 6.37 a: Healthy uninfected prothoracic tracheae. b: Melanisation of the trachea indicates infestation as seen on the right (of the photograph)

Clinical signs

In many hives there will be no or few clinical signs. The mites parasitise the adult bees generally over two weeks of age post emergence (five weeks of age). There may be some vague clinical indications, such as more bees having disjointed wings and being unable to fly. This is called K-wing syndrome. Clinically significant K-wing bees may be seen crawling around the entrance. But note that many bees may demonstrate a K-wing temporarily.

The stress caused by *Acarapis woodi* may reduce the lifespan of the workers and contribute to CCD.

If the hive is not producing a significant amount of honey, it is worth checking for *Acarapis woodi.*

The condition has also been associated with Acute bee paralysis virus (ABPV) and Chronic bee paralysis virus (CBPV).

Diagnosis

The mite is found in the prothoracic tracheal system of the honey bee. There may be 100s of mites in a single bee. They may also infest the abdominal and tracheal air sacs.

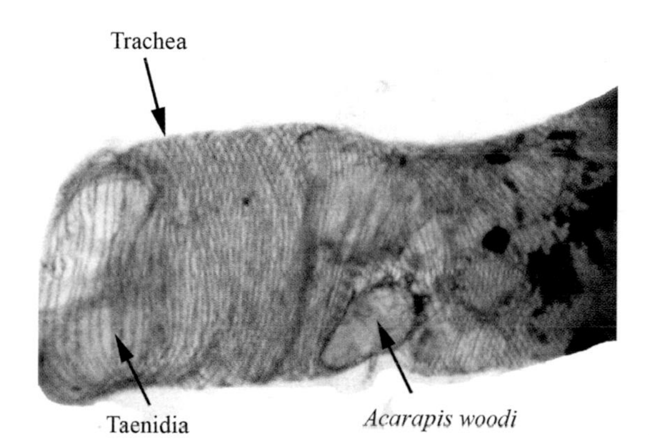

Figure 6.38 *Acarapis woodi* in the pro-thoracic trachea of a worker honey bee

Similar conditions

Note there are two other *Acarapis* spp affecting honey bees – *Acarapis externus* and *A. dorsalis.* These are non-pathogenic but look extremely similar to *A. woodi.* The mite needs to be found *in situ* in the trachea.

Treatment

Treatments for *Varroa* mite will control Tracheal mites *Acarapis woodi.*

Grease patties placed in the autumn through winter interfere with the ability of the mite to recognise the young bee.

A grease patty is made using vegetable shortening plus sugar (1:2 ratio). Add 4.5g peppermint flavouring to 300g of grease patty. Take a handful of patty, place it on waxed paper, top and bottom, and place between the two boxes. Remove the top layer of waxed paper. Note the oil may interfere with the bee's ability to thermoregulate.

Menthol fumes are toxic to the mites. Fume all equipment with the bees. Note it requires warmth to make the menthol crystals fume.

Formic acid can be used on the bottom boards. This can be useful to control tracheal mite. Administer 30ml every week for three to five applications in the spring, and then again in June. The air temperature needs to be above 22°C but below 30°C.

Amitraz may be used to good effect.

Management and control

Reduce drifting and robbing. Re-queen annually to keep strong colonies. Reduce other stresses. Maintain excellent biosecurity. Some Russian bees are claimed to be resistant. Breeding for hygiene behaviour will be beneficial.

Trauma to the queen

Feet balling

Many queens can be seen to have lost feet – this is called balling. This can affect the queen's ability to measure the cell to determine drone or worker cells.

Figure 6.39 A Queen honey bee who has lost her lower limb – balling. General and detailed view

Wing clipping

Wing clipping is another deliberate form of trauma. The queen's wings are deliberately clipped to stop her swarming. There appears to be no pain response from the clipping of the wing.

Figure 6.40 However, if the hive tries to swarm the queen can get caught outside as she cannot fly

However, wing clipping can lead to problems. If the queen gets outside the hive she cannot fly back. Be very careful not to shake the queen onto the ground when examining a brood frame. It can be a long way home and she is likely to get stood on or picked off by a passing hungry bird.

Trauma to workers

Trauma to an individual worker may not seem significant on the scale of the hive population but should be avoided. Worker bees trapped and killed between brood boxes or supers can spread pathogens such as Nosema.

Dead and dying bees inside the hive may also spread fungi and need to be removed. When handling the hive be careful and brush bees always from the top surface where replacing supers or brood boxes.

Figure 6.41 Workers crushed when the hive is remade

Figure 6.42 Carefully remove workers from the edge of the hive by a soft brush

Frames can be examined by having the top board in sections rather than one piece of wood. This can reduce the stress on the hive. Alternatively, examine the hive using a cover to reduce the flying of the nurse worker bees.

As the workers age their wings will become worn and damaged. This can be used as a rough guide to worker bees' ages. Especially useful in queen bumblebee identification.

Figure 6.43 Slatted top board reduces the exposure of workers to the environment and can reduce crushing workers when remaking the hive

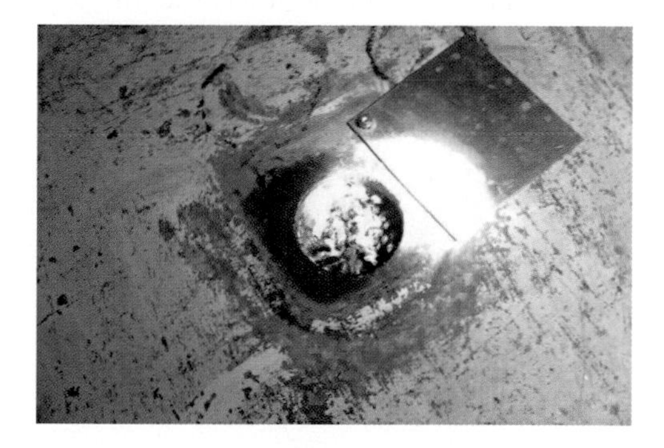

Figure 6.44 A viewing hole is especially useful in cold climates to reduce cold stress on the bees

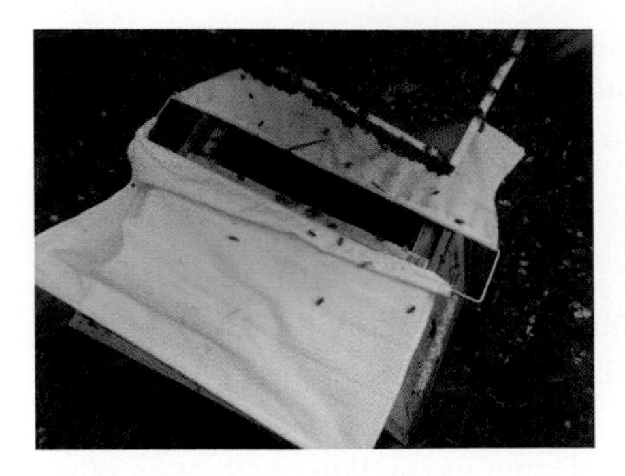

Figure 6.45 Control the number of workers flying around the hive during a hive examination using sheets or a special movable cover

Figure 6.46 A blanket can help to control the number of flying bees when manipulating the hive

Varroa destructor virus 1

Varroa destructor virus 1 is an Iflavirus (RNA) of *Apis mellifera*. The virus is related to Deformed wing virus. The virus may be identified more in the autumn. There is some association with *Varroa* mite infestation.

Wasp and other hymenoptera parasites

Syntretus splendidus and other wasps are members of the same family as bees – the hymenoptera. These wasps preferentially parasitise the later-nesting species of bumblebee – for example, *Bombus pascuorum*. Some may also parasitise the stingless bees (Meliponini).

The female lays her eggs in the thorax or abdomen of bumblebees – up to 70 in queens and 20 in workers. The eggs hatch and develop through five instars within the thorax and abdomen. The fully grown larva may be 6mm long. The bee forages and behaves normally, but the queen generally stops laying. The parasised queen's ovaries often become degenerate and egg laying ceases. The bumblebees may become very sluggish and weak immediately before the final fifth instar leaves the bumblebee. The larvae emerge from the abdomen. While the bee may not die from the emergence of the parasitic wasp it is weakened and starves to death.

Figure 6.47 Drawing of a *Syntretus ocularis* female, 3mm long

Mutilla europaea parasitises numerous species of bees including bumblebees and honey bees. The female is wingless and resembles a hairy ant. The eggs are laid in the pupa which is then consumed.

7

Managing the health of the hive as a whole

The hive needs to be considered as a living organism. This chapter looks at disorders which affect the hive itself.

The normal honey bee hive

Apis mellifera belongs to the group of *Apis* bees who build their hives in cavities – this includes *Apis cerana*. The obvious natural cavity would be an old hollow tree. If you look carefully in fallen logs or old trees you occasionally can see a 'wild' bee hive.

Figure 7.1 and 7.2 Natural hive in a log. Note the hanging 'frame' appearance.

Figure 7.3 and 7.4 An old tree with a bee hive

Anatomy of the comb

The main structure of the hive is the comb. The comb hangs from the shelf with the cells being made horizontal in the comb.

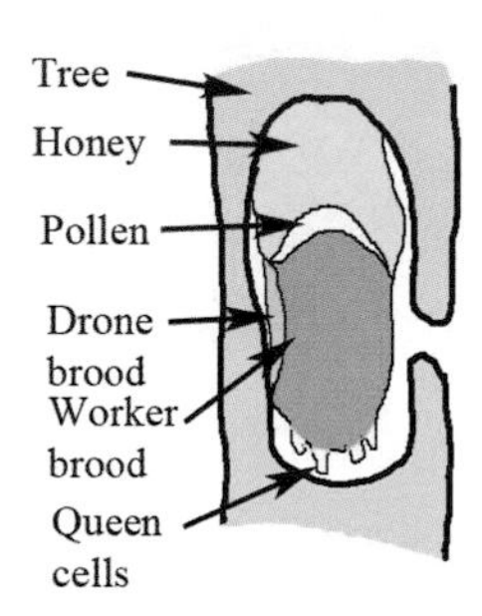

Figure 7.5 Drawing of the layout of the hive in a tree

Figure 7.6 The appearance of natural hive

Dissection of the hive demonstrates the relative position of the honey, pollen and brood-nest. The queen cells are produced at the very bottom of the brood-nest where the wax is newest. The brood of *Apis mellifera* is kept at a constant 36°C. The drone cells are on the edge of the brood-nest where it is cooler, at 35°C.

The same pattern is seen in the typical 'frame' with honey stored at the top of the frame and the brood in the warm centre. Between the honey (1) and the brood (3) is a narrow band of pollen (2). Note the cappings are uniform in colour and are convex (higher in the centre than at the edges).

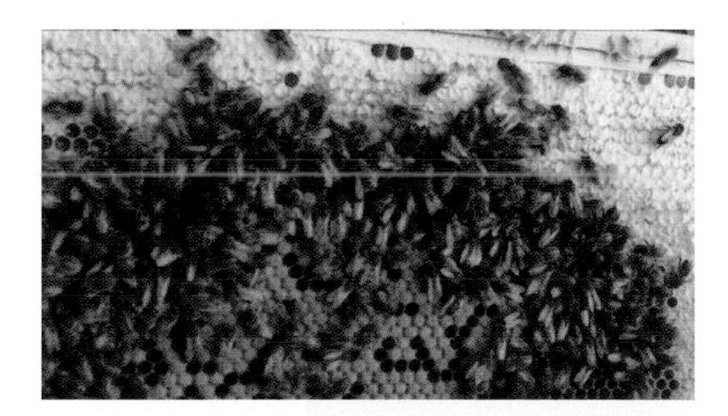

Figure 7.7 The frame with honey bridge

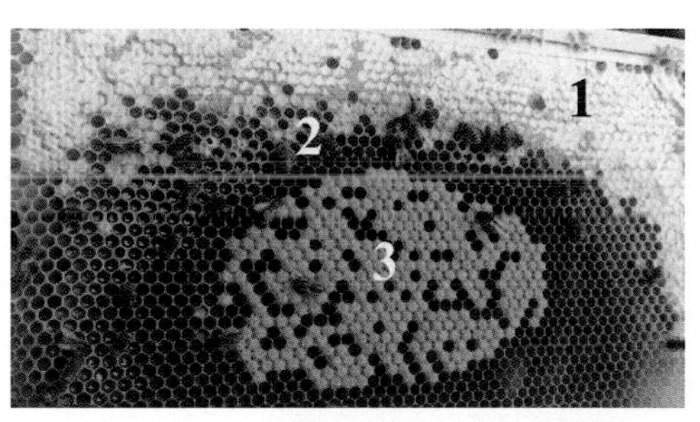

Figure 7.8 Bees shaken off into the hive, revealing the various zones

In the artificial hive – the honey stored in the brood box at the edge is going to be stored in the smaller worker cells (5mm).

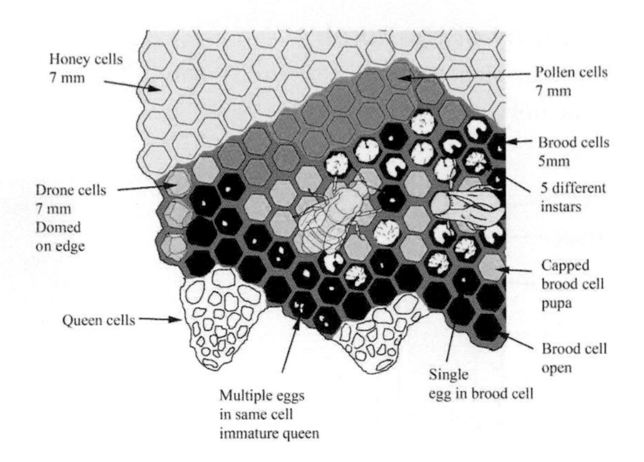

Figure 7.9 Drawing of the major parts of a comb

Anatomy of a cell

Figure 7.10 View into a worker cell. New wax is white

Figure 7.11 Cross section of a worker cell with eggs visible

Figure 7.12 As cells age they become darker

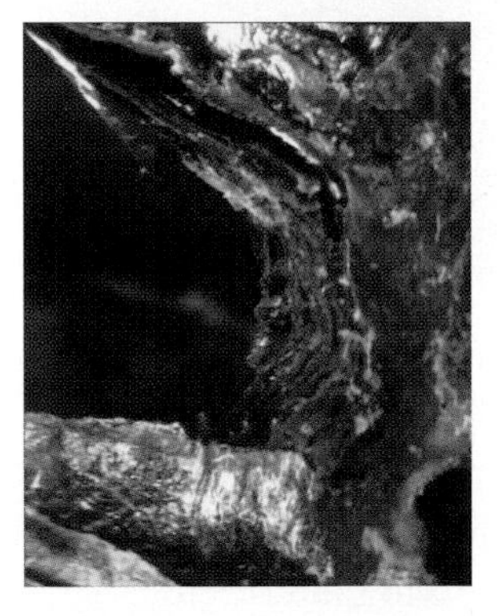

Figure 7.13 Detail of the wall of the cell showing the succession of layers laid down with each larva. The width of the cell progressively shrinks

Propolis

Propolis is a resin material collected from trees – leaves and buds. Bees use propolis to seal holes and provide a disinfectant to the hive. Propolis has natural antimicrobial properties and protects the hive and the bees from bacterial infections.

Figure 7.14 Propolis in the honey basket of a bee

Figure 7.15 Propolis used to seal holes and gaps.

Man-made bee hives

Log hives

One of the simplest hives that can be constructed can be made out of a log. A tree is felled and cut into cylindrical logs which are split into halves. The inside is carefully scooped out and the log hive put back together. Note to protect the hive from predators (humans for example), the hive is often hung in trees.

When the honey is collected, the hive is split open and the honey combs removed. The two halves can then be put back together for the start of the next honey crop.

A similar hive can be made from a tin can (carefully cleaned out).

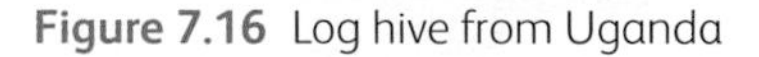

Figure 7.16 Log hive from Uganda

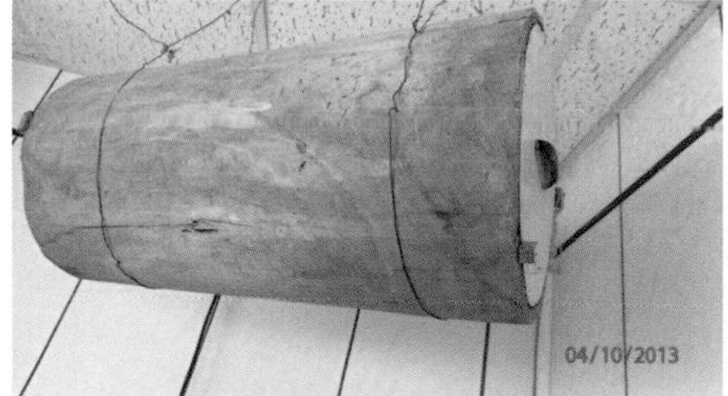

Figure 7.17 Log hive from China – the entrance is the slit at the front

Modified log hive – the basket hive

In central Africa, hives can be constructed from sticks, which are woven into a cylindrical shape. The outer wall is then smeared with wet soil or ideally cattle dung. The ends are sealed with banana fibres, grass or wood, leaving a few holes for the bees. There is only one entry point.

These tend to be a one-use only hive, as the hive is destroyed to get the honey and the bees are sacrificed or forced to swarm.

Figure 7.18 Weaving the sticks together

Figure 7.19 Starting to apply the mud outer casing

Figure 7.20 and 7.21 The outer-casing can be made from grass or straw

Figure 7.22 and 7.23 Entrance at one end with bee holes (left picture) other end sealed (right picture). Note the plastic cover (can be a metal plate) to protect from rain

Figure 7.24 Detail of the entrance for the bees

Figure 7.25 Hive suspended in a tree to protect from predators – in this case termites

Skep

In Europe a straw based hive has long been used for thousands of years. The dry straw is woven into a skep. These can be excellent for capturing swarming bees.

Figure 7.26 Skep

Figure 7.27 Skep making class

Improved hives

Kenyan top bar hive

This hive has great advantages as it allows the beekeeper to inspect the bees at any time without significantly disturbing the hive. It also allows easy collection of honey, bees, wax and other products as required.

The hive has to be carefully constructed with a sealed top to each top bar and the bee space below each top bar allows the bees to construct their hive hanging on the top bar.

Figure 7.28 Top bar hive in use. Note entrance holes 8–10mm wide

Figure 7.29 The top bar hive with the roof removed

Figure 7.30 Note no bee space visible between top bars

Figure 7.31 Free form on a Kenyan top bar

Figure 7.32 Drone cells on a Kenyan top bar

The Langstroth hive and related types – for example the National

The major advantage of the Langstroth hive is that the queen bee and her brood can be excluded from the supers carrying the honey. The reduction in traffic produces good clean honey and the honey can be easily extracted as there is little disturbance to the queen or the brood.

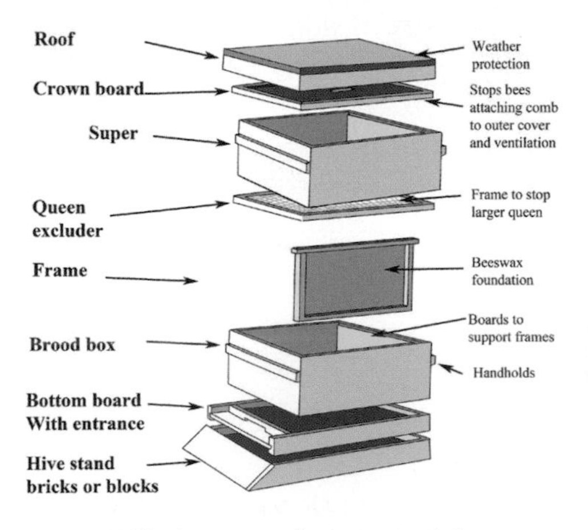

Figure 7.33 Anatomy of a Langstroth hive

Figure 7.34 Hive painted white in hot climates.

However, the hive is expensive and the honey is extracted from the frames using a centrifugal honey extractor.

Some of the terms change around the world. In North America the roof is called the outer cover. The crown board is called the inner cover. The brood box would be a brood super whereas the super would be just a honey super.

Figure 7.35 Supers under construction – note bee space

Figure 7.36 Frame with the foundation of beeswax

Figure 7.37 Hives in Australia

Figure 7.38 Startup hive in Uganda. Note no honey super

With prolific queens, two or more brood supers are often used. More honey supers can be added if there is a lot of food around.

Figure 7.39 and 7.40 Rows of hives in Canada. Note the hedge at the rear to protect the hives from the north winds

Figure 7.41 Hives placed to pollinate sunflowers in Ukraine

Figure 7.42 Hive in UK. Note electric fence to protect hive from sheep and cattle

Each frame will collect about 2–3kg of honey.

WBC hive

Figure 7.43 WBC hive

Figure 7.44 Inner and outer wall to provide insulation air gap

The WBC hive has two outer walls with the inner hive. This provides better insulation during the winter months.

Where to place the hive

Bee boles

Bee boles are recesses in a wall, often south facing in the UK to make it easier for the bees in the wintertime. The bole provides sufficient room to hold a skep hive.

Figure 7.45 Bee boles in the UK with two skeps in place

Detail of location

Bees need water, food, air (good ventilation) and good flooring (the hive).

Water

It is essential to place the hive near a good water supply – ideally a pond or river. Bees do not store water in their hive, so they need easy access. In the summer months providing water in a basin may be very beneficial. Place a leaf, brick or some new straw in the water so the bees can land and drink. Simple arrangements of gravel-, moss- or peat-filled dishes topped up with water are ideal. Place several around the hive and garden.

In cool months a hive requires about 150g of water per day to survive.

In the hot months this increases to 1kg water per day per hive.

Figure 7.46 Place the hives near a good water source

Figure 7.47 Support the hive in the hot summer months with additional water

Food

The food sources for bees are nectar, honeydew and pollen (see Chapter 2).

Energy

Bees can become very agitated after repeated rains. Rain washes nectar out of the flowers and it takes time for the plants to replenish their nectar supplies. Some plants also only produce nectar when the air temperature is high enough. For instance, with lime trees the temperature needs to be above 20°C to maximise their nectar flow.

Nectar is quite variable in quality but contains 5 to 80% sugars in water plus small amounts of proteins and other substances. The main sugars are fructose, glucose and sucrose. Nectar provides around 50 times the energy expended in its collection. A hive requires about 25kg of honey to overwinter.

Honeydew nectar originates from insects who suck the sap of plants. Honeydew nectar is secreted by the sap sucking insects, aphids for example and placed on the leaves of plants. Bees will then collect the honeydew.

Protein

The major source of protein to the colony is pollen. A colony requires about 20–30kg of pollen per year.

Pollen also provides minerals, lipids and vitamins. The protein content of pollen also varies, ideally bees require 20% protein content. Maize (*Zea mays*) pollen is low in protein (15%) whereas white clover (*Trifolium repens*) is good at 26%. A hive requires about 50–100g of protein a day, thus about 250–500g of pollen per day. Note if only poor protein content is available (<15%) this increases to 340–600g of pollen per day.

If supplementing protein sources, it is important to know the ideal aminoacid concentration. In bees the limiting aminoacid is leucine. All the other amino acids need to be in the correct ratio to the leucine concentration otherwise they are wasted. This is important when formulating a nutrition ration. Amino acids should comprise:

4.5%	Leucine
4%	Valine and Iso-leucine
3%	Threonine, Lysine and Arginine
2.5%	Phenylalanine
1.5%	Methinine and Histidine
1%	Tryptophan

Sunflowers pose a particular problem for bees, for while there may be plenty of pollen, the total protein content is low (17%) and the aminoacid ratios are incorrect, resulting in a food source that can be difficult to utilise.

Air (good ventilation)

The siting of the hive is important to both get the morning sun to avoid any excessive heat which can heat stress the hive. The bees will maintain the hive around 36°C irrespective of the outside temperature. They are able to survive –40°C to +45°C for several weeks.

In the southern hemisphere site, the hive should be facing south with protection from the afternoon sun (in the north).

A wire mesh floor assists hive ventilation. It will not allow cold air in – as cold air will fall not rise.

The hive is cold stressed below 14°C and heat stressed above 35°C. At both tolerance levels the protein requirements increase to ensure hive survival. Below 14°C bees are not keen on flying and may only undertake cleansing flights.

Figure 7.48 Hive placement in Australia to avoid overheating the bees in the hot afternoon sun

Figure 7.49 Ensure good distance between hives to stop cross over of flight paths and drifting.

The outer cover can be angled to provide additional ventilation especially in the hot summer months. This should angle backwards, not to pour water onto the landing pad. Note this is not a top gap through which the bees can leave.

Watch the behaviour of the bees at the entrance as this can be an excellent indication of events inside the hive.

Figure 7.50 A productive hive in Canada. Note the sloping roof to allow run off of rain water, but it should run backwards not towards the landing pad

Figure 7.51 Bees on good flight paths with guard bees at entrance. Good hive temperature

Figure 7.52 Bees crowding around entrance – indicating the hive may be too hot. This is called bearding

Ensure the frame is correctly made and square

When you make the hive ensure that the hive is square and that the brood box is not at an angle to the hive stand or you may create another entrance or exit for the bees and their pests.

Similarly the top board and roof need to be correctly placed so that rain does not gain access to the brood or super boxes.

Figure 7.53 Ensure the brood box is square onto the hive stand. The presence of bees in unusual locations normally indicates problems (arrow)

Figure 7.54 Ensure that the roof is properly placed. The presence of bees in unusual locations normally indicates problems (arrow)

Angle of the frames to the entrance

If the hive is in an exposed site or in the winter months, set the frames parallel to the entrance. This will reduce the air flow in the hive and protect the brood from chilling draughts.

To maximise the air circulation, in sheltered positions or in the summer, set the frames 90° to the entrance.

Figure 7.55 Summer time – frames right angles to the entrance

Figure 7.56 Winter time – frames parallel to entrance

Security

Bees are robbed of their honey by a number of animals – including man. Place the hive in a secure location. The hive stand could be constructed to stop pests from entering the hive. The hive stand should stand in oil without odours – which can get into the honey. Ensure that vegetation is cleared around the hive stand.

Figure 7.57 Bee hive in tree – good general security – but difficulty in accessing the hive for observation

Figure 7.58 Vegetation well cleared around hive stand. But hive stand not in any protective oil well to reduce ant attack

Figure 7.59 Vegetation well cleared around hive stand

Figure 7.60 Vegetation growing around the hive stand

Figure 7.61 Hive stand not providing any protection against arthropod pests

Propolis – keeping the hive safe

Propolis is a resin material collected from trees – leaves and buds – using the same honeybasket as pollen. Bees use propolis to seal holes and provide a disinfectant to the hive. Propolis will be seen sealing gaps between the wood and other small holes.

It is interesting that the bees on their own cannot remove the propolis from their honey-basket, they require help of their sisters.

Figure 7.62 Honey bee with propolis on her hind leg

Figure 7.63 Bees trying to fill in the gaps with propolis (black material). The gap was left by man when rushing to rebuild the hive

Wild *Apis* hives

The other *Apis* species can be divided into the type of hives they make.

Other wild cavity bee hives

Apis mellifera; A. cerana; A. koschevnikovi; A. nuluensis and *A. nigrocincta* are known as cavity bees as they make their hives inside cavities.

Figure 7.64 *Apis cerana* in a cupboard

Figure 7.65 *Apis cerana* hive detail of bees on outside of colony

Mini species of *Apis*

The other *Apis* are open-nesting. With the mini species (*A. florea* and *A. andreniformis)* the hives hang with the branch within the nest.

Figure 7.66 *Apis florea* and the micro bees hang their nest around the branch

Giant species of *Apis*

The giant bees *A. dorsata* and *A. dorsata laboriosa* hives hang underneath the branch with one massive comb.

Figure 7.67 *Apis dorsata* and the giant bees hang their nest underneath the branch of a durian fruit tree. The active hive is on the left; an abandoned hive is on the right

Figure 7.68 Detail of *Apis dorsata*

With *A. dorsata* the hive is kept at 34°C. If the hive temperature is above 36°C the brood can die. Below 30°C the brood develops slowly and is often deformed. *A. florea* maintains the nest between 33 and 35°C.

Bumblebee nests

There are numerous species of bumblebees (*Bombus*) each with its own peculiarities of life cycle.

Once the nesting site is selected an area is cleared by the queen to form a small chamber. Pollen is collected and moulded into a mass upon which the first wax cells are made. The queen then lays her first batch of eggs. The queen constructs another wax cell in front of the eggs which is filled with nectar – providing a honey pot.

The queen broods (incubates) the egg clump using her body and raises her brood at a temperature of 25°C, which is much cooler than *Apis mellifera* hives where the brood is at 36°C.

The nest in the autumn is abandoned by the older queen and workers. They may be seen foraging for themselves. However, they will not survive winter. Some young queens will start a new colony in their first and only summer.

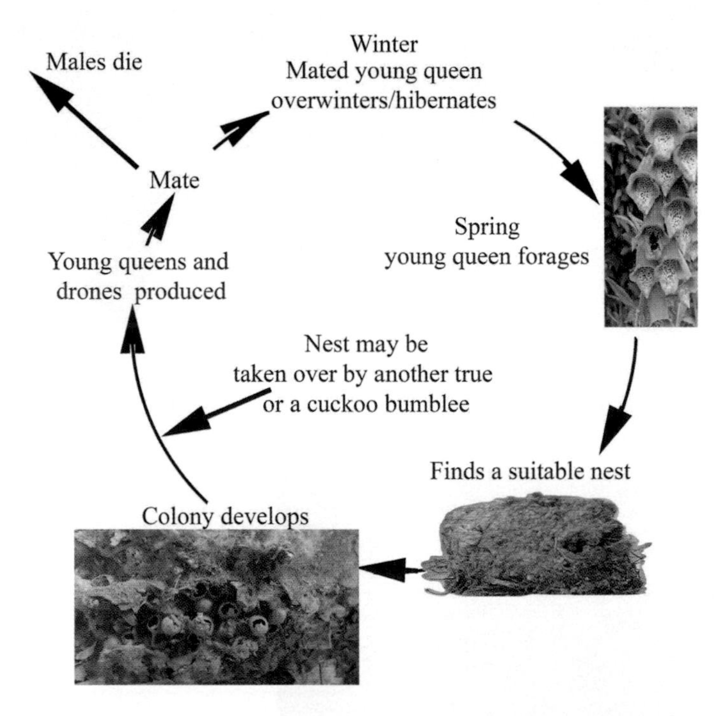

Figure 7.69 Drawing of the life cycle of a typical bumblebee. Note there are thousands of species of bumblebees and each has a slightly different life cycle to fill a different ecological niche

Artificial bumblebee nest box

Bumblebees are reared to assist the pollination of commercial greenhouse plants.

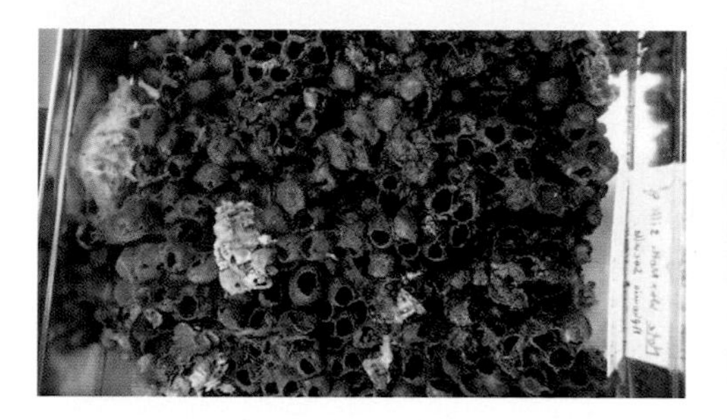

Figure 7.70 A bumblebee nest (*Bombus terrestris*) in a box

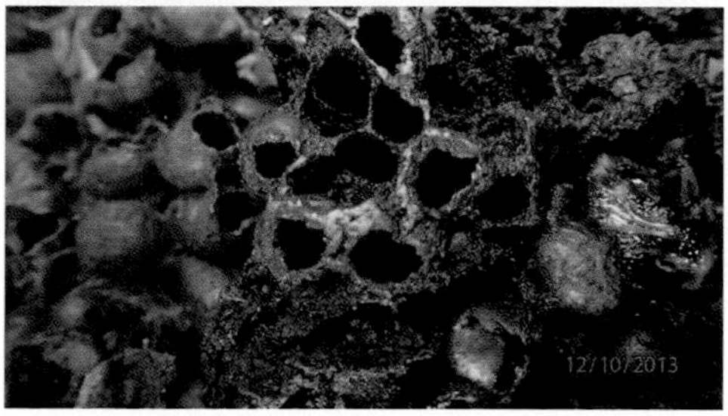

Figure 7.71 Detail of the brood cells

Stingless bees hives

Stingless bees, the Meliponini, are a large group of bees. They possess a sting apparatus but it is incapable of penetrating human skin. The stingless bee appears to have originated in South America and has then migrated into North America, East Asia, Australia and Africa. They are particularly common in the tropical zones. Their hives are quite diverse but many have the combs arranged in a spiral format.

Figure 7.72 Abandoned stingless bee hives being recycled in Borneo

Figure 7.73 Box wood hive for stingless bees in Borneo

Figure 7.74 Detail of guard bees in a stingless bee hive

Figure 7.75 Opened log hive of a stingless bee

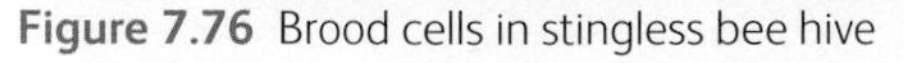

Figure 7.76 Brood cells in stingless bee hive

Figure 7.77 Water supply by newly established stingless bee hive

Disorders of the hive and nest

Alfalfa/lucerne stress (*Medicago sativa*)

While the alfalfa/lucerne plant has plenty of protein – 20–24%, the aminoacid ratio is incorrect for bees, being short of iso-leucine. The bees require 4% iso-leucine whereas lucerne has 3%. Therefore if the bees only are working lucerne the hive increasingly comes under stress. This will also happen when bees are working other mono-crops. Sunflowers are another stress inducing plant.

Amphibians

Toads and frogs will occasionally sit at entrances to *A. cerana* hives at dusk and eat bees.

Ants

All types of ants are the natural enemy of bees. It is essential to stop all ants from gaining entrance to the hive. Protect the legs of the hive with insect repellents or cover them in thick grease. Some beekeepers will stand each hive leg in engine oil. However, note odour which may get into the honey. Spreading wood ash or charcoal ash will help to keep the ants away.

Figure 7.78 Ant with a worker bee

Figure 7.79 Plant leaves resting against the alighting strip providing easy access to the hive by ants

Stop any branches or leaves that might come into contact with the hive. Grass growing up to the landing board allows easy access to the hive by ants. When doing your routine examination do not encourage ants by spilling honey or sugar around the hives. Do not just drop wax in front of the hive. Clean up carefully following each hive examination. Burn the hive debris as part of the hive biosecurity programme.

In Asia, ants are a particular concern to wild bee hives. Weaver ants, *Oecophylla smaragdina* will attack hives. These are large ants (10mm long) and attack dwarf bee species *A. florea* and *A. andreniformis*

They may also attack *A. cerana* but not as commonly. If the hive is targeted repeatedly this can lead to absconding.

Ants are also a natural enemy of stingless and bumblebee colonies.

Beekeeper

The greatest threat to the modern bee hive is the beekeeper. Stealing too much honey, pollen and other products puts enormous stress on the hive and will often lead to hive collapse. In many parts of the world, theft is a particular concern and bee hives have to be placed in areas difficult to reach.

Figure 7.80 The beekeeper is the biggest robber of the hive

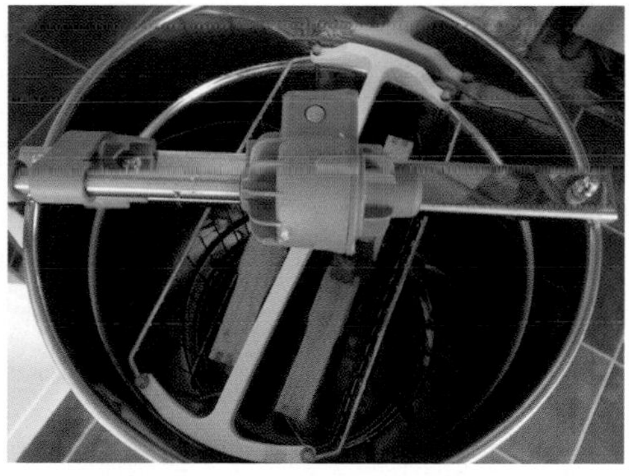

Figure 7.81 Honey being extracted from supers in the autumn

Beetles

Carefully examine any beetle found in the hive for the small hive beetle (*Aethina tumida*). Then remove any beetles found in the hive to assist the honey bees.

Beetles in bumblebee nests

Antherophagus nigricornis lives frequently in bumblebee nests. This is a small beetle 5mm long and 1.5mm in width. They are yellow brown in colour. *Antherophagus nigricornis* lives on the nest debris.

Trichodes ornatus (the ornate checkered beetle) from North America is known to eat bumblebee larva and pupa and needs to be avoided.

Hive beetles in honeybee hives

Small hive beetle (*Aethina tumida*)

Figure 7.82 The small hive beetle. Note the club antenna

Figure 7.83 Detail of the adult small hive beetle – *Aethina tumida*

The major beetle of note in honey bee production is the small hive beetle (*Aethina tumida).* This is a notifiable condition in many parts of the world where it is still absent. Where the beetle is present, its clinical signs are relatively mild, but it does add to the pathogen stress on the hive, increasing the risk of Colony Collapse Disorder (CCD). The beetle also will damage honey and other hive products. The beetles are small, brown to black and 5–7mm in length and 3–4.5mm wide. When looking at the beetle, note the clubs on the end of the antennae. Beetles will rapidly run from the light when the top board is removed.

The adult beetle lives for 4–6 months. The beetle is able to fly 5–15km from the hive looking for new hives. The beetle will deliberately follow migrating and swarming bees. In the evening it may follow returning foraging bees back to the hive. Note the beetle may also infest bumblebee colonies, making control extremely difficult.

Clinical Signs

The small hive beetle will be found throughout the hive, most commonly in the rear portions of the bottom board. There may be few obvious clinical signs, especially when the hive only has a few beetles present. This beetle is destructive as it feeds and defecates in the honey causing discolouration. It will uncap and eat bee brood, honey and stored pollen. The beetle will run fast when exposed to the air. Some bees may be in gaol, imprisoned by guard bees with propolis, but then fed honey and pollen by the same guard bees!

In Asia 2014, an infestation has occurred in the Philippines and poses a particular threat to the native bee population.

Hives may collapse under the pressure of a new infestation. Other species of Apis and stingless bees are potentially under threat.

In some of the stingless bee species (*Trigona carbonaria*) the beetle may be mummified by wax, propolis and mud.

Diagnosis

Recognition of the beetle in a hive. Noting the large club antenna helps to identify the beetle. Hive beetles move quickly but can be collected using a filtered aspirator. If you are concerned, collect the beetle and then ask your local bee inspector or veterinarian. The beetle may be confused with similar species notably *Cychramus luteus.*

Life cycle

The eggs hatch in 3–6 days.

The eggs are often laid in large clusters. The larvae are 10–11mm in size. The larvae have characteristic rows of spines on their back (absent in Wax moth larvae) and 3 pairs of tiny prolegs. Larval development takes 8–29 days depending on the environment. Larvae leave the hive and pupate in the soil for 2–12 weeks, again depending on the environment. High temperature and humidity is ideal for *Aethina tumida.* They also survive the harsh winters in North America safe within the hive.

Figure 7.84 The small hive beetle in a hive

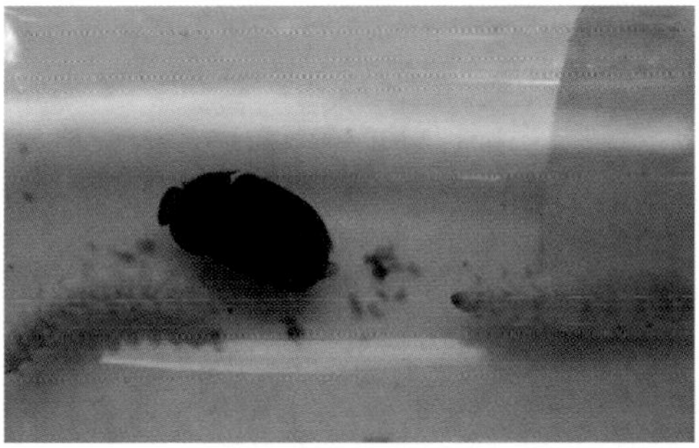

Figure 7.85 Adult and larva of the small hive beetle

Similar disease

Small/young wax moth larvae may be confused. The adults are small beetles and a lot of beetles can look alike. Look for the club antennae.

Treatment

Beetle traps are possible made from corrugated paper. Fungal biotreatments are being developed.

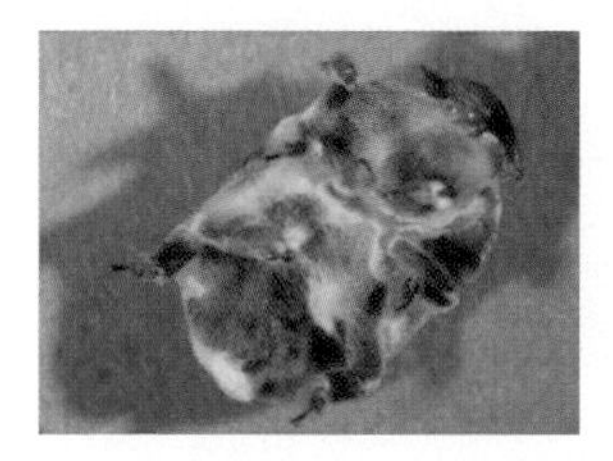

Figure 7.86 Biological control of the small hive beetle

Management and Control

- Store supers in an environmentally controlled room with a relative humidity below 50%.
- Store empty combs with moth crystals (paradichlorobenzene).
- Freezing (–20°C) combs overnight will kill the beetle.
- Fumigate the stored combs with 70% ethanol (eg methylated spirits).
- Keep the super storage area clean.

A more long-term programme is to select queens for hygienic behaviour (see Chapter 12). Purchase queens which are from high hygienic behaviour lines.

Zoonotic implication

None.

Large hive beetle – *Hoplostomus fuligineus,* also called the hive chafer

Hoplostomus fuligineus and *H. haroldii* is found in Sub-Saharan Africa. The large hive beetles have very hard carapaces, presumably to withstand any stinging action by the workers. If any of these beetles were found outside Africa they would constitute a notifiable condition. Biosecurity at ports and when purchasing queens from abroad is the key to controlling their

Figure 7.87 The large hive beetle inside the hive

Figure 7.88 The large hive beetle outside the hive under attack

Figure 7.89 Large hive beetle dorsal surface – with an African honey bee on the right

Figure 7.90 Large hive beetle ventral surface

introduction. They consume a lot of honey and can contribute to a weakened hives' collapse. Outside the hive, the beetle is attacked by large numbers of workers but inside the hive, appear to be unmolested. They are often seen just under the top board. They may emit a sound or chemical to disguise themselves within the hive. Once the workers can see the beetles they attack *en masse.*

Birds

Many species of birds will occasionally or preferentially catch and eat honey bees and bumblebees.

The shrike (*Lanius collurio*) will capture bumblebees. The shrike will either eat the bee immediately or store it for later in a larder with the bee impaled on thorns or spikes, barbed wire being particularly suitable. In some species of shrike, 10% of the feed intake is composed of bumblebees.

The great tit (*Parus major*) will attack hibernating bumblebees. The bird will peck at the abdomen to feed on the nectar in the honey stomach. Bumblebees (*Bombus lucorum* and *Bombus terrestris*) foraging on flowering lime can become drowsy with the nectar making them susceptible to attack, mutilation and death.

Bee-eaters are obviously designed to eat bees and hives. The green bee-eater *(Merops orientalis*) will attack hives of *A. dorsata* in mountainous regions of Asia. It has been seen that early in the morning a flock of 100 bee-eaters will cooperate in their attacks, some of them tormenting the hive until the bees leave the hive to attack the birds. In the cool of the mornings, the bees can succumb to the cold temperatures and become moribund. The birds then eat the slower bees.

The blue bearded bee-eater (*Nyctyornis athertoni*) initiates defensive behaviour from a hive. The bird then flies off with trapped bees in its plumage, the bees to be eaten later.
The blue throated bee-eater (*Merops viridis*) eats *Apis dorsata.* The bird will deliberately rub the bee's abdomen on a branch to release the alarm hormone. This in turn will attract other guard workers. The blue throated bee-eater then hops to one side and catches the guard coming to investigate the cause of the alarm hormone release.

The oriental honey buzzard (*Pernis ptilorhyncus*) specialises in attacking *A. dorsata* hives. It encourages the hive's guard bees to mount a defensive response and the buzzard attacks the guard bees. Once it has attacked the defence bees out of the hive, it then turns and rapidly returns to the undefended hive. But the bird takes a risk as 30 stings can kill.

Figure 7.91 Grey-backed fiscal (*Lanius excubitoroides*) a type of shrike.

Figure 7.92 White throated bee-eater (*Merops albicollis*)

The honeyguide (*Indicator indicator*) is a useful bird for man and for the honey badger, who will watch the bird and will follow it to the bee hive.

Green woodpeckers (*Picus viridus*) can damage hives and its parts and may even eat brood and bees.

If this is a problem build a cage from chicken wire around the hive – sufficiently separated from the hive to prevent the woodpecker's bill reaching the hive.

Figure 7.93 Hive attacked by a woodpecker

Cuckoo bumblebees (Psithyrus*)*

One of the interesting aspects of bumblebee health management is that other species of bumblebee may parasitise their nests. The parasitic bumblebees are called cuckoo bumblebees. Over time these have evolved from bumblebees but they do not produce wax and have no pollen basket. They tend to have a thicker exoskeleton presumably as a defence against stings from the resident bumblebee.

There are a number of species of cuckoo bumblebees, all with different survival strategies. In the late spring the female cuckoo bumblebee leaves its ground hibernation. By this time bumblebees have started to develop their nests with workers. The female cuckoo bumblebee enters the nest and may remain in the nest debris until she smells like the nest. She then may kill some of the workers, eggs and larvae. Sometimes even the queen bumblebee is killed. Note some workers and even older larvae are tolerated to provide 'workers'. If the queen bumblebee is not killed, she may be humbled to become a worker in behaviour. She now works for the queen cuckoo bumblebee.

The cuckoo bumblebee makes her own cells from the materials of the nest and then lays numerous eggs. The cuckoo eggs and larvae are then raised by their hosts. The adult newly mated cuckoo bumblebee female has a large fat body and hibernates over winter in the soil. The males die off over winter.

Some normal bumblebees may act like cuckoo bumblebees. *Bombus terrestris* workers may lay eggs in unrelated nests, for the larvae to be reared by the adopted parents.

Death's head hawk moth (*Acherontia spp.*)

The death's head hawk moth is named because of the obvious 'human' skull on its dorsal thorax. This is a large moth. It is able to enter a hive unmolested as it can emit a sound similar to the queen bee, so the workers remain calm. This is very unusual as moths do not normally make squeaking noises. The moth also produces a scent which is believed to help to mask its

Figure 7.94 Death's head hawk moth of Europe, *Acherontia atropos*

Figure 7.95 *Acherontia lachesis*

Figure 7.96 *Acherontia styx*

presence. The moth then walks around the hive consuming honey without interference. The moth then leaves the hive and flies away. The moth (*Acherontia atropos*) normally lives in Northern Africa and only migrates into Northern Europe occasionally.

In Borneo other cavity bees can be robbed by another hawk moth (*Acherontia styx*).

The hawk moth (*Acherontia lachesis*) is able to steal honey from all *Apis* – including the Giant Asian bee – *Apis dorsata.*

Dogs, cats, honey badgers, badgers, raccoons and other carnivores

These mammals will overturn hives and gain access to the honey. Place electric fencing around the hive. There are a number of mammals who specialise in being predators of bees.

Sunbears are also called 'Honey bears' (*Helarctos malayanus*) and have long tongues so they can access the honey without breaking open the tree.

It has been reported that pine martins (*Martes flavigula*) in Nepal will open an *Apis cerana* hive and then urinate on the hive, killing or paralysing the bees, before eating the hive.

Tigers (*Panthera tigris*) are able to attack low hanging hives of *Apis dorsata.*

Honey badgers (*Mellivora capensis*) are a major predator in Africa and one of the reasons the hives are hung in trees. Honey badgers will watch honey indicator birds (*Indicator indicator*) to lead them to a hive. Skunks may eat bees and spit out a cud of chewed up bees. The bee remains may also be seen in the faeces as their exoskeletons cannot be digested.

Figure 7.97 Sunbears (*Helarctos malayanus*) are particularly keen on eating honey

Figure 7.98 Tigers are able to attack low hanging *Apis dorsata* hives

Drifting

Bees may fail to return to the hive they left. This is called drifting. Older forager bees are more often guilty of drifting. Drones may also drift between hives. Prevailing winds may drive bees into a different hive. The bees will drift downwind.

Environments which are featureless make drifting easier. Hive entrances should be clearly marked with different shapes or colours to assist bees to recognise which hive they belong to. Drifting is a major biosecurity problem. Having each bee hive face a slightly different direction helps to reduce drifting.

Figure 7.99 Hive in a straight line encourages drifting

Figure 7.100 Hives facing different directions reduces drifting

Earwigs (*Forficula auricularia*)

Regularly removing earwigs is a good biosecurity preventative measure. Earwigs may damage cappings. Standing the feet of the hive in containers of water or disinfectant prevents earwigs from gaining access. They are extremely common in hives particularly around the top board.

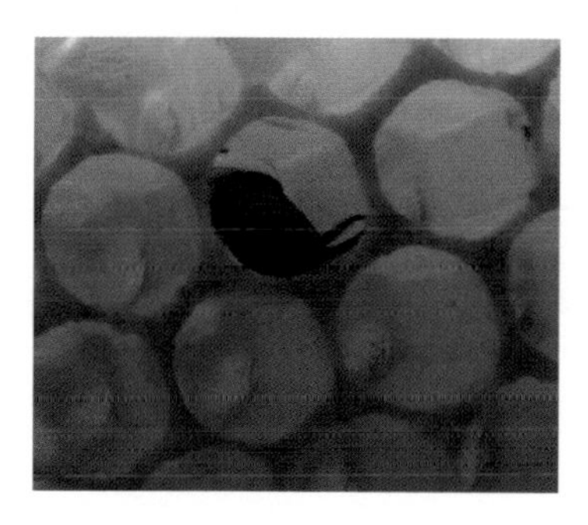

Figure 7.101 Earwig in a honey cell

Environmental stress

If the beekeeper fails to maintain or manage the hive the bees will be subjected to stressors which will reduce productivity and may even result in the complete collapse of the hive. Some of these environmental factors can be very simple mistakes.

- Hive under an overhanging tree – more likelihood of wasps and branches falling on to the hive.
- The hive being built incorrectly – the top roof the wrong way round.
- Hive sloping backwards – so water runs into the hive rather than out.
- Hive floor upside down – a haven for wax moths which will not be seen by the beekeeper until the hive almost collapses.
- Poor management of the ventilation system.

Farmed animals: sheep, horses, cows and deer

These animals find bee hives excellent rubbing places and can then destroy the hive. Note the bees may become sufficiently annoyed to swarm and kill the offending animal.

Figure 7.102 Protect hives with more sturdy fencing

Figure 7.103 Ankole cattle in an apiary. They can rub against hives and cause destruction

Flies

Several species of flies will cohabit bumblebee nests and live on the nest debris. *Fannia canicularis* adults, which albeit are smaller than *Brachicoma devia* (the serious parasitic fly of bumblebees) can be easily mistaken. They both resemble house flies (*Musca domestica*). Their presence can be an indication of poor hygiene as they will thrive in nests where the floor has become wet or fouled with bumblebee faeces.

Figure 7.104 *Fannia canicularis* male, 6mm long

Honey bound hive

When a brood becomes split by a frame of honey the queen will not visit the isolated brood. The isolated workers can start raising a queen in the isolated brood.

Hornets

Hornets (*Vespa crabro*) can attack an apiary taking bees in the air or from their alighting board. The hornet then alights on a nearby tree and eats the worker honey bee. If a hornet is terrorising your hive, it may be necessary to find the hornets' nest and destroy it. Some hornets terrorise the hive to such a degree that the workers will not forage. In such cases it may be necessary to move the apiary.

The Japanese hornet (*Vespa mandarina*) will prey on foraging bees. However, at the end of the season if they locate a hive they will mark the hive with a pheromone from their van der Vecht gland (these are on their metasomal sternum VI). This pheromone will attract other nest mates of the hornet. Up to 30 hornets may then attack the hive. *Apis cerana* will actively remove traces of this pheromone from their hive if they realise they have been targeted.

Figure 7.105 Hornet (*Vespa crabro*)

Apis cerana will also attack these hornets. A hundred bees will surround the hornet and smother the hornet raising its temperature to a fatal 47°C. *Apis cerana* can survive this temperature for short periods. *Apis cerana* will also try and confuse the hornet by 'shimmering' at the entrance.

Vespa multimaculata will hover in front of the landing area, intercepting incoming bees.

Vespa affinis flies to drone congregation areas of *Apis cerana*. This hornet mimics the actions of the virgin queens, which attracts the drones to the hornet. They are then captured by the hornet and eaten!

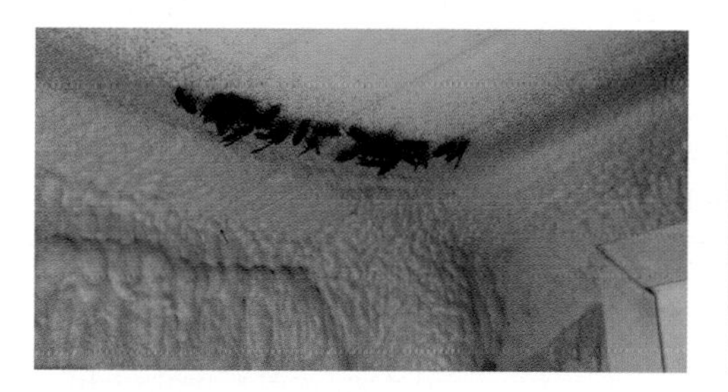

Figure 7.106 Asian giant hornet (*Vespa mandarinia*)

Figure 7.107 Hornet trap in Korea

Mice and shrews

When you examine the hive, note if pieces of wax comb are seen on the alighting board. This is a good sign that there may be a mouse in or visiting the hive.

If you suspect a mouse, raise the hives onto hive stands. Especially in the autumn, mice and shrews can gain access to the hive and destroy combs and bees. Remove any mice and shrews found in the hive. Fit a mouse guard/excluder in the autumn. Dead mummified mice may be found in the bottom of a hive as the workers killed the mouse, but were unable to drag the dead mouse from the hive.

Lucerne stress

Mite pests of bumblebees

The hives of wild bees and the nests of the bumblebee and the bee's body is home for a host of mites most causing minimal inconvenience. Those that live on detritus are referred to as saprophagous. If they live on feed stores they are called kleptophagous.

In wild bee populations *Neocypholaelaps* and *Afrocypholaelaps spp.* are commonly found. In *Apis cerana* if the worker returns with a *Neocypholaelaps indica* mite it may perform a shaking dance to encourage other bees to remove the offending mite from the rear of its thorax!

In the bumblebee, *Parasitus fucorum* is just one example where the immature stages live in the hive debris. The adult female attaches itself to the body of the bumblebee and will fly around with the bee. In this manner she can leave the bumblebee at a suitable flower waiting for another bee to infest or overwinter with the queen as she makes a new nest.

Some of the mites of the bumblebee may even have parasitic mites of their own.

Monkeys and primates

Several species of monkeys will steal bee hives, including the greatest bee predator – man (*Homo sapiens*). Macaque monkeys will shake *A. florea* and *A. cerana* hives until the adult bees abandon the hive.

Moulds in general

If the hive is damp after prolonged rains a variety of moulds and fungi can grow inside the hive. This can produce a foul taste in the honey. Improve the ventilation and air passageways.

Plants

Insectivorous plants may capture a variety of bees. Stingless bees in particular can be often be found in insectivorous plants. Frogs will sometimes sit in insectivorous plants waiting to eat unwary bees.

Pollen mites

These may be mistaken for *Varroa* at times. If found, remove and destroy old combs which contain a lot of pollen.

Figure 7.108 Pollen mites may be seen when you examine the hive debris

Reptiles

Various reptiles will eat bees. Geckos and skinks may feed on returning foragers.

Rhododendron

The rhododendron family of plants does not specifically cause any issue to the bees. But the honey that is produced has significant and potentially fatal effects on people. It is referred to as 'mad honey' and obviously should be avoided.

Robbing

Note the risks when collecting honey from the hive. When you are robbing the hive, others wasps and bees may all be attracted to the party.

Figure 7.109 Fighting at the entrance

Spiders

Many spiders may live in bee hives. They will interfere with the ventilation of the hive and regular inspection and brushing of webs from crevices and removing spiders are all good preventative measures.

Figure 7.110 Spider web within the nest

Figure 7.111 Bees trapped in a web spun by the entrance to the hive

Starvation

There are two forms of starvation:

Honey
Pollen

Honey starvation classically will occur in the winter or wet season. The hive should have four or five frames of honey at all times and a lot more going into the winter.

Pollen starvation can occur at any time of brood rearing.

Clinical signs

With honey starvation, the hive stops brood rearing. The worker bees may be slow moving and may demonstrate trembling. Examination of the frames will indicate dying and dead bees. The bees often die head first in cells.

With pollen starvation there is a lack of pollen stores. There can be eggs in the cells but no uncapped cells. Day old larvae may be seen without food. This will extend to no brood. The larvae and pupae may be found on the bottom board or ground around the hive. Some of the pupae may be partially eaten.

Termites

The termite does not attack the bees, but will destroy the hives, hive stand and equipment.

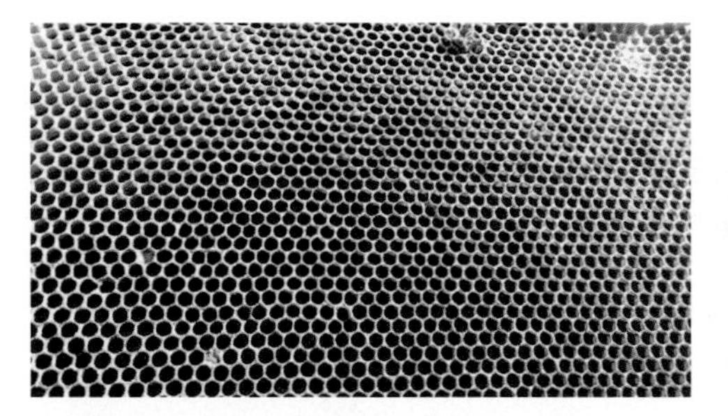

Figure 7.112 Complete lack of honey stores

Figure 7.113 Mass of dead worker bees which have also developed mould

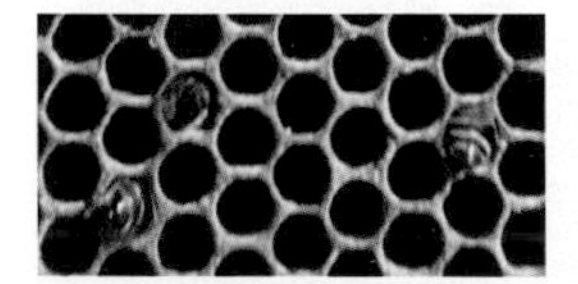

Figure 7.114 Starved bees with the classical head first in cell death pose

Volucella flies

Volucella flies lay their eggs in bumblebee nests. These larvae develop within the nest and scavenge below the comb, feeding on nest debris. Volucella are hover-flies which mimic bumblebees in general body size and colour patterns and can emit a bee-like buzz when disturbed.

Interestingly if the female volucella is killed by the bumblebee upon entry into the nest, she will still lay her eggs.

Wasps

Each country has its own significant wasp species. For example, in the UK three species of wasp are of note: *Vespula vulgaris, V. germanica* and *Dolichovespula saxonica.* Wasps are significant as they will rob and kill weak colonies.

When lifting the top cover check for wasps. *Dolichovespula arenaria* (North America) can be a particular problem when harvesting honey. They can gain access to the open hive causing considerable distress to the hive bees.

Figure 7.115 Wasps and robber bees assisting the beekeeper to rob the hive of its honey stores. Note the unusual bee with the black body. This could be a bee suffering with ABPV or CBPV

Wasp pirates

These can be a problem in Africa. These wasps will catch flying bees and can be a significant problem terrorizing the hive. The yellow bee pirate (*Philanthus diadema*) catches bees as they forage on flowers so is very difficult to control. The banded bee pirate (*Palarus latifrons*) ambushes forage bees as they return to the hive.

Management and control

Set up wasp traps – a jar with a 10mm hole in the lid and half filled with jam/water mixture. Do not use honey – this will attract the bees.

Wax moths – *Aphomia sociella* and *Galleria mellonella*

Aphomia sociella is the wax moth of bumblebees and can cause abandonment of the nest. In *Apis mellifera* (honey bees) *Galleria mellonella* (greater wax moth) and *Achroia grisella* (lesser wax moth), they can cause serious problems in honey production. The wax moth may also cause damage to the wood of the hive reducing the life of the hive.

The larvae may leave sufficient silk making it difficult for emerging bees to escape the cell – this is referred to as Galleriasis.

Galleria mellonella are similar in size to the honey bee. The female is 20mm long whereas the male is 16mm. The moth is usually stopped from entry into the hive by the guard bees. However in a stressed hive the guard bees are not as diligent.

Achroia grisella are slightly smaller and are grey compared with *Galleria mellonella*. *Achroia grisella* females measure 12mm long while males measure 10mm. The damage to the hive has been termed 'tubular brood'.

Benefits of wax moths

There is another side to wax moths. The larvae are highly valued by anglers. Wax moths can also be a food source in protein-deficient diets.

Figure 7.116 Galleriasis in a Kenyan top bar hive resulting in abscondment/collapse of the hive

Clinical signs

When the hive is examined, the caterpillar phase of the wax moth may be seen. Careful examination may reveal the adults flying from the hive. Examination of the hive debris may reveal dead wax moth caterpillars.

Wax moth development requires brood combs or brood cell cleansings. The life cycle takes between one and six months. Adults live 3– 28 days. Females will fly into the hive and lay eggs. If the female cannot gain access to the hive, she will lay eggs in the ground around the hive and the larvae will gain access into the hive themselves. The caterpillars take 18 days before becoming pupae. It is the larvae that cause most disruption to the hive.

The larvae eat the cappings revealing bald brood. Despite having the capping removed, the growing pupae will develop normally.

Diagnosis

The presence of the caterpillar stages and the webbing covering the frame. Check the hive debris for caterpillar stages.

With *Achroia grisella* the caterpillars burrow straight across the base of the cell creating a 'tubular brood'. They will rarely pass through the centre of the frame. Note wax moth larva have three prolegs at the front and paired prolegs at the back. There are no spines along the back (unlike in the small hive beetle larvae).

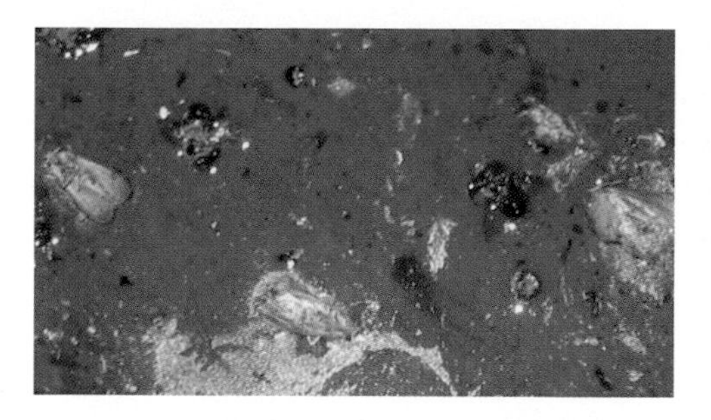

Figure 7.117 Adult wax moths – similar in size to workers

Figure 7.118 Wax moth larvae leave behind webbing – which the bees find very difficult to remove

Figure 7.119 Wax moth larvae – off white soft bodied caterpillar – see dead in hive debris

Figure 7.120 Cocoons of wax moth pupae are very tough and extremely difficult for bees to remove

Figure 7.121 Wax moths can damage hive insulation boards and woodwork.

Treatment

Remove larvae from hive.

Management and control

Note you must kill the larva as it can walk/crawl to another hive within 50 metres. Freeze old combs for at least 24 hours. Apply *Bacillus thuringiensis* spores (Bt spores).

Control with chemicals that do not get into the honey. Do not use 'moth balls' with paradichlorobenzene (PDB), as this is carcinogenic and can leave residues in wax and honey. After treatment, stack supers outside for at least a week before use. There must be no presence in honey sold in the EU.

Stacked dirty honey or brood supers over winter are an ideal environment for wax moths to multiply. Alternatively, place combs into a sealed box and fumigate with vapours from 80% ethanoic acid (acetic acid). Note cover any metal on the boxes with Vaseline to protect from the acid. Fumigate for at least ten days. Air well before use.

Increase ventilation to the hive. Wax moths do not like light and fresh air.

Wax moths in wild *Apis*

Wax moths can be extremely problematic in wild Asian bee hives. The tropical climate assists the rapid build-up of wax moths. *Apis florea, A. cerana* and *A. dorsata* hives can be particularly affected. It should be noted, however, that wax moths may play an important natural role in destroying and recycling abandoned hives.

Wax moths in bumblebees

Aphomia sociella is a destructive wax moth. The adult female enters the nest in the summer and lays a group of shiny white eggs. There are often over a hundred *Aphomia* larvae in the nest. As they grow the group of larvae eat the food stores and cells and occasionally even the bumblebee larvae. They leave the destroyed nest as a group and pupate in the ground and overwinter. They are more destructive than honey bee wax moths *Galleria mellonella* (greater wax moth) and *Achroia grisella* (lesser wax moth). Adult bumblebees may become trapped by the web spun by the moths, in their own nest, and then starve to death.

The adult *Aphomia sociella* are whitish yellow in colour with dark brown markings. The adults have a wing span of 30–35mm.

Fig 7.122 Wax moth web in a bumblebee nest

There are other species of moth, such as *Vitula edmandsii* (North America) or *Endrosis sarcitrella* (Europe), which are relatively harmless feeding on wax, pollen and nest debris. It is important therefore to correctly identify any species found in the bumblebee nest or honey bee hive before taking drastic actions.

White box stress (*Eucalyptus albens*)

White box has poor protein content of its pollen at 17% and the iso-leucine content is only 3.8%, reducing the digestible protein content further. The plant flowers in the winter months. Bees are therefore cold stressed and protein stressed and this can lead to hive collapse. Note that other plants produce identical symptoms.

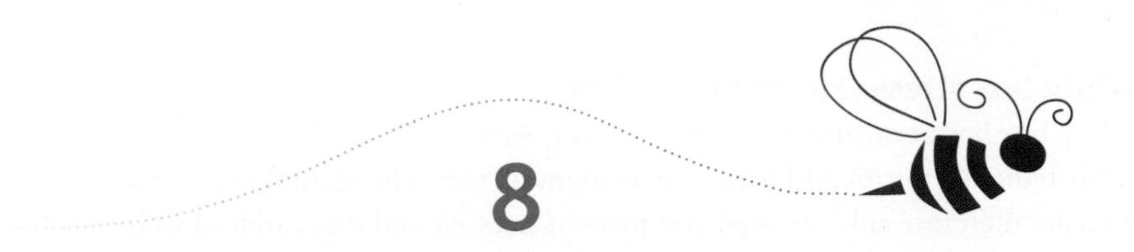

8

OIE and international bee health

The World Organisation for Animal Health is the international body which is responsible for monitoring the presence of animal pathogens around the world. In 2003 it changed its name from Office International des Epizooties but kept its historical acronym 'OIE'. For more details visit the OIE website at: www.oie.net.

Most countries are part of the OIE. The published list consists of transmissible pathogens with the following characteristics:

- Have the potential for very serious and rapid spread, irrespective of national borders.
- Have a serious socio-economic or public health consequence.
- Are of major importance in the international trade of animals and animal products.

Generally these pathogens do not occur in many regions of the world because they have never been present or because they have been eradicated and are kept out by government control measures. If any of these pathogens are suspected they must be reported immediately to the local authorities.

When a new condition is recognised across the planet it is generally added to the OIE list until its impact is understood.

Free, fringe or enzootic

Bee producing regions of the world can be classified as free, fringe or enzootic for any given pathogen. These terms are defined as follows:

'Free' describes those regions, countries or continents which are free from a pathogen. An exotic pathogen (disease) is a pathogen which does not occur in the region or country of your apiary. A pandemic disease is a pathogen which is widespread throughout a region or the world.

'Fringe' refers to areas which are generally free but which are always under threat of re-infection from outside (bordering an area where pathogens are common). Fringe areas may suffer sporadic outbreaks, which then have to be controlled and stamped out. Such outbreaks may occur several times a year, or several years apart.

'Enzootic' (endemic) refers to a disease that is permanently present in a population. In relation to diseases of bees, the population may be a hive or the apiaries in a region, a country or a continent. Strictly speaking, 'endemic' should only be applied to populations of people ('demic' is from the Latin *demos* meaning people/democracy) and 'enzootic' should be used for animals but in practise the two are interchangeable.

'Epizootic' (epidemic) refers to a disease which spreads, usually fairly quickly, to a large proportion of the bee population.

Which countries are free from which pathogen?

It should be noted that the global situation is extremely fluid and subject to rapid change.

Countries might be free of a pathogen in commercially farmed bees, but feral (wild) bees or related species may remain endemically infected.

Also, whereas the information in some countries is fairly reliable, in others it is not and in some there is very little information. The most reliable information is that in the countries listed in North and South America, the Antipodes and the EU. For up-to-date information, see the OIE website.

OIE list of honey bee pathogens, 2015

- *Melissococcus plutonius* (European foulbrood)
- *Paenibacillus larvae* (American foulbrood)
- *Acarapis woodi* (tracheal mite)
- Infestation with *Tropilaelaps spp.*
- Infestation with Varroosis (*Varroa spp.*)
- Infestation with *Aethina tumida* (small hive beetle)

Global situation

Melissococcus plutonius – European foulbrood

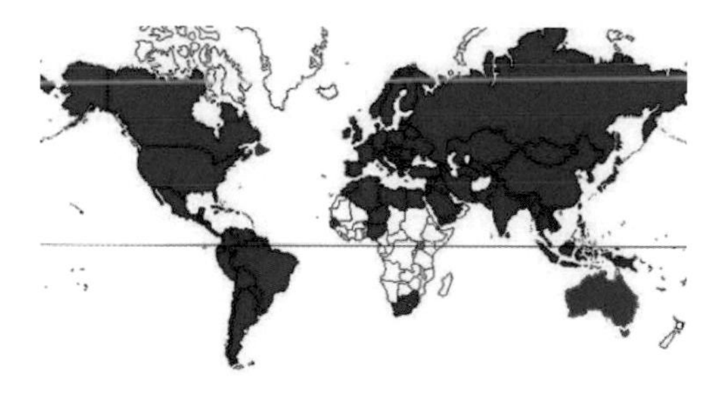

Figure 8.1 Global spread of *Melissococcus plutonius*

European foulbrood is associated with the bacteria *Melissococcus plutonius.* This is a Gram positive cocci (round) bacteria which can occur in chains. Occasionally the bacteria can be pleomorphic (have a variety of shapes). The bacteria can be difficult to grow, it may not appear to 4 days and requires anaerobic (oxygen free) atmosphere to grow.

Paenibacillus larvae – American foulbrood

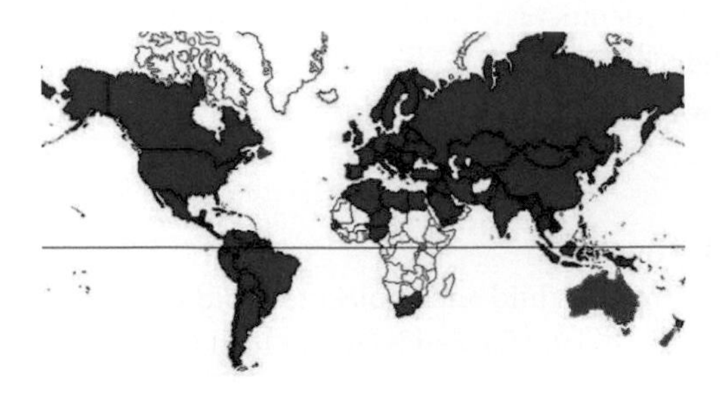

Figure 8.2 Global spread of *Paenibacillus larvae*

American foulbrood (AFB) is caused by the bacteria *Paenibacillus larvae.* This is a slender Gram positive rod with slightly rounded ends with a tendency to grow in chains. The bacteria produces spores making it very resistant. The bacteria does not grow on nutrient agar, it requires Difco brain-heart infusion. There are four types of AFB recognised: ERIC I worldwide, ERIC II Europe, ERIC III Chile, ERIC IV uncommon. Another bacteria has also been recognised, *Paenibacillus pulvifaciens,* but this may just be a variant of *P. larvae*

American foulbrood can cause problems at any time of the year. On opening the hive an unusual smell may be noticed, which is like a decayed glue pot – hence 'foulbrood'. The workers may also be unusually aggressive towards the hive examination. Examination of the brood frames will indicate that the capped brood distribution will be spotty and not even. The worker bees are trying to remove the dead and dying pupae. The bacteria infect the 1–2 day old larvae, but the clinical signs are seen in the pupae which die at 12–16 days of age, thus after capping. The bacteria cannot affect larvae older than three days of age (six days after lay). The larvae are infected through the honey they are being fed within the first couple of days of life. It may only require ten spores to infect a larva and on dying they produce billions. Close examination of the remaining capped brood may reveal some of the cappings are discoloured and convex (sunk inwards). Some may be punctured by workers who recognise a problem and have started to remove the dead pupa. Examination of the remaining pupae reveals that they are dull and often light brown to black instead of pearly-white. In some cells the pupa has decayed to become only a dried black scale, which may be stuck in the cell and difficult to remove. One characteristic is when a toothpick is put into the affected pupa the dead pupa sticks to the matchstick and the resultant 'mucus material' appears ropey. Interestingly, the tongue (glossa) of the dead pupa may remain as a fine thread.

Acarapis woodi – tracheal mite

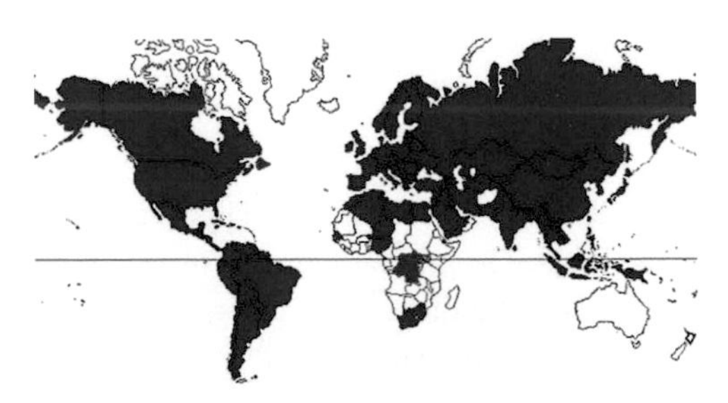

Figure 8.3 Global spread of *Acarapis woodi*

Tropilaelaps spp.

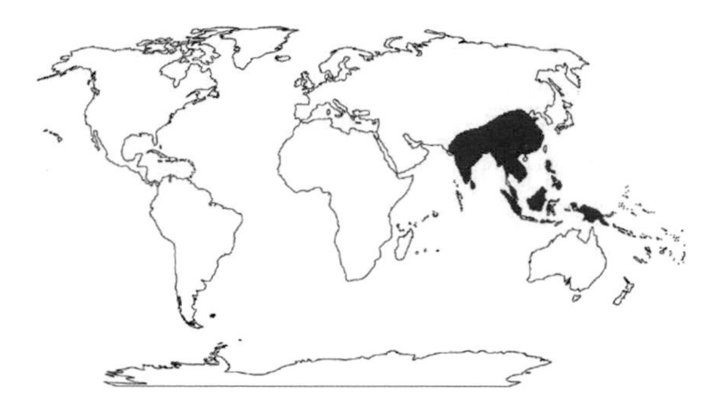

Figure 8.4 Geographical spread of *Tropilaelaps spp.*

Tropilaelaps mites are 1mm long and 0.6mm wide. There are two major species: *T. koenigerum* is slightly smaller than *T. clareae.*

Tropilaelaps is a parasite of *Apis* species, the natural host is probably *Apis dorsata* (the giant honey bee).

There are four recognised species of *Tropilaelaps*:

Tropilaelaps clareae infesting *A. mellifera* in the Philippines.
Tropilaelaps mercedesae infesting *A. dorsata, A. laboriosa* and *A. mellifera.*
Tropilaelaps koenigerum infesting *A. dorsata* and not *A. mellifera.*
Tropilaelaps thali infesting *A. dorsata laboriosa.*

The mite can live on a number of *Apis* species –*A. cerana* and *A. florea* but has not been seen to replicate. *Tropilaelaps mercedesae* is probably the greatest threat as it has a wide species association.

The adult female mite infests an L5 larva as its cell is about to be capped. The larva pupates and the mite feeds off the pupal haemolymph. The mite lays eggs. The first egg which hatches is a male and the subsequent mites are female. The male mates with his sisters. The life cycle is much faster than in *Varroa*, by a rate of 25:1. There is no preference for worker or drone infestation, unlike in *Varroa.*

The adult female mite cannot feed off adult bees and will die within three days without brood to feed upon. Therefore, the phoretic risk is much lower than with *Varroa.*

Varroa destructor – Varroosis

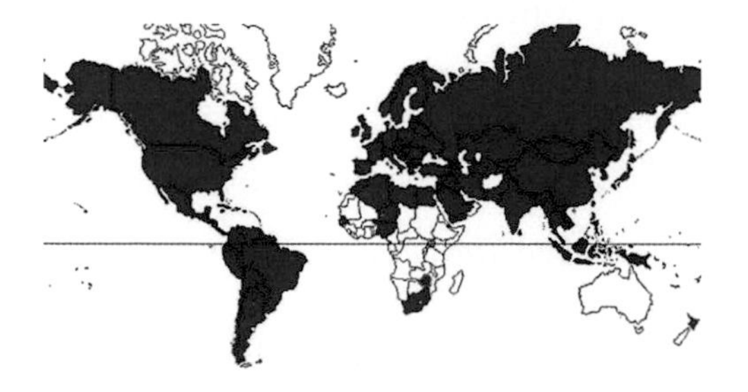

Figure 8.5 Global spread of *Varroa destructor*

The problem has arisen because *Varroa destructor* recently moved from its natural host, *Apis cerana*, into domesticated *Apis mellifera* hives.

The *Varroa* mites can also infest other native bee species – bumblebees (*Bombus pennsylvanicus*), for example. However, while they do not multiply on other bees, they can act as a transportation system (phoretic) which is important as it complicates the control of *Varroa*.

Varroa infestation of a hive can result in almost no clinical signs or in a combination of other issues, contributing to severe stress on the hive. *Varroa* mites may also infect *Apis mellifera* with a range of secondary viruses and may play a significant role in Colony Collapse Disorder.

Most hives are infested with *Varroa* with no or few clinical signs. If the mite remains at a low level the hive can be extremely productive and successful. Absence of the mite is extremely rare. Once an area is infested, beekeepers can only reduce the infestation, they cannot eliminate the mite. The mites move between hives by robbing and the drifting of honey bees and bumblebees can act as a means of transport – called the phoretic phase of the mite's life cycle.

Aethina tumida – small hive beetle

Figure 8.6 Global spread of *Aethina tumida. Note Hawaii (USA) is positive. Italy (pink) indicating its infestion in 2014*

Different countries

Each country may have its own requirements to protect the local bee population. For example, the United Kingdom and Western Australian policies are illustrated below. New Zealand is interesting as the island is free of European foulbrood – *Melissococcus plutonus*. New Zealand may also be free of Israeli acute paralysis virus. Many countries, especially island nations and territories, have active biosecurity programmes to protect their bee populations.

United Kingdom: The Bee Diseases and Pests Control (England) Order 2006

American foulbrood	*Paenibacillus larvae*
European foulbrood	*Melissococcus plutonus*
Small hive beetle	*Aethina tumida*
Tropilaelaps mites	*Tropilaelaps clareae* and *Tropilaelaps koenigerum*

Any foreign species, such as the asian hornet, would also be considered a reportable pathogen.

Western Australia

Note the bees in Australia (and Western Australia) are particularly pathogen free and the beekeepers and government work extremely hard to try and maintain this excellent health status. For this reason, their list of notifiable pathogens is longer than most countries.

Africanised honey bee	
American foulbrood	*Paenibacillus larvae*
Asian honey bee	*Apis cerana*
Braula fly	*Braula coeca*
European foulbrood	*Melissococcus plutonius*
Small hive beetle	*Aethina tumida*
Tropilaelaps mite	*Tropilaelaps clareae*
Tracheal mite	*Acarapis woodi*
Varroa destructor	Varroasis

Any pathogen or foreign species would also be considered a reportable disease.

Reporting a suspected notifiable condition

Before treating any condition it is essential to make a diagnosis. If the diagnosis is a possible notifiable condition do not treat. You must consult the government authorities, bee inspector or your local veterinarian.

If you suspect you have a notifiable condition then follow the following recommendations:

- Write down the GPS location of your hive.
- Note the phone number and email address of your local bee inspector.

When there is a suspected problem:

- Put the hive back together.
- Phone the bee inspector.
- Photograph the problem
- Email the photographs of the suspected problem.
- Do not seal the hive. This allows the flying bees to return to the hive.

Figure 8.7 Sealing the entrance after dark

Figure 8.8 Securing the hive with webbing

If the bee inspector has not arrived by evening, once all the forage workers have returned home, block the hive entrance. This utilises the fact that bees do not fly at night.

- Do not allow other beekeepers or interested parties to come and inspect the hive.
- Do not remove anything from the whole apiary site, including your clothing and equipment.
- Secure the hive with webbing.
- However, if the bee inspector has still not arrived, ensure that the hive remains sealed in the morning.
- Phone your veterinarian.

9

Poisoning of bees

Bees are in quite a unique position as they cover a lot more area than the normal farmed animal. Their territory will take them through farmland, households and commercial land. They can therefore be exposed to a wide range of potentially toxic compounds from the pollen of exotic plants to insecticides and pesticides.

The clinical signs of poisoning will vary depending on the toxin. The ones that are going to be noticed by the beekeeper are going to be:

- Acute – typically colony collapse, neurological disorders or piles of dead bees. Bees may be refused entry into the hive and therefore a lot of fighting will take place at the entrance, which may look like robbing. Bees may be found dead with their proboscis (tongue) extended.
- Chronic – deformities of the young.

Toxic chemicals – man-made

Poison	Stages affected	Adult	Brood	Colony
Toxic chemicals	Adult	Dead foragers. Perhaps nurse bees affected. Queen may be ok.	Usually few	Colony collapse. Weakened colony. Lots of dead bees in entrance.

Figure 9.1 Piles of dead bees outside the hive. Some of these may be in the ground around the landing pad as shown below. This was associated with local fruit trees being sprayed with insecticide for the autumn. Bees were also seen wandering around outside the hive

Figure 9.2 Bees were dead and dying on the landing board. Many were also trembling and showing signs of paralysis

Common poisons

Botanicals
Carbamates
Chlorinated hydrocarbons
Growth inhibitors
Microbials – *Bacillus thuringiensis (BT)*
Neonicotinoids
Organophosphates
Pyrethroids – synthetic
Spinosyns

Natural poisons – plants

The poison plants listed here are largely from North America. Each area will have plants which may poison the bees. Note the bee is not native of North America and therefore did not evolve in the presence of these plants and has no natural protection from their poisons.

Plant poisons	Stages affected	Adult	Brood	Colony
Aesculus californica	Young brood Forage bees	Deformed workers hairless with neurological problems. Can look like chronic paralysis. Queens may become infertile.	Drone brood produced. Workers die after emergence. Brood becomes uncapped.	Progressively weakened hive. Queen becomes superseded. Dead workers at entrance.

Aesculus flower stem

Figure 9.3 Californian buckweed

Aesculus leaves

Figure 9.4

Plant poisons	Stages affected	Adult	Brood	Colony
Astralagus miser		Young nurse bees		Death and collection of young workers at the hive entrance
Cyrilla racemiflora	Larvae	No effect	Many blue or purple larvae	Slight to severe weakening of the hive

Figure 9.5 Cyrilla Southern Leatherwood

Plant poisons	Stages affected	Adult	Brood	Colony
Gelsemium sempervirens	All	Young adults die. Older adults normal	Pupae die in cells and become mummified	Hive slightly weakened

Figure 9.6 Yellow Jessamine, Woodbine

Plant poisons	Stages affected	Adult	Brood	Colony
Kalmina spp. (Laurel)	None	None	None	Produces poisoned honey containing a deadly poison – andromedotoxin

Figure 9.7 *Kalmia latiflora*

Plant poisons	Stages affected	Adult	Brood	Colony
Loco plants. These are plants that produce the toxin swainsonine – actually made from a fungus living with the plants. Thus many species of plant can be locoweeds.	Pupae Adult – queen	Forage bees die away from hive. Those who do return have neurological issues. Queen may die	Many cells contain dried pupae	Colony collapse. Dwindling colony

Oxytropis in North America

Astragalus in North America

Swainsona in Australia

Figure 9.8 Locoweeds

Plant poisons	Stages affected	Adult	Brood	Colony
Rhododendron, especially *R. luteum* and *R. ponticum*	None	None	None	Produces "mad honey" through grayanotoxins

Figure 9.9 and 9.10 *Rhododendron luteum* and *Rhododendron ponticum*

Plant poisons	Stages affected	Adult	Brood	Colony
Tilia cordata	Foraging adults	May become very dozy and easy prey for birds. Bumblebees are particularly prone	None	None

Figure 9.11 Lime tree or Linden tree detail of flowers and leaves

Plant poisons	Stages affected	Adult	Brood	Colony
Veratrum californicum	Adult	Forage bees die away from hive. Adults die in a curled state. Queen normal	No effect	Forage bee loss – reduction in production

Veratrum full plant

Veratrum detail of flowers

Figure 9.12 and 9.13 False Hellebore

Plant poisons	Stages affected	Adult	Brood	Colony
Zigadenus venenosus (deathcamas)	All	Paralysis and death	Death	May kill a large portion of the hive

Figure 9.14 *Zigadenus venenosus*

Pollen

If pollen contains more than 2% ash (minerals) there is a demonstrated decline in brood rearing. If the pollen contains 8% or more ash (minerals) brood rearing ceases.

Pollen needs to have 4% iso-leucine to provide sufficient essential amino acids for bees to succeed.

Physical problems

While not a 'poisoning' in itself ,the bee may be physically incapacitated by plant or other animal materials.

Galleriasis

Wax moth (*Galleria mellonella*) infestation may be sufficiently high that the developing wax moth larvae create so many silk threads, it can make it difficult for emerging bees to escape the cell. The emerging workers are thus found dead trapped by the wax moth silk threads (see Chapter 5).

Milkweed pollinia (*Aslepias spp.*)

The pollen is 'coherent'– in pairs connected by a slender filament (this is called the pollinia) – which looks like a bird's wishbone. When foragers collect the pollen they can become ensnared and cannot free themselves from the plant. Those that do have difficulty getting back to the hive with the pollinia still attached to their bodies. The pollinia will then be hopefully removed by other worker bees.

Figure 9.15 *Aslepias syriaca*

Figure 9.16 Aslepias pollinia attached to the bee's tongue

Propolis

Propolis is interesting in that the honey bee collecting the propolis cannot remove it from its honey basket – it requires the help of the other workers within the hive.

Figure 9.17 Propolis in the honey basket of a bee

Action if you believe your bees have been poisoned

- Stay calm.
- Write down everything you note – particularly behaviour of the bees.
- Note time of day and date.
- Take photographs.
- Note the direction the bees are coming from.
- Note the colour of any pollen on the bees, especially the ones affected.

- Estimate the number which are affected.
- Take a sample of at least 200 dead bees.
- Alert the local bee inspector – who may already be investigating other cases.

In the UK, contact APHA.

Treatment options

There may be few specific treatment options. In most cases the hive will recover in a couple of weeks. But support in terms of protein patties and sugar supplements may be required to assist the hive.

The hive may require being combined with another hive.

Get your neighbours and local farmers to inform you before they spray

Ask your neighbours and local farmers not to spray chemicals which can affect the bees. There is much more awareness today over the use of chemicals.

However, if you cannot stop them using chemicals, they will generally give you 24 hours warning prior to using any chemicals. This gives you the chance to close up the hive the night.

10

Bees as neighbours

The general public is quite anxious and nervous about bees. Everyone loves what bees do but there is a sting in the tail. Parents particularly do not want a hive around their children. A swarm fills the imagination with particular fear, which is unwarranted as swarming bees tend to be quiet as they have stocked up on honey prior to swarming.

You are serious about getting a hive

If you are going to place your hive in the town you need to consider your neighbours, their children and pets. Even in a country village be considerate.

Being a good neighbour

Some simple rules to consider:

- You do not want your bees to be a nuisance.
- Your neighbours have a right to feel safe and comfortable in their gardens.
- Provide a shrubbery or fence about 2 metres tall by the beehive. This forces the bees to rise in their flight path and over people's heads. The shrubbery may also provide a windbreak for the hive. Properly placed it may protect the hive from direct sunlight at times as well, essential in hot climates.
- Consider the design of the entrance area – this can be directed upwards. Note that bees normally live in a hollow tree and the nest can be quite far from the entrance point.
- Place the beehive out of direct sight.
- Provide a good supply of water. Your bees are going to need water and if you fail to provide it they will source their water from your neighbours' ponds and even swimming pools.
- Keep your hive strong.
- Do not place the hive where a neighbour's night security light is beaming on the hive: this can disturb the hive even at night.
- Minimise examining your bees at times when they are likely to get annoyed and start following people.
- Do not examine the hive when the neighbours are mowing their lawn or having a garden party and absolutely not when the kids next door have a birthday!

- Try not to have the flight path over the neighbour's car park area. The cleansing flights can leave unsightly marks on the car, which is unpleasant to wash off.
- If the hive is inclined to sting or be aggressive, re-queen with a quieter type.
- Collect swarms that occur locally. Manage your own hive to reduce the risk of swarming. But remember this is normal behaviour.
- Let the neighbours see the bees. Have a spare suit with gloves.
- Especially let the children see the bees and explain the use of the various parts of the hives – even without bees in them.
- Provide a little honey for your neighbours.

Flight path

The bees leave the hive in an organised manner. This is a direct line and is called the flight path. The bees are moving very rapidly in the flight path. The flight path wants to be above 2–3m so they do not run into people. It is common to get stung as a worker bounces off your head and gets caught in your hair if you walk into their flight path. Placing a high hedge or trees in front of the beehive will encourage the bees to rise before setting off on their tasks. Note, bees do not always follow your best efforts!

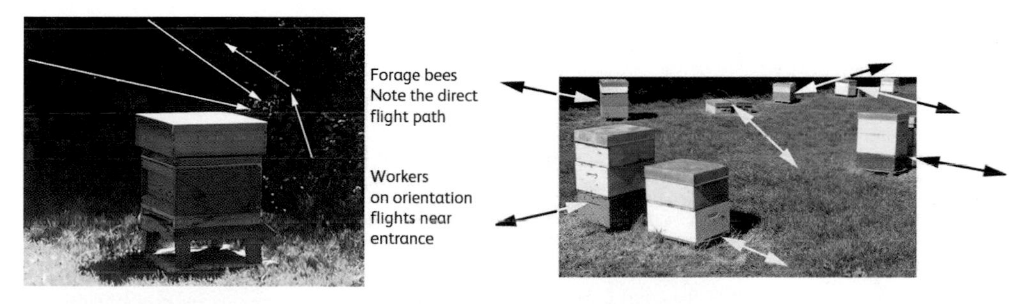

Figure 10.1 and 10.2 The flight paths of bees from the hive. In garden hives, have the bees climb to over two metres before they set out on their flight paths

Figure 10.3 The hive can be placed high on the roof so that its flight path is out of the way of the public

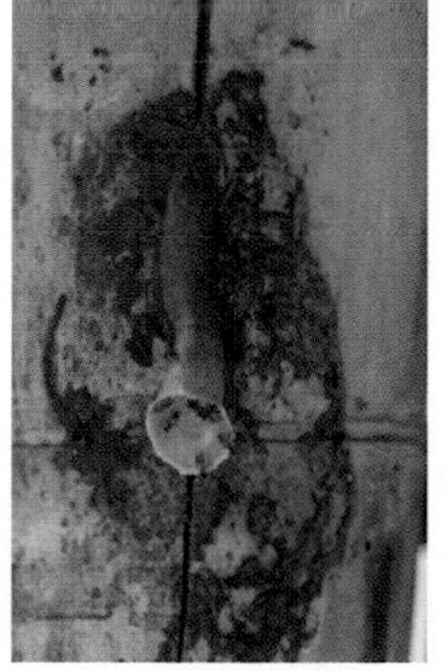

Figure 10.4 The entrance to the hive can be down a tube placed over two metres above the hive. This is illustrated by these stingless bees

Figure 10.5 Design of the flight path to reduce risk of stings by placing a high hedge or trees in front of the hive

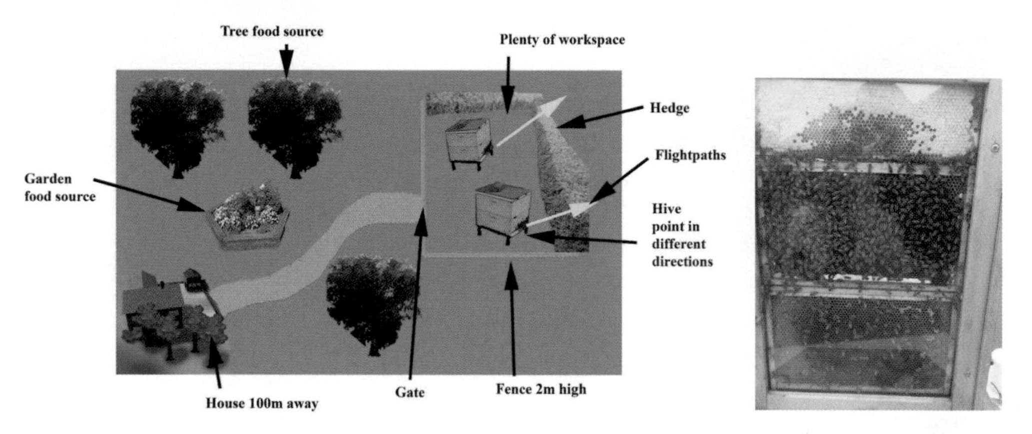

Figure 10.6 Plan your apiary

Figure 10.7 This public hive has an entrance outside the building via a polypipe tunnel leading from the roof to the observation hive inside a school

Where to place the hive

Hives need a good food source close by in spring – apple, for example, in the UK – and in the autumn –ivy for example in the UK. There are plenty of flowers in the middle of summer. Bees need a good source of nectar within 2km (12500 hectares).

Local large fields of monocrops may provide lots of food, but only temporarily. The hive may need to be fed afterwards. With rape the honey may need to be removed (before it crystallises) and sugar supplemented to make up what has been taken. Hives may be moved temporarily to the crop area and then returned home. Note the biosecurity risks involved.

Figure 10.8 Hive placement in Australia

Where not to place a hive

Do not place hives where they are obviously going to be a nuisance.

- Do not place near a public footpath.
- Do not place near a public swimming pool.
- Do not place near a public playing field or school (note the school may have its own bees).
- Near a police station!

Figure 10.9 Hive placed by the public footpath and bridleway. This is asking for trouble

Swarm management

Do not capture and use swarms of unknown health. They can prove to be a serious risk to your own apiary.

However, this can be harsh and beekeepers have a duty to protect the general public from the harassment caused by swarms. From a neighbourly view, capture the swarm, and then isolate and quarantine the bees. Monitor their health using separate equipment including clothing. Wear gloves when handling equipment and the bees.

Ideally have a set of equipment designed for swarm control, a skep is very useful for this purpose. Wash your hands, clothing, boots and equipment.

If the swarm hive has significant pathogens, which you cannot eliminate humanely, destroy the bees. Note that brood pathogens are left behind by the swarming bees.

Figure 10.10 Get the swarm out of the tree

Figure 10.11 Getting bees to walk into the hive box allowing them to be moved

Managing a sting

1 Remove the sting by scraping out with your thumbnail. Do not squeeze out the sting – this will empty more toxin from the venom sac.
2 Take pain killers.
3 Be aware of sensitisation and antihistamines may be required.

Figure 10.12 The extended sting in a honey bee

Figure 10.13 The effects of a sting under the right eye with swelling

Bees as deterrents

In some parts of the world the honey bee can be used as a deterrent to other animals. In Uganda the villages use bees to protect their crops from marauding elephants. Beehives are placed around the edge of villages or important crops.

Figure 10.14 and 10.15 Beehives can be situated around the village or vulnerable crops to protect them from marauding animals – elephants for example in East Africa

The best time to examine the hive is when:

- Most of the forage workers are out collecting nectar
- Good nectar flow
- Hive not under stress from wasps
- Good sunlight onto hive
- Good temperature 25°C or more
- When there is no party going on next to the hive
- Avoid the middle of the day
- Avoid windy days
- Avoid cold days
- Avoid thundery days

Neighbours can be good to you

Ask your neighbours not to spay chemicals which can affect the bees. Today there is much more awareness and sympathy with the plight of bees and helpful insects in general. However, if you cannot stop people using legal chemicals, they will generally give you 24 hours warning prior to using any chemicals. This gives you the chance to close up the hive the night before (once all the foragers have returned home). Keep the hive closed for the next day – assuming it is not too hot.

Figure 10.16 Stop the entrance to reduce the risk of exposing your bees to any toxic chemicals

11

Maintaining health

Integrated pest management (IPM)

Pest and parasites can be managed in an organised manner, without having to resort to over-use of harsh chemicals. IPM is primarily aimed at *Varroa* control. There is no method of *Varroa* elimination, but *Varroa* control can be obtained. See Chapters 5, 6 and 7 for more details on *Varroa* and other parasites.

Table 11.1 Review of possible IPM techniques over the year to control *Varroa* and other mite parasites (note these need to be legal in your area)

Technique	Winter	Spring	Summer	Autumn
Open mesh floor sticky paper				
Drone brood removal				
Drone comb trapping				
Thymol products				
Fluvalinate/flumathrin				
Coumaphos				
Oxalic acid				
Icing sugar dusting				

Maintenance of hive health

While integrated pest management (IPM) is a great idea, it misses the whole point of a hive. It only concentrates on the pests and parasites and even then mainly *Varroa* infestation. But it is a great starting point in looking after the health of the hive. This book wants to encourage a more holistic approach to health. The aim of managing the hive is to maximise the productivity of the hive to benefit the beekeeper and the health of the resident bees.

Health management involves a more holistic approach of

- Good bee keeping
- Keeping all pathogens at a level where the health of the bees is not compromised

- Minimising the greed of the beekeeper
- Maximising the productivity of the bee
- Looking after the wellbeing and welfare of the bees.

Tools and set up

Have a dedicated toolbox for the apiary. For example:

- Straight sided plastic bucket to collect wax and debris, cut out queen cells, propolis
- Smoker and wind-proof matches/lighter
- Hive tool coloured for each hive
- Brush
- Scraper
- Capping tool
- Frame holders
- Blanket
- Spare fuel
- Lighter
- Match box to collect bees
- Sealed jar to place the match box in so bees do not escape in the car on the way home.

If you do not have a dedicated hive tool for each hive, have at least more than one and a jar of disinfectant – such as surgical spirit; 0.5% bleach or sodium hypochlorite.

After examining a hive, place the used hive tool into the disinfectant and use the alternative hive tool on the next hive. Colour each hive tool differently to make the identification easy.

Figure 11.1 Straight sided bucket – so wax and debris are not spilled on the ground around the hive

Figure 11.2 Hive tool

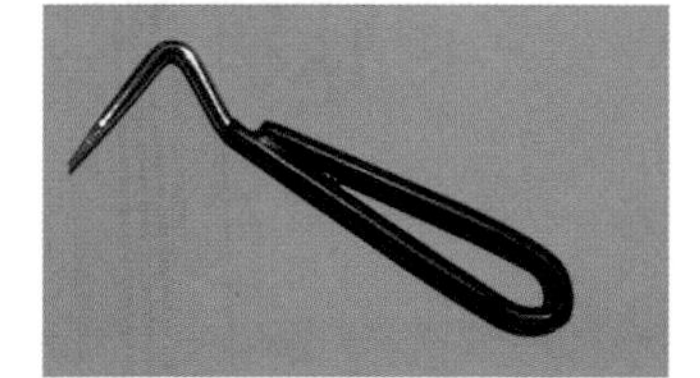

Figure 11.3 Scraper

Figure 11.4 Capping tool

Figure 11.5 Frame holder

Figure 11.6 Blanket

Figure 11.7 Smoker

Hive set up

- Number all the hives, brood boxes, supers and frames.
- Keep records of where all the numbered brood and super boxes belong and have been.
- Place the hive entrance in a southerly direction in the northern hemisphere to help the bees get warm in the morning or a northerly direction in the southern hemisphere.
- Protect the hive from the afternoon sun where the sun's heat can become intense. In hot climates this might require the hive to be painted white or heat reflective.
- Do not place the hive under an overhanging tree where birds, hornets and wasps may stand and stalk the hive. This also stops branches and leaves from breaking or smothering the hive.
- Place the hive on a hive stand – this allows easy access to the hive debris and reduces back strain when examining the hive.

Use of a dummy frame

Dummy frames are part of the hive maintenance programme as they reduce the likelihood of squashed bees when the hive and frames are being manipulated. It will reduce the hive's productivity, but the health benefits outweigh this disadvantage. Note, the main predators of bees are mankind and greed.

Figure 11.8 Dummy board gives the bees a little more space and reduces squashing bees when manipulating the hive

With or without gloves

Ideally the hive should be manipulated by the beekeeper without gloves. It provides the beekeeper with more sensitivity toward the bees. The beekeeper is more aware of the hive temperature. It encourages the beekeeper to have a more gentle touch around the bees. It encourages the beekeeper to have quieter bees.

But, it can be daunting to have bees walking on your hands while manipulating the hive and sudden movements need to be restrained or the bees will sting. Once one bee stings, the others are all encouraged by the pheromones released. The beginner beekeeper is encouraged to progress from gloves to latex to nothing as their experience and bee handling skills improve.

Figure 11.9 Handling bees without gloves promotes gentle handling from the beekeeper

Figure 11.10 Handling bees with gloves provides the beekeeper with more confidence

Smoke

Use natural materials to produce smoke and learn to use a minimum amount of smoke if possible although sometimes keeping the smoker going can be an art in itself. Egg cartons make good smoke along with some dried grass.

The smoke makes the bees worry about possible fire in the location. The bees instinctively will then fill their honey stomach with honey making them quieter. The same happens when they are swarming.

Only use absolute minimum smoke on the supers as the smell of the smoke can get into the honey.

Figure 11.11 Lighting and getting the smoker to stay lit is an art

Figure 11.12 Smoking a Kenyan topbox

Hive records

Hive Record Card						Year					
Location						Colony #			Origin of queen		
Date	Q	QC	Brood	Stores	Health	Varroa	Temper	Feed	Super	Weather	Notes
		Frame #	E Pattern Frame #	# Frames	OK	Drop count	1-10	Litre given	# change	F, C, S, R	

Note: Q = Queen seen.

Figure 11.13 Hive records being completed after each visit

Figure 11.14 Keep the hive records with the hive

How to examine the hive on a normal day

The visit should follow a basic plan. On all visits:

Pre-visit – prepare.
Observe the entrance from a distance.
Approach the hive avoiding the flight path.
Open the hive.

Pre-visit

1. Ensure that your bee suit is clean.
2. Wash bee suit in non-perfumed washing products. Wash on a high temperature wash to kill as many pathogens as possible. Note angry bees will deposit pheromones and alarm substances on your bee suit and equipment.
3. Avoid (if possible) manipulations of the hive on cold, wet, thundery or windy days.
4. Take wind proof matches so you can light your smoker.
5. Avoid being hot and sweaty – body odour issues.
6. Avoid having perfumed products – this can include hair shampoo in the morning you plan to examine the hive.
7. Avoid having any alcohol – this can disturb bees.
8. Avoid banana as the smell mimics the alarm pheromone.
9. Handle the combs gently. As with any stock avoid sudden movements.
10. Do not discard ash from your smoker carelessly. Hot ashes are a fire risk.

Observation of the hive entrance and surrounds

Good signs

1. Bees on active flight path. Lots of bees with full pollen sacs.
2. Many drones flying – normal in afternoon in late spring to summer.
3. Hive stand is free of grass and objects which may allow ants into the hive.

More observation is required if:

1 Small pieces of wax at entrance – bees uncapping stores.
2 Bees fanning and exposing Nassanov glands.
3 Bees fanning/bearding – colony too hot.
4 Bees issuing from hive in a swirling and ascending mass – colony is swarming.

Bad signs

1 Bees covering entrance – too many bees in the hive, colony too hot – bearding in the summer afternoon – swarming in the spring morning.
2 Bees walking aimlessly around the front of the hive – possible disease.
3 Dead larvae actively being carried out – disease in the hive.
4 Mummified larvae – Chalkbrood (*Ascosphaera apis*).
5 Dying bees – dead at bottom and live on top – paralysis virus.
6 Bees unable to fly and staggering around, bright black bloated abdomens.
7 Bees fighting at entrance – robbing.
8 Faecal spotting – Nosema – dysentery (*Nosema apis* and *Nosema ceranae*).
9 Large pieces of wax at entrance – mice in the hive.
10 Foul smelling – death of the hive.

Figure 11.15 Bees outside hive

Figure 11.16 Aimless bees on landing area or adjacent plants

Figure 11.17 Dead bees around the hive or close by

Figure 11.18 Unusual bees in the hive – bee on the right with the black body for example

Figure 11.19 Fighting at entrance

Figure 11.20 Faecal spotting

Approach the hive

- From the side.
- Do not stand in the flight path of the bees.
- Measure the infra-red temperature of the hive.
- Observe the number of dead bees in the ground around the entrance.
- Check that the top cover is properly placed.
- Listen to the hive.

Examine the hive debris

If the hive has an open bottom, the hive debris will fall onto a tray. This can be an excellent opportunity to observe the health of the hive.

- Place a sticky sheet into the hive floor 7 days prior to the visit.
- Place the hive debris into windscreen washing liquid or methylated spirits.
- *Varroa* mites will float to the surface-count the number.
- Divide the number of mites by the number of days to get a mite/day estimation.

Figure 11.21 Chalkbrood mummies in the hive debris

Figure 11.22 The hive debris indicates where the bees are in the middle of winter

Basic hive examination when nothing is suspected as a problem

Basically try to not disturb the bees. The examination should be the minimum required to determine if the bees are happy and productive.

Remove the roof

Look for wax moth – greater and lesser – and also any wasp presence.

Listen for the presence of the bees in the hive. Note the distribution of the hive debris on the debris board. This will give you an indication of where the brood is placed.

To examine the bees the use of a slotted crown board is recommended so that only one or two frames are examined at a time. If this is not available use a blanket or special examination cloth.

Smell the hive

The middle of the brood box is the place where the queen will be laying the eggs. Examine the brood to determine if:

- The queen is still present – there are eggs in cells. She was there 3 days ago at least.
- The cells only contain one egg each – workers will lay more than one egg in the cell.
- The cappings are evenly distributed.
- Check that drone cells are not in the middle of the frame.
- Check that drone cells are not in worker cells.
- That the L3 and L4 larva is pearly-white and are in a C-shape, filling the cell.
- That there are drones present but with workers.

Close the hive

How to clinically examine the hive in detail

Remove the roof

Look for wax moth – greater and lesser – and also any wasp presence.

Examination of the hive

1. Smell the hive.
2. How many bees are present, are there too many?
3. Measure the infra-red temperature of the hive.
4. Ensure each brood frame is numbered.
5. Place the dummy frame on the floor.
6. Place the frame holding tool on the right side of the hive (assuming right handed beekeeper).
7. Examine each frame right to left one side, and then the other side.
8. Place the first frame examined on the frame holding tool.
9. Examine each frame in turn.

10 Shake the nursery bees on the frame into the hive box.
11 For each brood frame note the number of frames with brood on them.
12 Is there an active frame replacement programme – three frames a year per brood box (30% of the frames)?
13 Check that the hive is 'queen right'. Is there an active queen present?
14 If you see the queen – note her marking colour. Is the queen the same as the one supposed to be in the hive?

Figure 11.23 Too many bees are in the hive

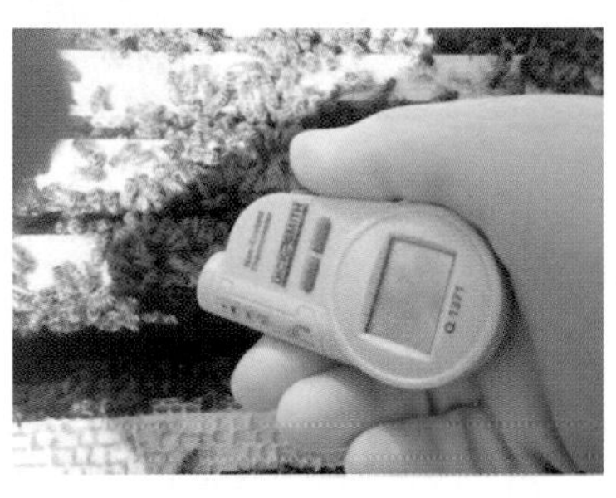

Figure 11.24 Measure temperature of the hive

Figure 11.25 Place frame holder tools on hive

Figure 11.26 Examine first frame

Figure 11.27 Examine frame right to left

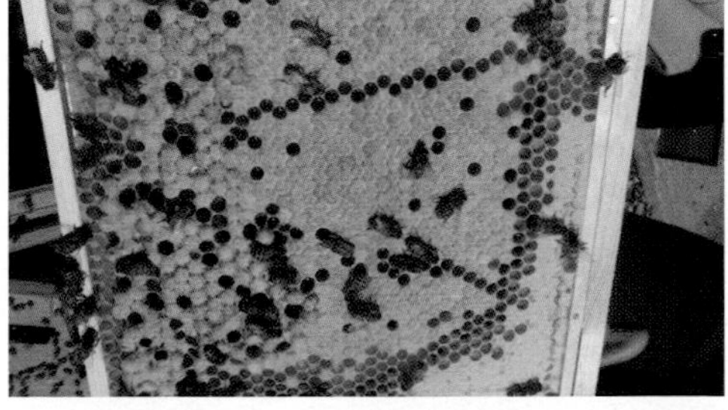

Figure 11.28 Examine the other side of the frame – right to left. Note the line of empty cells by the foundation wires

Figure 11.29 Place first frame on frame holder tool

Figure 11.30 Examine each frame in turn

15 Note the condition of the queen – especially note any thin queen.
16 What is the condition of the frames – evidence of wax moths or other pests?
17 Where are the drone cells? It is a problem if a lot are irregularly spread across comb.
18 How many eggs are in each cell? If more than one may be workers laying or new queen.

If you think there may not be a queen place a numbered frame from an adjacent healthy hive containing eggs. If the workers do not raise queens then the original hive has a queen.

19 Check brood larvae are pearly-white in colour.
20 If there are a lot (more than 10% of empty cells in areas with capped brood, are heater bees entering the cells trying to heat adjacent cells – may indicate a cold hive.
21 Ensure that the older frames are at the end of the brood box. Check condition of the combs. If you cannot see through the comb, replace.
22 Are there special frames – such as super frames in the brood box used to provide drone cells for *Varroa* control?
23 Estimate the amount of stores present in the brood and super.

Figure 11.31 Marked frames

Figure 11.32 Brood super for drone removal

24 A hive weight scale can be extremely useful to provide information on the health of the hive.
25 Avoid crushing bees when manipulating the colony.
26 Avoid placing the frames on the ground or spilling any honey around the hive to minimise attracting wasps or other bees.

Examination of the bees

Equipment required
Fine tweezers
Wide mouthed funnel
Hand lens
Wire mesh #8 or #12
White cotton cloth
Starter engine spray – ether or washing-up liquid 500 ml
Ultraviolent light 3100 to 4000 A – for foulbrood
Lactic acid
Corkboard
Pins
Wind matches – also matches to examine larvae
Formalin saline (10% formalin)
Icing sugar – to assist *Varroa* identification
Infrared thermometer

1 Remove a frame and knock down around 500 bees into a container with a lid. Brush the bees off a comb through a large mouthed funnel into the container. Ensure the queen is not present.
2 Add 500ml of windscreen washing liquid – ethanol, or alternatively methylated spirits.
3 Collect 30 bees from the landing board area using a matchbox. Foraging bees are older but note deformed bees may be unable to fly.
4 Examination of brood may also be necessary.

Varroa (Varroa destructor)

Shake the bottle for 1 minute to dislodge the mites. Examine for evidence of any *Varroa* mites using a hand lens. This can be assisted by passing the bees and alcohol through a wide screen #8 or #12 mesh to remove the bees and then sieving again through cotton cloth.

Nosema – dysentery – (Nosema apis *and* Nosema ceranae)

Remove 10 bees remove the digestive tract.
Examine the digestive tract, especially the ventriculus for evidence of Nosema spores.

Acarina – tracheal mites (Acarapis woodi)

Remove 10 bees and section across the thorax.
Examine cross-section for evidence of Acarinae.

Viruses

Examine each bee for evidence of deformity – particularly note the wing structure.

Place any bees of interest into the refrigerator for 20 minutes and then into chilled 10% buffered formaldehyde. Place a small piece of tissue paper on top of the bees to ensure they are immersed into the formalin.

Protein content

Remove the remaining bees and submit for total crude protein concentration as a general indicator of health.

Examine the hive debris

1 Place a sticky sheet into the hive floor 7 days prior to the visit.
2 Place the hive debris into windscreen washing liquid or methylated spirits.
3 *Varroa* mites will float to the surface count the number.
4 Divide the number of mites by the number of days (7 for example) to get a mite/day estimation.

The hive through the year

Design a hive calendar for your region. Each part of the world tends to have honey flow periods and high hive activity. At other times hive activity may be low. This may be associated with cold weather or dry weather, wet or dry seasons. Bees will tend to rest in the cold or wet seasons. Joining a local bee club will help you to design your calendar of events.

A similar calendar of events can be constructed to monitor your bumblebee nest or stingless beehives.

The dates are suggested, the day has to be suitable to examine the hive and minimise the discomfort of the hive.

Suggestion of hive examination

Break the year into 12 periods of examination.

1 Beginning of the season.

Cold northern	1st April	*Temperate northern*	1st March	*Equatorial*	Middle wet period
Cold southern	20th August	*Temperate southern*	1st August		

Examination of the hive

Look at the hive in general and its surroundings.
Look at the activity at the hive entrance and landing area.
Look at the hive debris for an indication of the location of the main brood area.
Remove the crown board.

Listen for the presence of the bees.
Smell the hive.
Do not open the hive unless the air temperature is over 16°C.
If warm enough recentre the brood frames by moving frames on either side – look at the distribution of the hive debris.
Move the brood frames as a single unit using the hive tool as a lever.

Examination for the stores

Observe using the bee space the presence of capped honey stores.
It may be necessary to distribute the frames in the super to ensure the capped honey is central.
Ensure that there is an absolute minimum of 4 full frames of capped honey.
Feed if necessary.
Weigh the hive to provide a baseline.

Maintenance of health

Check the spare stored supers for wax moth infestation.
If there is any evidence of wax moth, the supers can be placed dry into the freezer for 7 days.
Start *Varroa* examination of hive debris.
Apply mite treatment if required – note this will also control tracheal mites.
Follow all manufacturer requirements and remove treatments as required.
Grease patty with methanol may be a useful alternative to medical chemicals.
Carry out a detailed post mortem on all dead colonies.
If you have concerns consult the local bee inspector or veterinarian.
Repair all fences and large animal protection barriers.
Trim the grass around the hive stand.

Hive records

Complete the hive record sheets.

2 Preparation for the brood season

Cold northern	20th April	*Temperate northern*	1st April	*Equatorial*	End of wet period
Cold southern	20th September	*Temperate southern*	1st September		

Examination of the hive

As before but also . . .

Turn the brood boxes so the frames are perpendicular to the entrance. This will maximise the ventilation through the frames.
Remove any winter wraps.
Remove any mouse guard but monitor and adjust the entrance space.

If there are two brood boxes exchange the top for the bottom brood box.
The bottom box will probably be empty of brood. By putting the empty brood box on top, this will help to reduce swarming by providing the hive with more space to grow into.
Make sure the queen is present – make sure you see eggs.
Check all the brood frames especially the brood frames in the middle of the brood.
Check that the capping is even with few skipped cells.
Ensure drone cells are on the edge of the brood area.
Remove any queen and play cells noted.
Re-queen if required.
Check that the queen present is the same as the one in the autumn. Note the previous queen's colour marking.
Place an additional super/brood box if required.

Examination of the stores

Depending on the weather supplement with sugar and/or pollen as required. Note the wet weather can significantly limit nectar availability and may also be associated with cold weather.

Maintenance of health

Ensure your spare supers are dry and clean. Watch for any evidence of wax moth. Freeze for one week to kill any wax moth eggs or larvae.
Initiate control of *Varroa*, especially if you are in an area with lots of hives.
Place super in the brood area to start drone brood removal to control *Varroa.*
Use grease patty with menthol to assist control of tracheal mite.
If Nosema is a problem use fumigillin if legal within your area.
Check for Chalkbrood. Hive must be dry and well ventilated.
Discuss pathogens in the area with the local bee inspector.
In some parts of the world tetracycline is advised at this time if your hive is in an area susceptible to foulbrood diseases. This must be coordinated with your veterinarian and local bee inspector.
Monitor the size of the entrance area. Remove the mouse guard.
Carry out a detailed post mortem on all dead colonies.
If you have concerns consult the local bee inspector or veterinarian.
Repair all fences and large animal protection barriers.
Trim the grass around the hive stand.

Hive records

Complete the hive record sheets.

3 *Beginning of the brood season*

Cold northern	1st May	*Temperate northern*	1st May	*Equatorial*	End of wet period
Cold Southern	1st October	*Temperate southern*	1st October		

Examination of the hive

Estimate hive strength.
Combine hives to maximise hive strength.
If you see the queen, check she is the same as previous known queen – is she marked?
Do not move brood cells above the queen excluder.
Review possible swarming.
If too many bees add brood box or brood super.
Reverse brood boxes if lower brood box has lower amount of brood.
Remove brood frames with capped brood and place in weaker colonies. Note this will also move *Varroa*.
Consider producing your own queens and possibly split strong colony.
Add super as required. Note place capped honey super above new super.

Examination of the stores

Monitor the need for sugar and pollen supplementary feeding in order to stimulate brood development.

Maintenance of health

Remove all insecticide medicines in preparation for the honey flow.
Monitor *Varroa* numbers. Carry out capped drone pupae removal. Incinerate the removed drone brood.
Check brood for brood diseases.
If you have concerns consult the local bee inspector or veterinarian.
Repair all fences and large animal protection barriers.
Trim the grass around the hive stand.

Hive records

Complete the hive record sheets.

4 Growth of the hive

Cold northern	20th May	*Temperate northern*	20th May	*Equatorial*	Dry period
Cold southern	20th October	*Temperate southern*	20th October		

Examination of the hive

Thoroughly inspect the hive, especially if the hive appears quiet with few foragers.
Evaluate queen. Ensure that there are eggs present in singles in cells.
Evaluate brood capping patterns. Look for minimal skipped cells.
Re-queen if required. If re-queen, will need to re-examine hive in 7 days.
Remove all queen cells and play cells.

Examination of the stores

There should be no requirement to feed unless the weather is particularly inclement.

Maintenance of health

Monitor *Varroa* numbers. Carry out capped drone pupae removal.
Check brood for brood diseases.
If you have concerns consult the local bee inspector or veterinarian.
Repair all fences and large animal protection barriers.
Trim the grass around the hive stand.

Hive records

Complete the hive record sheets.

5 Honey flow and growth of hive

Cold northern	1st June	*Temperate northern*	1st June	*Equatorial*	Peak flow
Cold southern	1st November	*Temperate southern*	1st November		

Examination of the hive

Look at the hive in general and its surroundings.
Look at the activity at the hive entrance and landing area.
Look at the hive debris for an indication of the location of the brood mass.
Remove the crown board.
Listen for the presence of the bees.
Smell the hive.
Thoroughly inspect the hive, especially if the hive appears quiet with few foragers.
Evaluate queen. Ensure that there are eggs present in singles in cells.
Evaluate brood capping patterns. Look for minimal skipped cells.
Re-queen if required. If re-queening, you will need to re-examine hive in 7 days.
Remove all queen cells and play cells.
Ensure queen excluder in place to ensure queen at bottom of hive. Obviously not for top box or natural hives.
If honey supers are full it may be necessary to remove, collect the honey quickly and return supers to the hive for cleaning and reuse. Place empty super under working super.

Examination of the stores

There should be no requirement to feed unless the weather is particularly inclement.

Maintenance of health

Regularly inspect any unused/stored supers for wax moth and freeze for at least one day if any evidence found.

Ideally place all supers on active strong hives.
Keep your equipment and tools exceptionally clean.
Practise excellent biosecurity at all times.
Monitor *Varroa* numbers. Carry out capped drone pupae comb removal as required.
Check for small hive beetle.
Check brood for brood diseases.
If you have concerns consult the local bee inspector or veterinarian.
Repair all fences and large animal protection barriers.
Trim the grass around the hive stand.

Hive records
Complete the hive record sheets.

6 Honey flow and growth of hive

Cold northern	20th June	*Temperate northern*	20th June	*Equatorial*	Peak flow
Cold southern	20th November	*Temperate southern*	20th November		

Examination of the hive
Look at the hive in general and its surroundings.
Look at the activity at the hive entrance and landing area.
Look at the hive debris for an indication of the location of the brood mass.
Remove the crown board.
Listen for the presence of the bees.
Smell the hive.
Thoroughly inspect the hive, especially if the hive appears quiet with few foragers.
Evaluate queen. Ensure that there are eggs present in singles in cells.
Evaluate brood capping patterns. Look for minimal skipped cells.
Re-queen if required. If re-queen, will need to re-examine hive in 7 days.
Remove all queen cells and play cells.
Ensure queen excluder in place to ensure queen at bottom on hive. Obviously not for top box or natural hives.
If honey supers are full it may be necessary to remove, collect the honey quickly and return super's to the hive for cleaning and reuse. Place empty super under working super.

Examination of the stores
There should be no requirement to feed unless the weather particularly inclement.

Maintenance of health
Regularly inspect any unused/stored supers for wax moth and freeze for at least one day if any evidence found.

Ideally place all supers on active strong hives.
Keep your equipment and tools exceptionally clean.
Practise excellent biosecurity at all times.
Monitor *Varroa* numbers. Carry out capped drone pupae comb removal as required.
Check for small hive beetle.
Check brood for brood diseases.
If you have concerns consult the local bee inspector or veterinarian.
Repair all fences and large animal protection barriers.
Trim the grass around the hive stand.

Hive records

Complete the hive record sheets.

7 Honey flow and growth of hive

Cold northern	1st July	*Temperate northern*	1st July	*Equatorial*	Peak flow
Cold southern	1st December	*Temperate southern*	1st December		

Examination of the hive

Look at the hive in general and its surroundings.
Look at the activity at the hive entrance and landing area.
Look at the hive debris for an indication of the location of the brood mass.
Remove the crown board.
Listen for the presence of the bees.
Smell the hive.
Thoroughly inspect the hive, especially if the hive appears quiet with few foragers.
Evaluate queen. Ensure that there are eggs present in singles in cells.
Evaluate brood capping patterns. Look for minimal skipped cells.
Re-queen if required. If re-queen will need to re-examine hive in 7 days.
Remove all queen cells and play cells.
Ensure queen excluder in place to ensure queen at bottom on hive. Obviously not for top box or natural hives.
If honey supers are full it may be necessary to remove, collect the honey quickly and return supers to the hive for cleaning and reuse. Place empty super under working super.

Examination of the stores

Same steps as previous.

8 *Mid season*

Cold northern	20th July	*Temperate northern*	20th July	*Equatorial*	Peak flow
Cold southern	20th December	*Temperate southern*	20th December		

Examination of the hive

Same steps as previous.

Examination of the stores

There should be no requirement to feed unless the weather is particularly inclement.
Note there may be a shortage of flowers in midsummer in temperate Europe.

Maintenance of health

Regularly inspect any unused/stored supers for wax moth and freeze for at least one day if any evidence found.
Ideally place all supers on active strong hives.
Keep your equipment and tools exceptionally clean.
Practise excellent biosecurity at all times.
Monitor *Varroa* numbers. Carry out capped drone pupae comb removal as required.
Check for small hive beetle.
Check brood for brood diseases.
If you have concerns consult the local bee inspector or veterinarian.
Repair all fences and large animal protection barriers.
Trim the grass around the hive stand.

Hive records

Complete the hive record sheets.

9 *Honey collection and mite stabilisation*

Cold northern	20th August	*Temperate northern*	20th August	*Equatorial*	End of dry period
Cold southern	1st January	*Temperate southern*	1st January		

Examination of the hive

Move brood frames around to ensure that the brood is even in the middle. Check that the hive is not splitting internally with honey between brood boxes.
If honey supers are full it may be necessary to remove, collect the honey quickly and return supers to the original hive for cleaning and reuse. Place empty super under working super.
Move supers frames around to make even honey deposition.

Collect pollen for use as supplemental feeding later in the year. Place pollen in freezer.
May need to reduce the entrance, even using grass to block the entrance. This is particularly the case if robbing is possibly occurring.

Examination of the stores

There should be no requirement to feed unless the weather is particularly inclement.

Maintenance of health

Regularly inspect any unused/stored supers for wax moth and freeze for at least one day if any evidence found.
Ideally place all supers on active strong hives.
Keep your equipment and tools exceptionally clean.
Practise excellent biosecurity at all times.
Monitor *Varroa* numbers. Carry out capped drone pupae comb removal as required.
Treat hive for mites if required after honey is removed.
Check for small hive beetle.
Check brood for brood diseases.
If you have concerns consult the local bee inspector or veterinarian.
Do not mix supers between hives. This reduces pathogen spread and robbing.
Repair all fences and large animal protection barriers.
Trim the grass around the hive stand.

Hive records

Complete the hive record sheets.

10 Honey collection prepare for winter

Cold northern	20th September	*Temperate northern*	20th September	*Equatorial*	Beginning of wet period
Cold southern	1st February	*Temperate southern*	1st February		

Examination of the hive

Remove honey fall crop but leave sufficient honey and pollen stores. Make the top super larger than the lower super.
Some drones are forced from hive may be seen dead around the hive.
Turn the brood box frames so they are parallel to the entrance. This will reduce any draughts into the hive, making it easier for the cluster to stay warm.

Examination of the stores

Start feeding with strong sugar solution to balance the stores you have removed.
Place one or two frames of pollen and capped honey in the middle of the upper brood box. These can be taken from the sides of the brood box.

Maintenance of health

Regularly inspect any unused/stored supers for wax moth and freeze for at least one day if any evidence found.
Ideally place all supers on active strong hives.
After honey removal clean all supers by replacing on the hive for two days.
Place bee excluder board for one day prior to removal of supers, once clean.
Keep your equipment and tools exceptionally clean.
Practise excellent biosecurity at all times.
Monitor *Varroa* numbers. Carry out capped drone pupae comb removal as required.
Treat hive for mites if required after honey removed.
Check for small hive beetle.
Check brood for brood diseases.
If you have concerns consult the local bee inspector or veterinarian.
If you have a weak colony, consider recombining to produce a strong colony.
Repair all fences and large animal protection barriers.
Trim the grass around the hive stand.

Hive records

Complete the hive record sheets.

11 End of the season preparing for winter or for rainy season

Cold northern	15th October	*Temperate northern*	15th October	*Equatorial*	Wet period
Cold southern	15th March	*Temperate southern*	15th March		

Examination of the hive

In cold climates the hives may be moved to their winter yards.
Look at the hive in general and its surroundings.
Look at the activity at the hive entrance and landing area.
Look at the hive debris for an indication of the location of the brood mass.
Remove the crown board.
Listen for the presence of the bees.
Smell the hive.
Do not open the hive unless the air temperature is over 16°C.
Reduce entrance space.
Install mouse guard.
Note drones driven out of hive may be seen dead surrounding hive.
Tilt hive 2–3° forward to allow water to drain.

Examination of the stores

Feed hive which has insufficient stores.

Maintenance of health

Regularly inspect any unused/stored supers for wax moth and freeze for at least one day if any evidence found.
Keep your equipment and tools exceptionally clean.
Practise excellent biosecurity at all times.
Monitor *Varroa* numbers.
Treat hive for mites if required. Follow the manufacturers' recommendations.
If brood still is being produced may consider drone comb removal.
If Nosema is a problem, treat with fumigillin if legal.
If you have a weak colony, consider recombining to produce a strong colony.
If you have concerns consult the local bee inspector or veterinarian.
Repair all fences and large animal protection barriers.
Trim the grass around the hive stand.

Hive records

Complete the hive record sheets.

12 Winter/rainy season survival

Cold northern	15th November	*Temperate northern*	15th November	*Equatorial*	Wet period
Cold southern	15th April	*Temperate southern*	15th April		

Examination of the hive

Prepare hive for winter. In cold climates the hives may be moved to winter yards.
Insulate and wrap hives as required.
Look at the hive in general and its surroundings.
Look at the activity at the hive entrance and landing area.
Look at the hive debris for an indication of the location of the brood mass.
Listen for the presence of the bees.
Do not open the hive unless the air temperature is over 16°C.

Examination of the stores

Only feed if absolutely necessary. The hive's stores should be already in place.

Maintenance of health

Regularly inspect any unused/stored supers for wax moth and freeze for at least one day if any evidence found.
Keep your equipment and tools exceptionally clean.
Practise excellent biosecurity at all times.
Remove all mite treatment programmes.
If you have concerns consult the local bee inspector or veterinarian.

Repair all fences and large animal protection barriers.
Trim the grass around the hive stand.

Hive records
Complete the hive record sheets.

Signs that additional super/new brood boxes are required

- Bees and brood filling both brood chambers
- White wax deposited on bottom and top of crown board
- Lot of bees at entrance

Figure 11.33 Hive too full: too many bees in the hive

Figure 11.34 Hive too full: freeform wax between the crown board and roof

Figure 11.35 Hive too full: freeform cells expanding into free space

Figure 11.36 Freeform cells are beautiful in their own right

Removal of workers from honey supers

When the supers need to be removed or the supers need to be cleaned use a bee escape worker chamber (Porter Bee Escape, for example). Place the bees' escape board between the super boxes and the brood frame for 24 hours.

A simple hive's appearance over the year

This photomontage illustrates a simple National hive's change in appearance over a season in the UK.

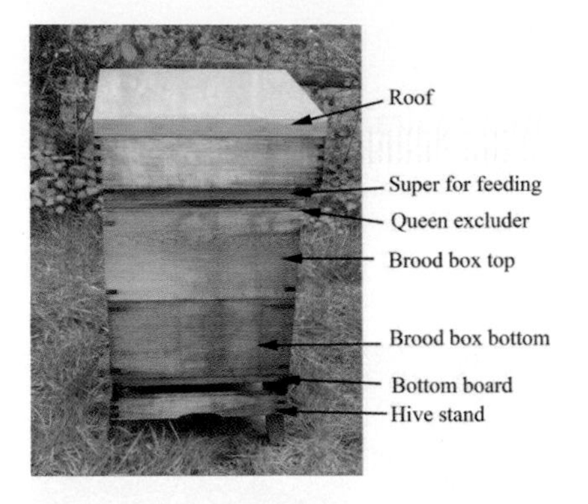

Figure 11.37 Over winter – two brood boxes and a super with feed

Figure 11.38 Early spring – brood boxes exchange. Super removed replaced with an eke under the roof

Figure 11.39 Spring – super added

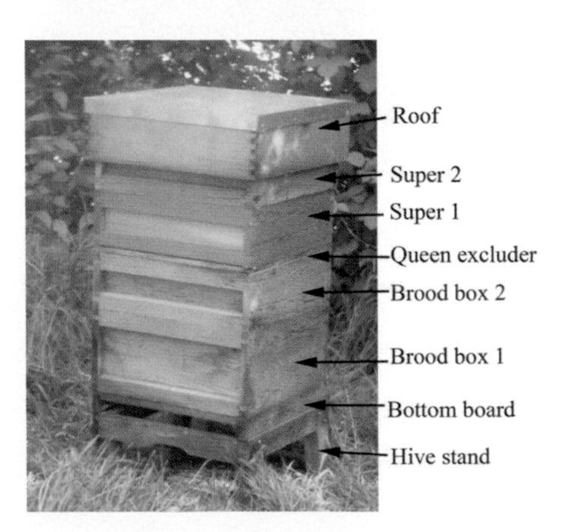

Figure 11.40 Summer – second super added. More brood boxes or reserve supers could be added increasing the height of the hive depending on the food supplies available.

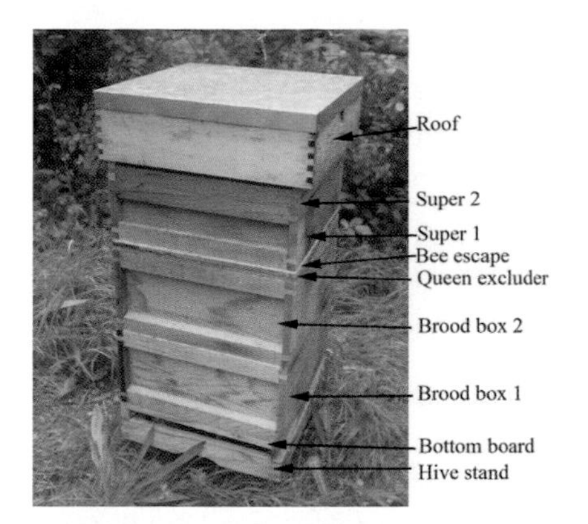

Figure 11.41 Late summer. A bee escape added to help removal of supers

Figure 11.42 Late summer day after supers removed –honey being collected

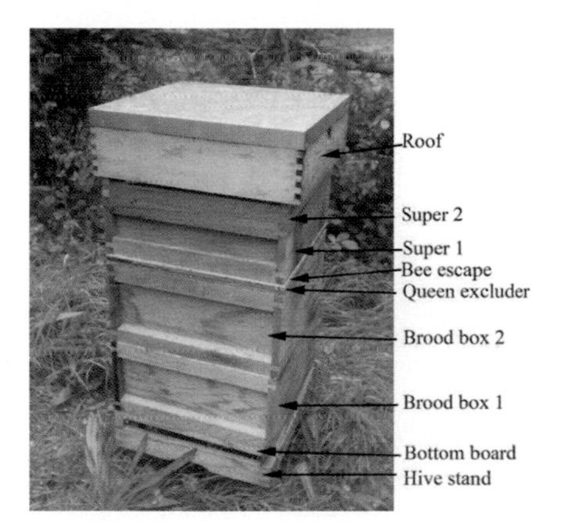

Figure 11.43 Late summer – later that day. Supers returned with bee escape to clean supers

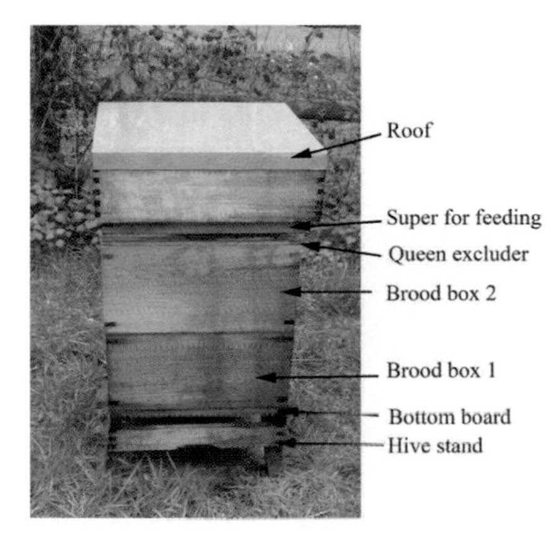

Figure 11.44 Back to overwintering

Prevent robbing

- If the hive has died, seal the entrance to the hive before you deal with the dead hive and its frames. Do not leave dead hives lying around.
- When a colony is weak, reduce the entrance to only one bee space to allow the guards time to examine each bee entering the hive.
- After removing honey, replace supers onto the same original hive for cleaning. This is to reduce giving other hives a taste of a different hive's honey.
- Monitor hive weight. A hive increasing its weight at a time when there is no honey flow is robbing.
- A hive losing weight rapidly is likely to be robbed.
- Avoid having leaves and other material against the landing board which may encourage robbing by ants and other insects.

- Observe the hive entrance for signs of fighting. This can be difficult to detect.
- Act quickly on a queenless hive, especially at the beginning and end of the season.
- When there is no honey flow, keep manipulations to a minimum and make them quick.
- Avoid *Varroa* treatments which have a strong smell as this may confuse the guard bees.
- Watch for wasps and hornets staking the hive.
- Observe for early signs of disease or distress in the hive.

Swarming

Swarming is natural in the spring. It is almost the primary desire of all active beehives.

Do not capture and use swarms of unknown health. They can prove to be a serious risk to your own apiary.

However, this can be harsh and beekeepers have a duty to protect the general public from the harassment caused by swarms. From a neighbourly view, capture the swarm, and then isolate and quarantine the bees. Monitor their health using separate equipment including

Figure 11.45 Swarm in a tree

Figure 11.46 Detail of the swarm

Figure 11.47 Place a brood box under the swarm. Have a couple of frames of honey in the swarm box to encourage the bees

Figure 11.48 The bees will walk into the brood box. Close the top of the box.

Figure 11.49 Hive about to swarm – the hive is full of bees and they are very active

Figure 11.50 Clipped queen on the landing board returning to the hive

clothing. Wear gloves when handling equipment and the bees. Ideally have a set of equipment designed for swarm control, a skep is useful for this purpose. After collecting the swarm, wash your hands, clothing, boots and equipment.

If the swarm hive has significant pathogens, which you cannot eliminate humanely, destroy the bees. Note that brood pathogens are left behind by the swarming bees.

Swarming tends to occur in the morning, from 10:00 to 14:00, in the late spring. The hive will be seen to be restless.

Problems with swarming may occur with queens who have clipped wings. They cannot swarm as they cannot fly. This particular hive was extremely lucky the beekeeper noticed the swarming behaviour and was able to redirect the clipped queen back into the hive. The hive was then managed by false swarming.

Bearding

Bearding occurs in the late afternoon in the hot summer months when the temperature is above 35°C. The behaviour of the hive may be very similar to swarming with lots of bees outside the hive. Some will be fanning the hive trying to cool it.

Biosecurity considerations

1. Age of hive and other hives in location.
2. Have new hives been introduced recently?
3. Does the apiary practise migration pollination?
4. Have new queens been purchased and from where?
5. Has a swarmed hive been collected and introduced recently?
6. Location of hive in relation to other hives – Google Earth can be useful in this manner.
7. Location of hive in relation to public path.
8. Location of hive in relation to access by livestock.
9. Note general appearance.
10. Water source.
11. Available food sources – note some food like rape may need frequent collection of honey.
12. Security measures to stop pests and people – electric fencing.

Figure 11.51 Hot hive bearding on the outside of the hive

13 If electric fencing used, are the wires clear of vegetation?
14 Is the hive placed on a secure hive stand?
15 Is hive stand clear of vegetation?
16 Is hive stand prepared to stop ants and other insects from entering the hive?
17 Note source of bee equipment – note source of wax for frame foundation.
18 Other beekeepers been recently – especially with their own equipment.
19 Hygiene of bee examination equipment. Sterilise hive tools and frame scrapes after cleaning by soaking in 0.5% sodium hypochlorite for 20 minutes. Do not allow other beekeepers to bring their own equipment.
20 Wear disposable gloves.
21 Provide boots for all visitors.

Biosecurity introduction of new stock – the new queen and her attendants

1 Keep good records and a calendar of events.
2 Enquire about the health of the area that is the source of the new queen.
3 Enquire if the source provides specific health information regarding their queen bees.
4 When bees arrive keep them separate from your hives and any bee equipment.
5 Examine bees in detail – looking for any parasitic conditions – Varroosis, Tropilaelaps and small hive beetle – and for any deformities, especially in wing structure.
6 Sacrifice some of the workers and post mortem them. Dissect bees and look for Tracheal mites (*Acarapis woodi*), and test for European foulbrood (*Melissococcus plutonius*) and American foulbrood (*Paenibacillus larvae*) by lateral flow. Macerate bees and examine for Nosema (*Nosema apis* and *Nosema ceranae*).
7 Calculate the crude protein content of the worker bees as an indication of general health.
8 Set up the new hive as far away as possible from your current, established apiaries.
9 After one month, examine hive in detail. Review the health of the new hive. If no problems are noted, relocate hive to your apiary.

Examples of simple poor hive set ups resulting in health issues

Here the roof was incorrectly applied, allowing rain to run onto the landing board making the board wet and reducing the foraging ability of the hive.

Fig. 11.52 Roof placed incorrectly and rain run off patterns illustrated

Here the hive's *Varroa* floor was put in upside down. Wax moths were able to escape the vigilance of the guard bees and were able to nearly destroy the hive.

Figure 11.53 Wax moths hiding under an upside down *Varroa* floor

Fig 11.54 The hives' ventilation blocked with spiders and debris reducing air flow in the hive

Figure 11.55 Hive covered in feed on a farm where feed overspilled from a feed bin this swamped the hive

Ageing the queen

Queens can last four years but to maintain good hive health are generally replaced in their third year. While distressing, replacing your faithful queen early optimises her hive for the future.

However, visual examination of the queen does not allow for the age to be easily ascertained. Placing a simple mark on her thorax allows beekeepers to know the year the queen was born. A simple internationally recognised code is used:

Marking	Year ending
White (grey)	1 or 6
Yellow	2 or 7
Red	3 or 8
Green	4 or 9
Blue	5 or 0

11.56 Queen marked white – from 2011

11.57 Queen marked red from 2013

Naming your queen

In small apiaries it is advised to name the queen and number the hives.

Here is Alice, our first queen:

Figure 11.58 Queen Alice

Here are a few suggested names for your queen:

Alice	Boadicea	Cleopatra
Daenerys	Elisabeth	Frederica
Guinevere	Hera	Isabella
Josephine	Katharine	Larena
Margaret	Nefertiti	Orla
Panthea	Queenie	Regina
Sheba	Titania	Una
Victoria	Wallis	Xandra
Yolanda	Zenobia	

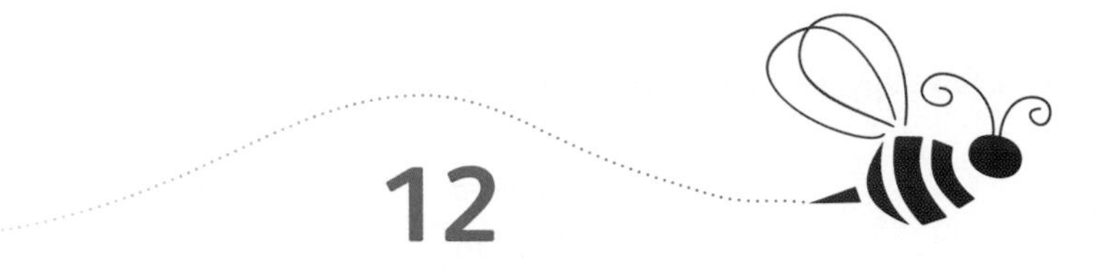

12

Techniques

To enjoy and maintain the health of your bees a few techniques may be employed. This chapter presents various practical techniques which may be employed to enhance the health of your bees and also increase your knowledge base of their needs.

Note that for some of these procedures, it is encouraged that the reader discusses these with their local beekeeping club, bee inspector and/or veterinarian. Each country has different rules and regulations regarding bee health, especially with regard to notifiable diseases for which medications are allowed and how they can be applied.

Conversion tables

Distance			
1 inch	2.54 centimetre	1 millimetre	0.03937 inch
1 foot	0.3048 metre	1 centimetre	0.3937 inch
1 yard	0.9144 metre	1 metre	1.094 yard
1 mile	1.6093 kilometre	1 kilometre	0.6214 mile
1 nautical mile	1.8532 kilometre	1 kilometre	1.539 nautical mile
Surface or area			
1 square inch	6.4516 square centimetre	1 square centimetre	0.1550 square inch
1 square foot	0.0929 square metre	1 square metre	10.76 square feet
1 square yard	0.8361 square metre	1 hectare	2.4711 acre
1 acre	0.4047 hectare	1 hectare	1000 square metre
1 square mile	259.0 hectare		
Volume/capacity			
1 cubic inch	16.387 cubic centimetre	1 cubic centimetre	0.061 cubic inch
1 cubic foot	0.0283 cubic metre	1 cubic decimetre	0.035 cubic foot
1 cubic yard	0.7646 cubic metre	1 cubic metre	1.308 cubic yard

Measures			
Imperial		US	
1 pint	0.5506 litre	1 pint	0.473 litre
1 quart	1.136 litre	1 quart	0.9463 litre
1 gallon	4.546 litre	1 gallon	3.785 litre
1 bushel	26.369 litre	1 bushel	35.24 litre
Avoirdupois weight			
1 ounce	28.35 gram	1 gram	0.035 ounce
1 pound	453.59 gram	1 kilogram	2.205 pound
1 hundred weight	45.36 kilogram	1 tonne	0.984 ton
1 ton	1.016 tonne	1 bushel (eg corn)	56lb (25.4kg)
Temperature			
°Celcius – ((Fahrenheit–32)*5)/9		°Fahrenheit – ((Celcius)*9)/5 + 32	
Velocity			
1 mile per hour		1.6093 kilometre per hour	
1 foot per minute		0.3048 metre per minute	
Energy			
1 Joule	0.239 Calorie	1 Calorie	4.184 Joule
Pressure			
Pounds per square inch	0.069 Bar	Bar	14.5 psi
Pounds per square inch	6.8949 kPa	KPascals	0.145 psi

How big is big?

Metric numbers	
1 Picogram 10^{-12}	= 10^{-12} gram
1000 picogram	= 1 nanogram or 10^{-9} gram
1000 nanogram	= 1 microgram or 10^{-6} gram
1000 microgram	= 1 milligram or 10^{-3} gram
1000 milligram	= 1 gram
1000 gram	= 1 kilogram
ppm	= mg/kg or g/tonne

RELATIVE SIZES OF THE INFECTIOUS AGENTS	Can be Seen	Approximate Size
Viruses	Electron microscope only	20–300nm
Spiroplasma	High power microscope	0.5μm
Bacteria	High power microscope	0.5–30μm
Fungi	Low power microscope	5–80μm
Protozoa	Low power microscope	6–12μm
Mites	Low power microscope	0.5–4mm
Varroa	By eye	
Metazoan parasites	Low power microscope or by eye	0.5–1mm

Comparative	
1	1cm metazoan
3	3mm Varroa mites
200	50μm fungi
1000	10μm protozoa
10,000	1μ bacteria
1,000,000	100nm viruses

Reviewing the health of the hive

How many bees should I take to check the health of the hive?

The following is a guide as expectations change with hive size and the prevalence of a pathogen at any given time.

Consult with your bee inspector and/or veterinarian to provide a guide to prevalence.

To achieve a 95% confidence that *one* bee will be positive in the sample you need to test:

Group size	Prevalence (%)								
	50	40	30	20	10	5	2	1	0.5
1500	5	6	9	14	29	58	142	271	493
10000	5	6	9	14	29	59	148	294	581
20000	5	6	9	14	29	59	148	296	589
30000	5	6	9	14	29	59	148	297	592
50000	5	6	9	14	29	59	149	298	595
80000	5	6	9	14	29	59	149	298	596

If the investigation requires a 99% confidence that *one* bee will be positive in the sample you need to test:

Group size	Prevalence (%)								
	50	40	30	20	10	5	2	1	0.5
1500	7	10	13	21	44	88	212	395	687
10000	7	10	14	21	44	88	226	448	878
20000	7	10	14	21	44	88	227	453	898
30000	7	10	14	21	44	88	228	455	905
50000	7	10	14	21	44	88	228	457	911
80000	7	10	14	21	44	88	228	457	914

Using a matchbox to collect a sample of bees from the landing board collect about 30 bees. This sample would then be useful to look for pathogens with about 10–20% prevalence such as *Varroa* but would not be useful to look for rare pathogens such as some of the viral diseases. Even *Varroa* may not be on 20% of the foraging bees – which are the ones likely to be collected on the landing board.

As an example of using these tables – if *Varroa* mite infestation in October is estimated to occur on 2% of the foraging bees in a hive of 20000 worker bees to have a 95% chance of finding at least one *Varroa* mite you would need to check 226 foraging bees – slightly more than would be normally collected using a matchbox!

One frame of bees is around 500 bees. This level of investigation would provide sufficient power for most pathogens likely to be present in the hive.

Clinical examination of a hive

Pre-visit

1 Ensure that your bee suit is clean.
2 Wash bee suit in non-perfumed washing products. Note angry bees will deposit pheromones and alarm substances on your bee suit and equipment.
3 Avoid (if possible) manipulations of the hive on cold, wet, thundery or windy days.
4 Avoid (if possible) manipulation of the hive if the day is very hot.
5 Take wind proof matches so you can light your smoker.
6 Avoid being hot and sweaty – there should be no body odour issues.
7 Avoid having perfumed products – this can include hair shampoo in the morning.
8 Avoid having any alcohol – this can disturb bees.
9 Avoid banana as the smell mimics the alarm pheromone.
10 Handle the combs gently. Avoid sudden movements.
11 Do not discard ash from your smoker carelessly. Hot ashes are a fire risk.

Biosecurity considerations

Observation of the hive entrance and surrounds

Good signs

1 Bees on active flight path.
2 Lots of bees with full pollen sacs entering hive.
3 Many drones flying – normal in afternoon in late spring to summer.
4 Hive stand is free of grass and objects which may allow ants into the hive.

More observation is required if

1 Small pieces of wax at entrance – bees uncapping stores or a mouse.
2 Bees fanning and exposing Nassanov glands.
3 Bees fanning/bearding – colony too hot.
4 Bees issuing from hive in a swirling and ascending mass – colony is swarming.
5 Fighting at the entrance – robbing.

Bad signs

1 Bees covering entrance – too many bees in the hive, colony too hot – bearding in the summer afternoon – swarming in the late spring morning.
2 Bees walking aimlessly around the front of the hive – possible disease.
3 Dead larvae actively being carried out – disease in the hive.
4 Mummified larvae in hive debris and being removed – Chalkbrood (*Ascosphaera apis*).
5 Dying bees – dead at bottom of the hive and live on top of hive – paralysis virus.
6 Bees unable to fly and staggering around, bright black bloated abdomens.
7 Bees fighting at entrance – robbing.
8 Faecal spotting – Nosema – dysentery (*Nosema apis* and *Nosema ceranae*).
9 Large pieces of wax at entrance – mice in the hive.
10 Foul smelling – death of the hive.
11 No bees on landing area.
12 No noise from the hive.
13 Look for any wasps or hornets flying close to the hive.

Figure 12.1 Bees outside the hive

Figure 12.2 Bees walking around aimlessly on landing area or adjacent plants

Figure 12.3 Mummified larvae in hive debris or being carried out of the hive

Figure 12.4 Dead bees around the hive or close by

Figure 12.5 Unusual bees in the hive – bee on the right with the black body for example

Figure 12.6 Fighting at entrance

Figure 12.7 Faecal spotting

Figure 12.8 Wax at entrance

Examine the hive debris

If the hive has an open bottom or a removable tray, the hive debris will fall onto a tray or on the ground. This can be an excellent opportunity to observe the health of the hive.

1 Place a sticky sheet into the hive floor 7 days prior to the visit.
2 Place the hive debris into windscreen washing liquid or methylated spirits.

3 *Varroa* mites will float to the surface. Then count the number.
4 Filter through some white disposable kitchen towelling to assist identification.
5 Divide the number of mites by the number of days since last check to get a mite drop/day estimation.

Figure 12.9 Wax moth larvae in the hive debris

Figure 12.10 Chalkbrood mummies in the hive debris

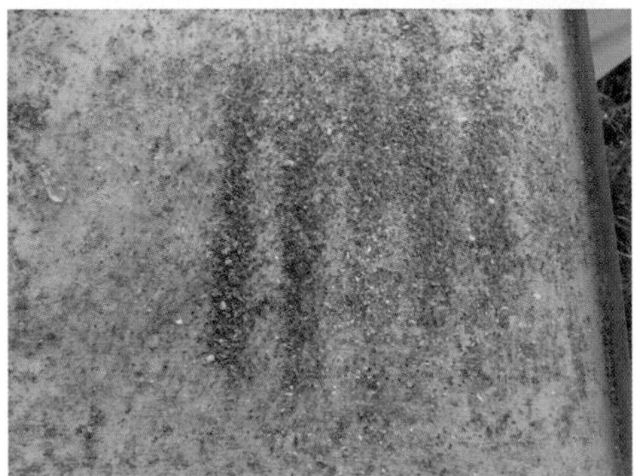

Figure 12.11 The hive debris indicates where the bees are in the brood box above in the winter time

Approach the hive

- From the side.
- Do not stand in the flight path of the bees.
- Measure the infra-red temperature of the hive from the outside.

Remove roof

- Look for wax moth and also any wasp presence.
- Observe again the number of dead bees in the ground around the entrance.
- Check that the top cover is properly placed.
- Look at the number of bees in the hive – are there too many?

Examination of the bees

1 Remove a frame and knock down around 500 bees into a container with a lid.
2 Brush the bees off a comb through a large mouthed funnel into the container. *Ensure the queen is not present.* Add 500ml of windscreen washing liquid – ethanol – or alternatively methylated spirits.
3 Collect 30 bees from the landing board area using a matchbox. Foraging bees are older but note deformed bees may be unable to fly.
4 Examination of brood may also be necessary.

Useful extra equipment

Infrared thermometer
Ultraviolet light 3100 to 4000Å – for necrotic tissue typical in foulbrood.

Indicators of the general health of the bees

Hive debris

If the hive has an open bottom, the hive debris will fall onto a tray. This can be an excellent opportunity to observe the health of the hive.

1 Place a sticky sheet into the hive floor seven days prior to the visit.
2 Place the hive debris into windscreen washing liquid or methylated spirits.
3 *Varroa* mites will float to the surface. Count the number.
4 Divide the number of mites by the number of days to get a mite/day estimation.

Figure 12.12 Examine hive debris with a hand lens, examine particularly for *Varroa*

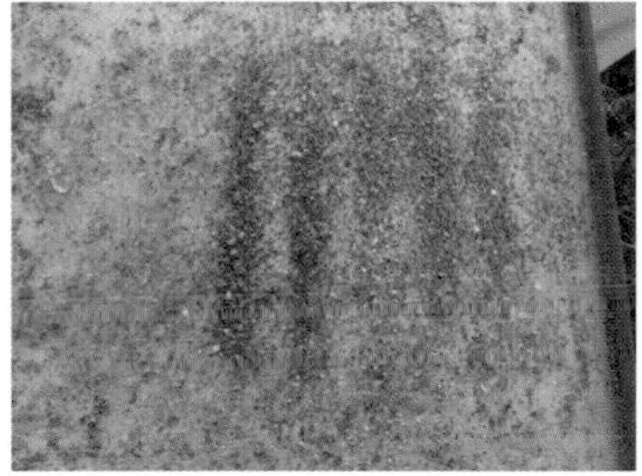

Figure 12.13 The hive debris indicates where the bees are in the brood box, above in the winter time

The weight of the hive

Having the hive on a weighing scale can provide invaluable information regarding the number of stores and general heath of the hive. This can also be integrated into a wireless system so that the hive can be monitored through the Internet. This can prove to be a deterrent to thieves as your computer/phone can alert the owners of any serious upset. The progressive loss of weight may also be a helpful early indicator of disorders such as colony collapse disorder or that the hive is being robbed.

Protein concentration

The crude protein concentration % can vary in working bees from 21–67%. It is important to maintain the workers with a crude protein concentration above 40%. If the protein concentration falls below 40%, the lifespan of the workers will fall from the normal 46–50 days to only 20–26 days. This significantly reduces their ability to forage and feed the hive. Foragers will not be able to fly for so long or for so far, further reducing the health of the hive.

If the protein concentration falls below 30%, the bees become very susceptible to European foulbrood (*Melissococcus plutonius)* and Nosema (*Nosema apis* and *Nosema ceranae*). This is particularly important to check in the autumn, as bees with low protein concentration will generally fail to overwinter.

Post mortem examination of the bee

Note you may need some degree of dexterity to perform some of these tasks.

Equipment required in 'lab back home'

Wind matches – also matches to examine larvae
Fine tweezers
Scissors
Wide mouthed funnel
Hand lens
Wire mesh #8 or #12
White cotton cloth
Starter engine spray – ethanol or washing-up liquid – 500 ml
Corkboard
Pins
Formalin saline (10%) – note health and safety issues
Lactic acid

Examination equipment

Hand lens
Dissecting microscope
Light microscope
Wax holding tray to hold about 1cm of wax.

Sample collection

Collect bees using a matchbox. One matchbox full of bees is around 30 workers. *Do not collect the queen.*

The agent being looked for will determine the type and number of bee(s) to collect. If the study wants to examine diseases, older worker bees can be examined by collecting flying bees from the entrance. *Varroa* is a pathogen primarily of pupae and there will only be low numbers on flying bees. Bees with viruses – for example, deformed wing bees – are not going to be collected from the entrance.

1 Block the entrance to the hive and leave the hive for a couple of minutes.
2 Open the hive and collect the flying bees.
3 The collected worker foraging bees will normally be older than 3 weeks post emergence.
4 Place the bees in a refrigerator for 20 minutes to anaesthetise, and then kill with washing up liquid.
5 For dry bees, place the bees in the freezer overnight prior to working on them.

External examination of the bee

1 Under a dissecting microscope or hand lens examine the bees.
2 Note the sex of the bees under examination.
3 Note the colour of the eyes.
4 Note the shape and structure of the wings.
5 Note any mites or other abnormalities.
6 Note the hair distribution and areas of hair loss.
7 Is the bee's shape, especially its abdomen, normal?

Obtaining a sample of haemolymph

1 Make a capillary tube from the tip of a Pasteur pipette.
2 Puncture the intersegmental membrane directly behind the coxa of the first leg.
3 The haemolymph will flow into the capillary tube.
4 Place the drop of haemolymph on a glass slide.
5 Smear on a glass microscope slide, stain and examine looking for haemocytes (see Figure 1.54).

Examination of the tracheal system

1 Pin the bee on its back.
2 Remove the bee's head and first pair of legs by pushing them off with a scalpel or razor blade using a downwards and forwards motion.
3 Cut a disk of thin transverse section of the prothorax.
4 Place the disk on a microscope slide and add a few drops of lactic acid. This makes the material more transparent and helps to separate the muscles.
5 With the aid of a dissecting microscope carefully separate muscles and examine trachea.
6 Remove suspect trachea and examine at x 40–100 down the microscope or dissecting microscope.

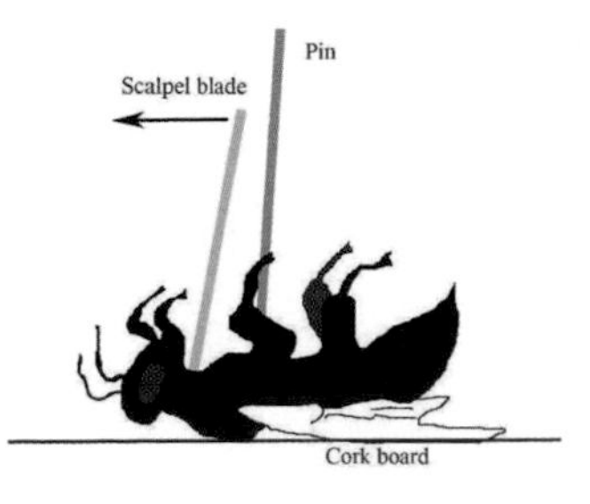

Figure 12.14 Drawing of scalpel placement to remove head and first pair of legs

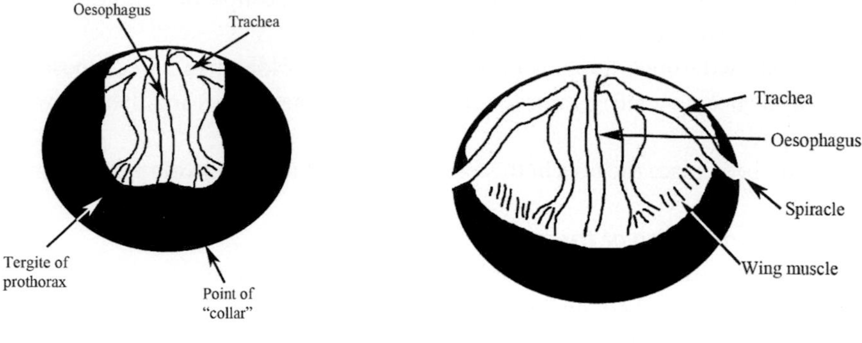

Figure 12.15 Drawings of the gross view of transverse section of the prothorax

7 Note with formal saline the trachea becomes the same colour as the wing muscles making identification of the trachea extremely difficult.

Examination of the digestive tract

The digestive tract can be easily obtained from a bee. The head has already been removed.

Grasp as much of the stinger as possible with a pair of fine tweezers and then with a steady, gentle pull withdraw the entire digestive tract from the abdomen with the stinger attached.

With a dissecting microscope arrange the digestive tract and separate from the stinger apparatus.

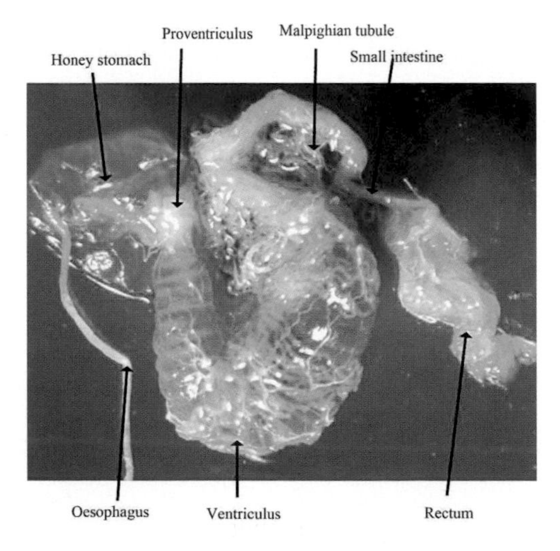

Figure 12.16 The gross anatomy of the digestive tract and Malpighian tubules

Examination of the Malpighian tubules

The Malpighian tubules are long threadlike projections originating at the pylorus. The pylorus is at the junction of the ventriculus and small intestine. The Malpighian tubules can be teased away from the digestive tract with a pair of fine tweezers. Place the tubule on a microscope slide.

Using a cover slip crush the tubule and examine under the microscope.

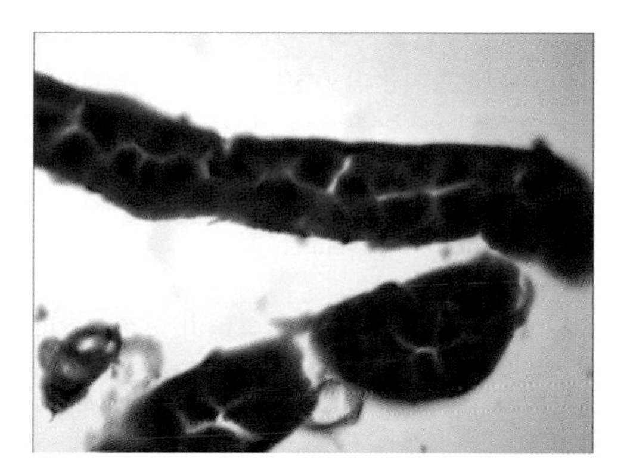

Figure 12.17 Normal Malpighian tubule in a queen H&E

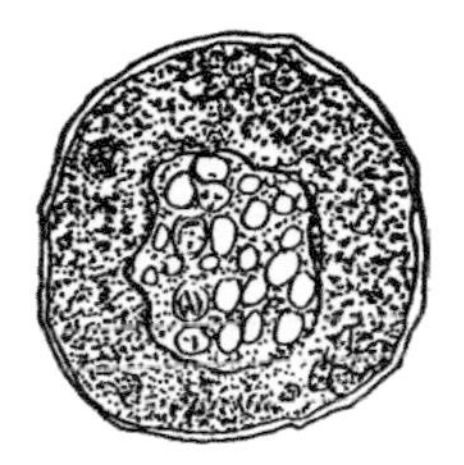

Figure 12.18 Drawing of the Malpighian tubule with amoeba cysts – the round circles- *Malpighamoeba mellificae*

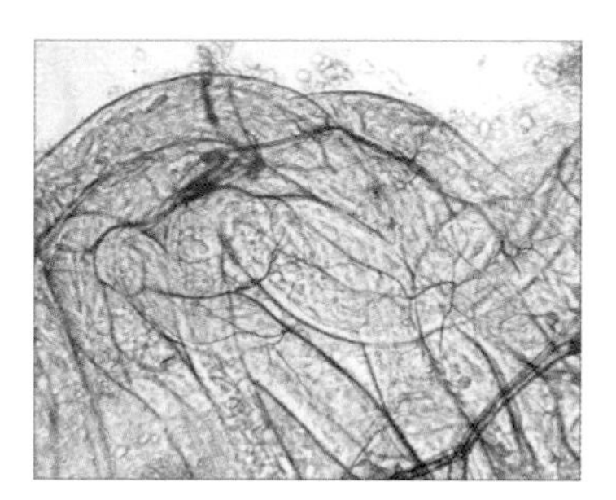

Figure 12.19 The normal Malpighian tubule worker under a low power light microscope

Examination of whole bee anatomy

Embed the bee in wax and dissect the bee using the dissecting microscope. Examine in detail the abdomen to reveal the layout of the general organs (see Chapter 1). This should be carried out in workers, drones and the older queen to make the beekeeper familiar with the anatomy and structure of the animals under his or her care. When embedding bees, place some on their backs to reveal the ventral surface of the organs. In others, remove the legs and embed them on their front to reveal the dorsal surface of the organs, especially the heart and then the nervous system.

The first few bees of the examination can be difficult and exasperating. But perseverance provides great rewards. To embed the bee, melt a small piece of wax with a bent nail and place the bee in the liquid wax. Wait until the wax re-solidifies before working on the bee. Immersing the bee in isoproplyene alcohol will improve the view down the microscope.

Examine also the larva and pupa

This encourages the beekeeper to understand more about the beauty and the workings and behaviour of the hive (see Chapter 1).

When the queen dies or has to be replaced

Examine the old queen in detail to become more familiar with a major player in the hive.

Post mortem of a hive

Conduct a post mortem on dead colonies to determine the cause of death before reusing the equipment. Check for diseased brood and scales. If you suspect a foulbrood disease, even in an abandoned hive, contact your local bee inspector or veterinarian.

If you are unsure about a diagnosis, send a sample to the laboratory. Discuss sample collection and submission with your local bee inspector or veterinarian.

1. View the hive from the side: do not approach the hive entrance directly. The hive may still be active. If you have any suspicions that the hive is still active you may need the smoker.
2. Look around the hive. Especially note the number and position of any dead bee bodies.
3. If there is a *Varroa* tray, remove it and examine the hive debris.
4. Look at the tray for evidence of wax moth.
5. Look under the hive at the hive debris.
6. Remove the roof and listen for the bees.
7. Remove the cover board and listen for bees.
8. Look under the inside of the cover board for wax moth infestation.
9. Open the super and examine each frame in turn placing them on the ground by the hive. Note this may attract passing bees and wasps to the smell of any still stored honey.
10. Examine the amount of capped and uncapped stores that are remaining in the hive.
11. Note any damage to the shape of the comb.
12. Open the brood box.
13. Examine each frame in turn and place it on the ground.
14. Note the position and characteristics of any capped and dried uncapped brood.
15. Note the amount of Chalkbrood or Stonebrood present.
16. Note any damage to the shape of the comb.
17. Examine the brood and super boxes.
18. Dismantle the hive stand.
19. Dispose of the old hive carefully or thoroughly clean and disinfect before re-use.
20. All frames and wooden items should be placed in a freezer (–20°C) for seven days to kill any wax moths or other arthropod parasites.

Figure 12.20 Chalkbrood and dead capped cells in an abandoned hive

Figure 12.21 Capped honey with mould in an abandoned hive

Figure 12.22 Wax moth infestation in an abandoned honey bee hive blocking the entrance to the hive

Figure 12.23 Wax moth infestation in an abandoned bumblebee nest

Figure 12.24 Do not leave old hives abandoned

Figure 12.25 Safe burning is an efficient method of disposal of old hives

Repair and reuse of an old hive

- All old hives should be viewed with caution, but hives are expensive.
- Hives should be repaired especially any holes or cracked wood.
- Foundations should be replaced.
- Old 'rescued' wax should be used for non-bee use only.
- Frames can be placed in the freezer for seven days to kill any arthropod larvae and eggs, in particular wax moth and *Varroa*.
- The entrance frame, hive stand, box and wooden parts of the frames can be sterilised using suitable disinfectants – H_20_2, peroxide, sodium hypochlorite and sodium hydroxide (bleach) would be examples.
- The surfaces can be flamed with the blow torch, but remember that wood burns.
- Put the whole hive out in the sun for a week and turn periodically so that all the wooden surfaces are exposed to sunlight – ultra violet rays.
- Ideally all hive components should be fumigated before reuse.

Bacterial identification

Gram stain

It is possible to start the classification of bacteria by using a stain, which illustrates the type of cell wall and the shape of the bacteria.

Always place organisms in centre of a glass microscope slide and always use 'frosted side up'.

1. Make a thin smear of the organism with a tiny droplet of water on a slide.
2. Dry – best if air-dried.
3. Fix by gently heating 2–3 times through the flame.
4. Flood smear with Crystal violet for five seconds, hold side perfectly level, then dump.
5. Wash with distilled water.
6. Flood smear with Gram's iodine for five seconds.
7. Wash with distilled water.
8. Decolorise with acetone-alcohol for about three seconds.
9. Wash with distilled water.
10. Counterstain with safranin for five seconds.
11. Wash with distilled water.
12. Blot slide dry, face down on a paper towel.

Catalase test

1. Dip a capillary tube into 3% H_20_2.
2. Touch a colony.
3. Observe the tube for bubble indicating a positive reaction.

Do not contaminate the bacterial colony with blood agar as red blood cells contain catalase thus resulting in a false result. Old cultures can lose their catalase activity.

Oxidase test

1 Hold a piece of oxidase test paper with forceps and touch onto an area of heavy growth.
2 Rapid purple reaction (within 10 seconds) indicates positive. If within a minute, it indicates a delayed positive.

Note an oxidised organism will also be catalase positive.

Diff-quick stain

This can be very useful for smears – for example, Haemocytes.

1 Make a smear of a drop of haemolymph and air dry.
2 Fix in Diff-quick solution A for 10 seconds.
3 Dip slides 25 times in Solution B. Do not rinse slide.
4 Dip slide 25 times in solution C.
5 Wash with phosphate buffered saline or distilled, deionized water.

Permanent fixation

Air dry the slides. Clear in Xylene and mount using a synthetic mounting medium.

Note only fixed and mounted slides may be taken into Australia – see appropriate customs regulations.

Lateral flow diagnostics

Lateral flow diagnostic test kits enable beekeepers, bee inspectors and veterinarians to test for pathogens and have the results immediately and easily in the apiary or in the laboratory.

Note the test will only detect one pathogen. You need to use the specific pathogen kit.

The classic pathogens to be diagnosed are *Paenibacillus larvae* (American foulbrood) and *Melissococcus plutonius* (European foulbrood). See Chapter 5 for more details.

How to use the lateral flow test kit

Whole or part samples can be used, but it is recommended that a whole, infected larva be used to obtain the best results.

1 Extract a larva from its cell showing suspicious symptoms with the spatula.
2 Unscrew the lid from the extraction bottle.
3 Use the spatula to deposit sample in the bottle.
4 Replace the lid and shake vigorously for about 20 seconds.
5 Beware of the potential human risk – the bottle contains a buffer and sodium azide.
6 Remove a test device from foil pack. Do not touch viewing window.
7 Unscrew the lid of the extraction bottle and use the supplied pipette to remove a sample of liquid from the bottle. For best results remove the sample immediately after shaking to prevent bacteria from settling out of suspension.
8 Hold the test device horizontally and gently squeeze two drops onto the sample well of the device.

9 Keep device horizontal until extract is absorbed (about 30 seconds) and a blue dye appears in the viewing window.
10 Wait until the control line appears (labelled C) and read the result (in about 1–3 minutes).

Interpretation of the results

A positive result (two blue/purple lines show up – both Test and Control, see below) indicates that the target pathogen is present in the sample.

A negative result (Control line shows up only as blue/purple, no Test line) indicates that the pathogen has not been detected in the sample.

Note the test has failed if:

- One blue/purple line (T only)
- No lines present
- Brown C or T lines

If the test has failed, carry out another test using a new sample from the same original comb with a new device.

Figure 12.26 Two possible results using the lateral flow device

Nosema detection

Examination of an individual bee

Remove the bee's head. Grasp as much of the stinger as possible with a pair of fine tweezers and then with a steady, gentle pull withdraw the entire digestive tract. Place the intestinal tract on a glass slide and locate the ventriculus. Crush the ventriculus with a glass slide and examine the contents down the microscope. The nosema spores can be recognised as little "rice grains".

Hive examination

Basic examination

Collect 30 bees (about a matchbox full). Place the bees in the refrigerator/freezer for 20 minutes to anaesthetise the bees. Place 15–30 ml of cold tap water with the bees in a crucible. Quickly kill the bees by maceration in a pestle and mortar with the 15–30ml of water. Examine one drop of fluid on a glass slide down the light microscope at x400 view. Note the shape and size of any Nosema spores present.

Results:

Number of Nosema spores seen per x400 view	Interpretation
0–20 spores	Mild infestation
20–100 spores	Moderate infestation
100 spores or more	Heavy infestation

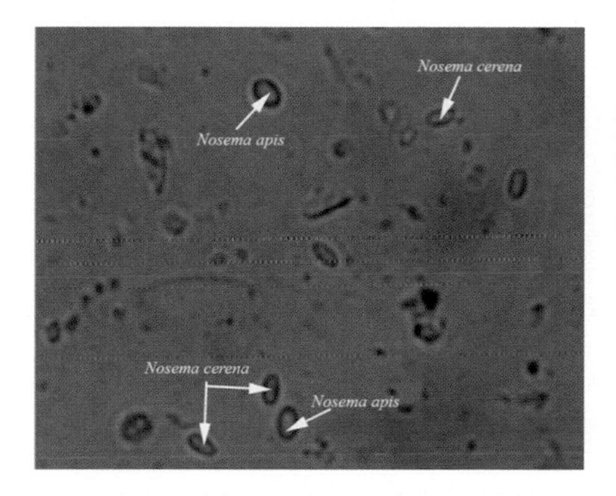

Figure 12.27 Mild infestation of mixed Nosema

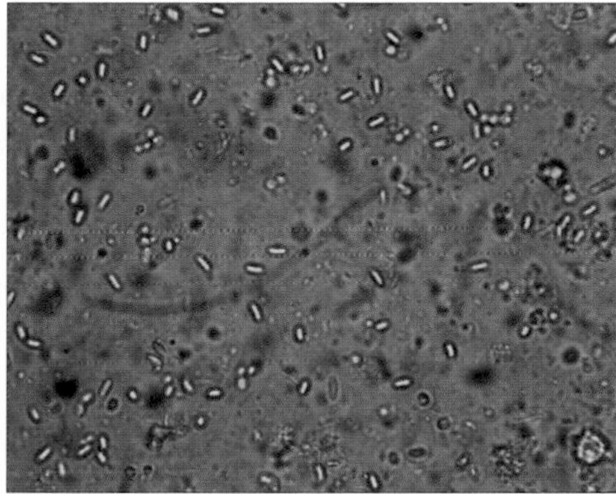

Figure 12.28 Heavy infestation of ***Nosema cerana***

Numerical examination by the use of a haemocytometer

This provides a guide to the number of spores per ml of ventriculus contents. Bees may have 10 million spores/ml. Read the instructions provided by the haemocytometer manufacturer.

Counting the number of spores per ml

1 Make a 1:100 dilution of the ventriculus sample by accurately pipetting 0.1ml of ventriculus contents sample into 9.9ml of distilled water.

2 Clean the glass haemocytometer and special coverslip thoroughly with a soft tissue. Press the coverslip onto the slide so that Newton's rings are clearly visible on the contact surfaces.

3 Ensure that the diluted semen sample is thoroughly mixed. A drop is then expelled gently into the chamber from a fine pipette so that the entire cavity is filled with diluted semen. The process is repeated for the second counting chamber. Excess diluted semen should not be allowed to flow into the grooves bounding the two chambers.

Counting the number of Nosema spores

1

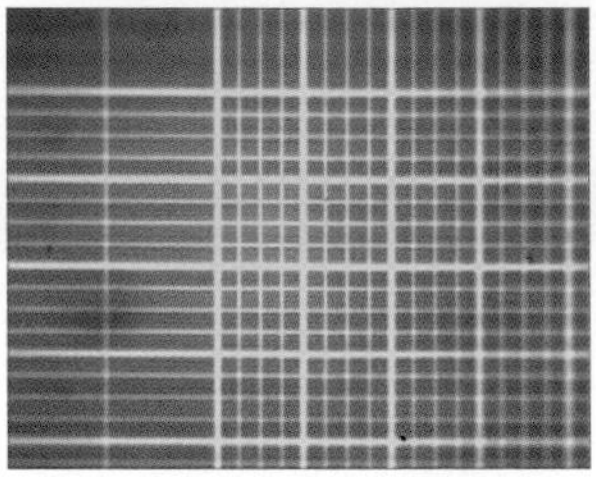

Figure 12.29 View of the haemocytometer down the microscope

View of the haemocytometer down the microscope under low power.

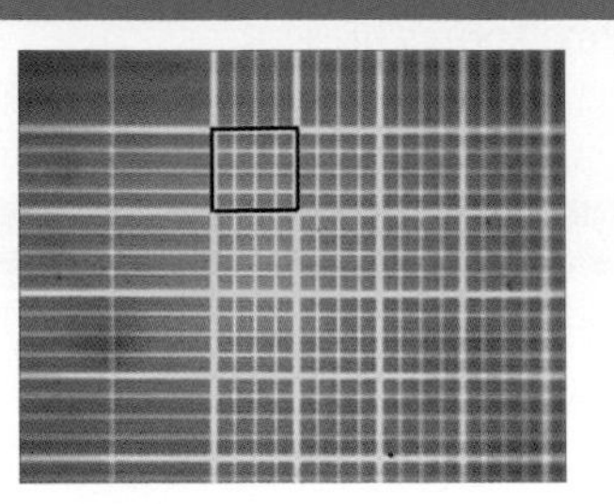

Figure 12.30 One counting cell in a haemocytometer

You need to identify the small squares with the 16 very small squares – as shown above.

2

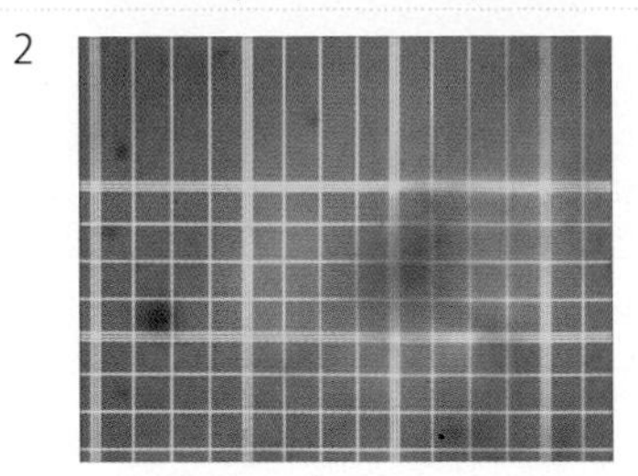

Figure 12.31 One counting cell in detail

Look at one of the small squares with the 16 very small squares (4 × 4).

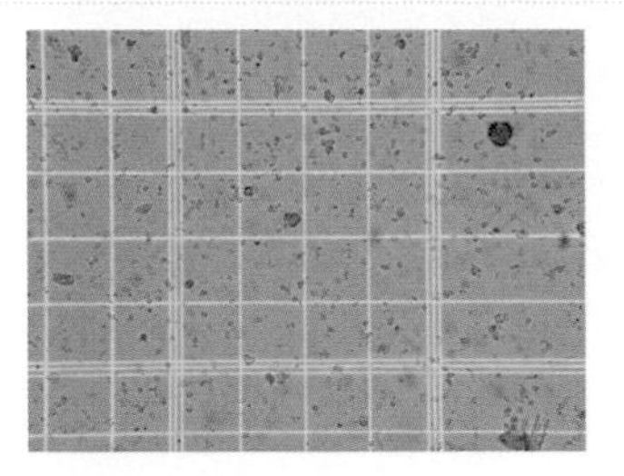

Figure 12.32 Nosema with the haemacytometer

Concentrate on to one of the squares with 16 very small squares so you will see the spores.

3 Count the number of spores in each one of the small squares (of 16 very small squares).

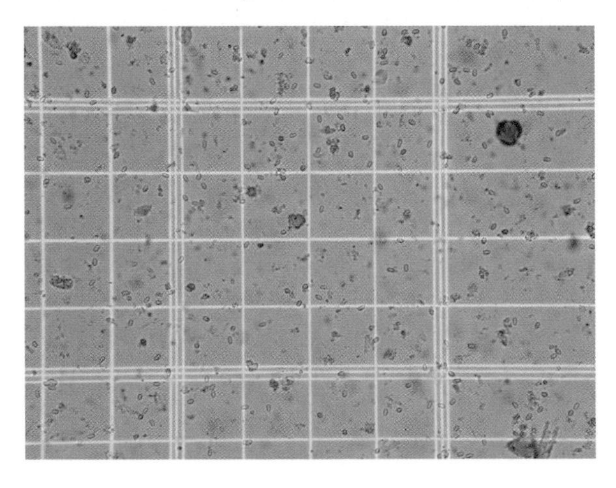

Figure 12.33 Counting nosema spores with the haemocytometer

4 Count the number of spores using a hand tally counter. Repeat the count on the second chamber and use the mean of the two counts for the calculation of the concentration.
One small square of 16 in the counting grid of a haemocytometer chamber is shown above. Count 25 of the small cells – this is one large square.

5

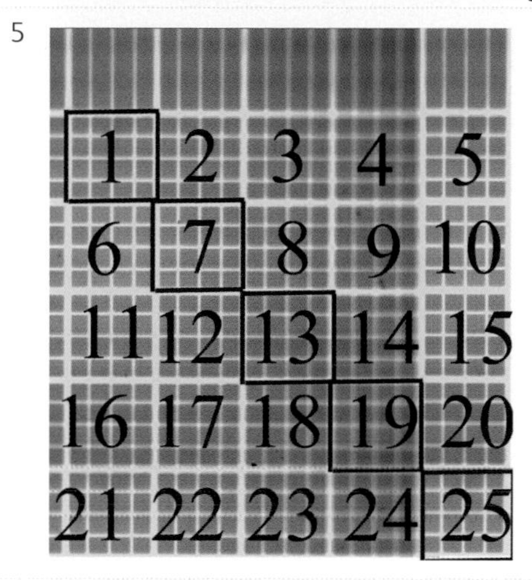

Figure 12.34 method of counting all 25 cells in a haemocytometer

6 The formula will then be that the number of spores in the 25 small squares = N.
Spore concentration = N × 10^6 nosema per ml.

Small hive beetle (*Aethina tumida*) examination

Collecting small hive beetles

The small hive beetle moves very quickly and will avoid the light, making it difficult to catch one for examination. It is possible to chase the beetle with a small container but the beekeeper must be persistent for a few minutes, chasing the beetles around and between the frames. Another option is using a mouth aspirator. But this involves beekeeper having the veil off which is potentially dangerous.

Identification of small hive beetles

Figure 12.35 Adult small hive beetle

Figure 12.36 Note these beetles are small and rapidly run from the light.

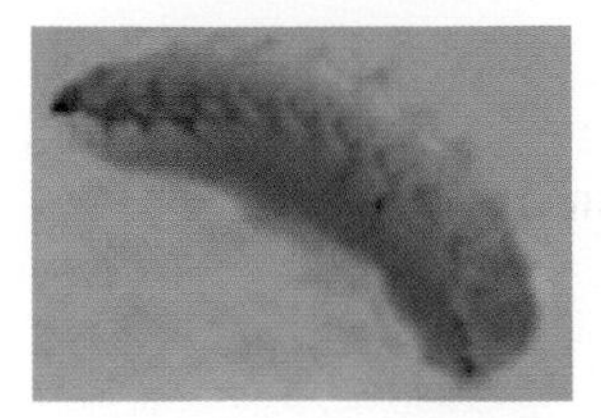

Figure 12.37 Small hive beetle larva with spines on its back

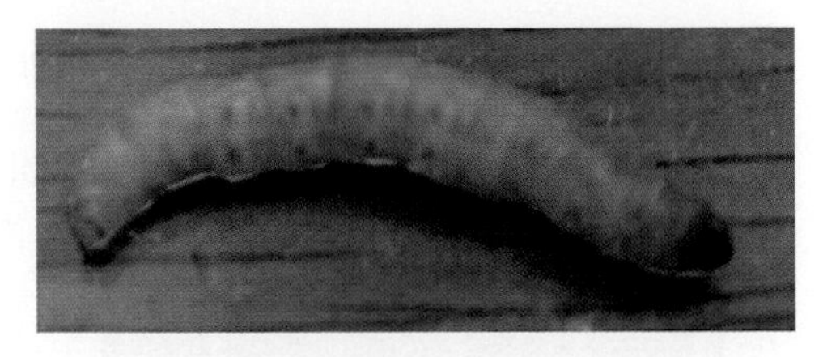

Figure 12.38 Wax moth larva smooth back, without spines.

It is a small and very active beetle, brown or black in colour. Note that the adult beetle has clubbed antennae. The larva has three prolegs and spines on its back. It may be confused with the small wax moth larva, but spines are not present on the back of the wax moth.

Sending a sample to the lab

Selection of specimens

Bees selected for laboratory analysis, ideally should be free from antimicrobial therapy and in an early or acute disease stage. Selected tissues should be collected as aseptically (cleanly) as possible. In addition, a meaningful history of the disease outbreak and a tentative diagnosis, based upon clinical evaluation, should be included. Laboratory tests results are directly affected by the selection, preparation, handling and shipment of selected specimens.

Identify tissue and samples with:

- Site of apiary GPS
- The address and post code of the location of the hive
- Hive identification number
- Type of material submitted

Preparation and collection of samples

Tissues – fresh

1 Collect 30 bees, place in the freezer for 60 minutes to cool down and die. (The number and type of bees required does depend on the suspected pathogen.)
2 Place in a plastic bag (e.g. whirlpacks).
3 Double bag in whirlpack bags.
4 Do not mix different hives in one single bag.
5 Transport with cold packs.

Histopathology

1 Bees should be placed in the refrigerator for 20 minutes to anaesthetise.
2 Place the 10% buffered formalin in the refrigerator for 20 minutes.
3 Place bees in chilled 10% buffered formalin.
4 Place a small piece of tissue paper over the bees so that the bees are completely in the formalin.

5 Ensure the bees are completely wet and immersed in the formalin.
6 Use ten times the volume of the tissues being fixed.
7 Replace back into the fridge for at least one hour before sending to the laboratory.

Packing specimens

To avoid leaking in transit, double bag the samples. Whirlpack bags work well for this purpose. Wrap sample bags and 2–4 ice packs in absorbent paper (e.g. newspaper) to absorb liquid in the event of leaking. Place the package into a styrofoam container. Completed submission forms should be inserted into the envelope on the inside cover of the cardboard box.

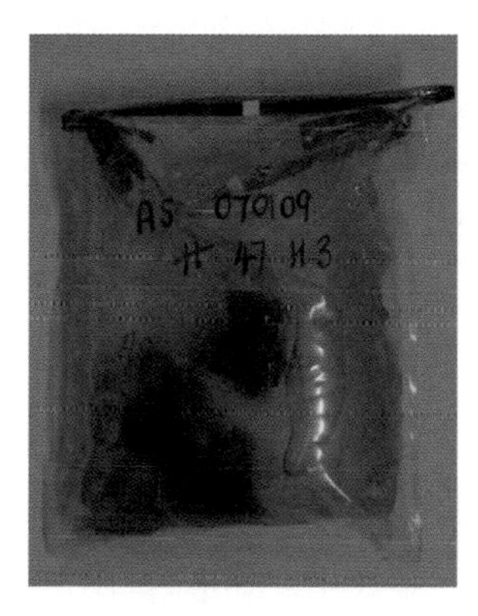

Figure 12.39 Samples ready to send

Mailing

Samples should be submitted by the fastest means possible to avoid deterioration of specimens. Next day or overnight delivery is preferred over other methods. Discuss with the mailing system selected any specific requirements. Ideally take the samples to the diagnostic laboratory personally or by carrier. Note, try to avoid Friday or Holiday samples.

Ensure that all samples are adequately identified and a suitable history is provided with the samples.

Administering medical treatments to the hive

Always read all the instructions provided with the medication. If you have any issues discuss application methods with your veterinarian and bee inspector. Keep all medications out of the reach of children and animals. Store all medications according to the manufacturer's requirements. Note, always dispose of all medication products appropriately. Even the medications you do not use. Keep records of all medicine used for five years, even if you stop having bees.

Feeding the hive

Artificial feeding options

Grease patty

Vegetable shortening plus sugar (1:2 ratio).

100g of vegetable shortening with 200g of sugar.

Add 4.5g peppermint flavouring.

Take a handful of patty, place on waxed paper, top and bottom, and place between the two boxes.

Remove top layer of waxed paper.

Energy in liquid form – sugar syrup

This is made from granulated sugar (sucrose) and should be presented in a concentration of more than 50% sugar so the bees can readily use the sugar. It is vital that the concentration is not too strong or the sugar will start to crystallise out.

There are two major concentrations which would be commonly used.

Autumn feeding – a 61.7% sugar solution.

This is made by adding 2.5kg of boiling water to 4kg of granulated sugar. The boiling water will help the sugar dissolve. Stir until the sugar is totally dissolved. Do not reheat the solution or you will create toffee. Add one drop of thymol per litre of fluid to help reduce fungal growth.

If there is no handy weighing scale available this concentration can be made by half filling any container with water and then filling to the top with sugar.

Spring feeding – a 50% sugar solution

This is made by adding 1kg of boiling water to 1kg of granulated sugar. The boiling water will help the sugar dissolve. Stir until the sugar is totally dissolved. Do not reheat the solution or you will create toffee. Add one drop of thymol per litre of fluid to help reduce fungal growth.

Thymol solution of preserving feeding sugar solutions

Fill a small bottle ⅓ with thymol crystals and top up with ethanol or surgical sprits.

Feeding protein

Pollen bee bread

Bee bread can be made to make an even more ideal supportive material for the bees.

Ingredient	Makes up to 1402g
Pollen	1000g
Honey	150g
Water clean	250g
Whey	2g

Artificial protein source

This is never as successful as natural collected pollen. Most artificial substitutes are actually low in protein. Soya products are generally used. Filter the products to get the particle size below 0.5mm. Once made the final product can be rolled and cut into dry biscuits or cakes and stored frozen.

Ingredient	Amount
Pollen	10–25%
Soy Flour	20–100%
Yeast	20–25%
Sugar/honey/water	20–50%

Properties of honey

Water	<18% (or fermentation starts)
pH	3.5–5.5
Enzymes	Saccharase, amylase (diastase), glucose oxidase
Water insoluble solids	The lower content, greater clarity
Osmotic pressure	2000 mOsmols/kg
Refractive index	1.55 at 13% water 1.49 at 18% water
Energy content	2.6 MJ/kg (1380 cal/lb)

Using a refractometer

Figure 12.40 Use of refractometer

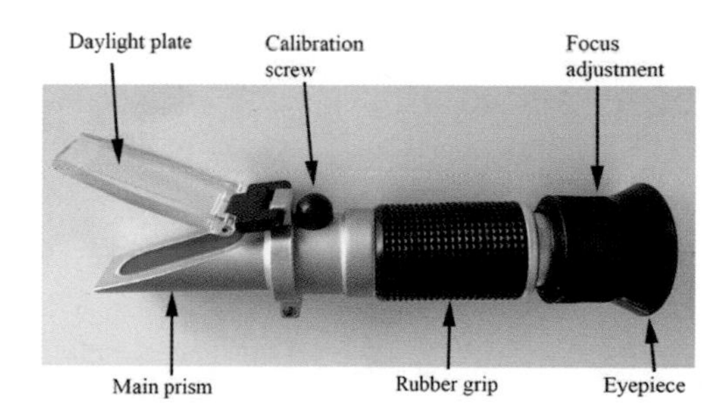

Figure 12.41 Refractometer

1 Open the daylight plate and place five drops of honey on the main prism.
2 Close the plate so that the honey sample spreads across the entire surface of the prism. There must be no air bubbles or dry spots.

3 Wait 30 seconds to allow settling.
4 Hold the daylight plate towards a light source and look into the eyepiece. You will see the gradation. Upper portion is blue the lower is clear/white.
5 Take the reading at the boundary. Depending on the machine you can obtain a water percentage and a sugar content measure from the different scales.

Good

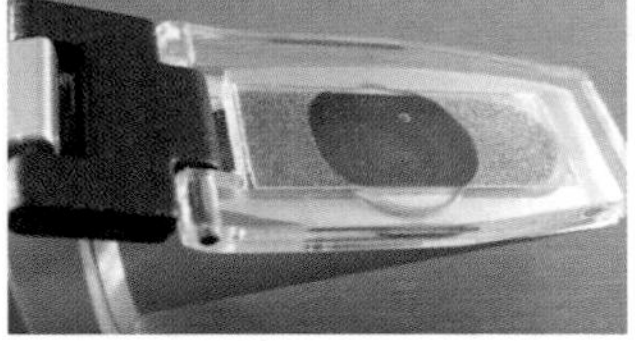

Poor – inadequate

Poor – air bubbles

Figure 12.42 Spread of honey on refractometer

If the water content is below 17%, fermentation is unlikely and the honey will have a long shelf life. Note that honey will absorb water from the air so its water content will increase with time.

Ideally harvest honey with a water content below 20%. If you wish to make mead, use honey with a water content above 20% to assist fermentation.

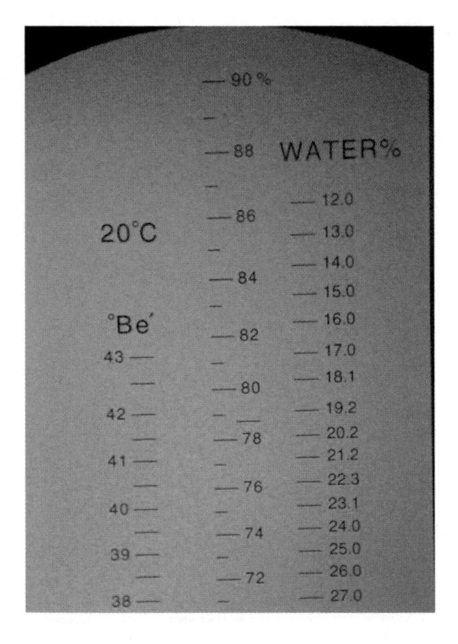

Figure 12.43 Normal scale, no honey

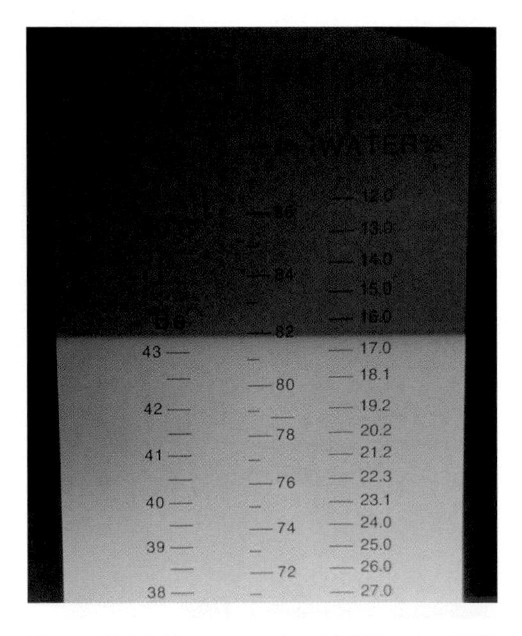

Figure 12.44 Honey sample – 16.8 % water, 82 % sugar

Note that the refractometer may require calibration from time to time. Additionally, ensure that the refractometer is designed for honey as veterinary and human refractometers are set for urine analysis and have a different scale. A honey refractometer is actually the reverse of this.

Examination of pollen

What type of pollen are my bees collecting?

This can be quite an interesting question.

Colour

The colour of the pollen can provide a guide but people see colours differently and shades in printed books can also vary. Pollen colours range from white, yellow, orange, green, brown and red to purple and can provide an excellent guide to the pollen being brought back to the hive. The bee only collects pollen from one type of plant at a time and then the nurse bees store each pollen type separately within its own cell – a filing system which is beyond most of us. Photographing the bees allows you to examine the type of pollen being brought to the hive.

Shape

Using a microscope the pollen can be further identified by its shape. Pollen can be collected from the plants themselves, from captured bees, using a pollen trap, from the frame and even from the honey itself.

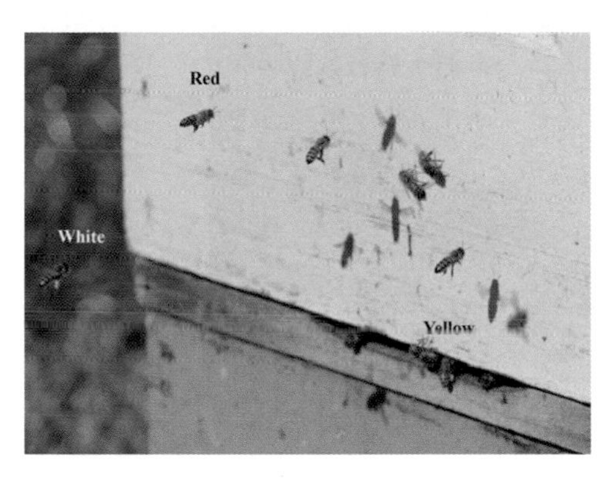

Figure 12.45 Forage bees carrying different types of pollen back to the hive

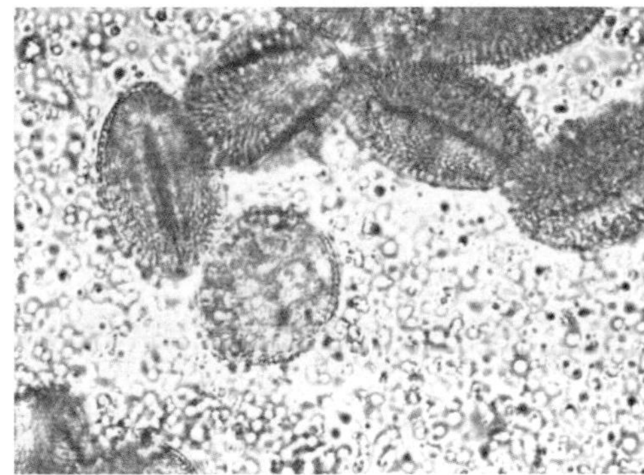

Figure 12.46 Pollen grains down the microscope

Technique

1 Place the pollen sample on a glass slide and wash with 95% ethanol to remove oils.
2 Allow the ethanol to evaporate, stain to make the pollen easier to see by adding a drop of basic fuchsin in some glycerol onto the dried pollen. Allow the slide to dry.
3 Place a small piece of glycerine jelly and a coverslip over the stained pollen area. Gently warm the slide with a small flame until the glycerine melts. The aim is to fill the space under the coverslip.

4 Seal the edges of the coverslip with nail varnish to prevent the jelly from drying out and set aside to dry.
5 Observe under the microscope the wonderful world of pollen.

Pollen shapes can be described as round, oval (flattened or elongated), triangular, long, boat shaped, multisided or irregular.

Manipulating the queen

Queen cells

Three types of cells can be recognised which may be associated with the future queen of the hive. Understanding these cells provides an insight into the current politics of the hive.

- The queen play cell – the worker bees will practise developing queen cells with a structure called a 'play cell' which is an enlarged cell outside the main frame structure but without an egg or developing larva inside.
- The swarming queen cell is constructed at the bottom of the frame.
- The superseding queen cell – a cell which is hanging from the middle of the brood is often an emergency superseding cell.

Figure 12.47 Play cell – generally at the edges of the frame

Figure 12.48 Swarming cell – at the bottom of the frame

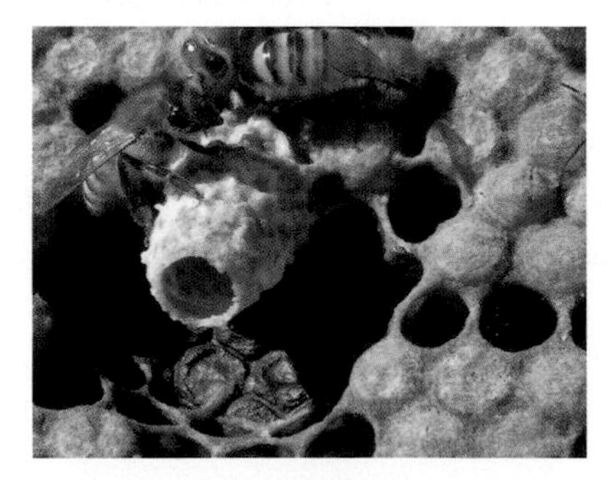

Figure 12.49 Superseding cell – in the middle of the frame note flattened worker cells surrounding the developing queen cell

Cleansing behaviour

Bees with a high hygienic behaviour appear to be able to protect themselves against a range of pathogens including *Varroa* and Chalkbrood. Utilising a hygienic behaviour examination can be a significant step to controlling Colony Collapse Disorder.

It is possible to examine a hive for hygienic behaviour with a 24 hour test:

1. Identify a frame with a solid area of capped worked pupae (pink eye stage) 15 days post lay.
2. Shake the nursery bees from the frame.
3. Place the frame horizontally on a solid surface.
4. Using a 6cm circular ring kill all the pupae using a sharp pin penetrating the cappings and pupae. The pupae may be killed by freezing with liquid nitrogen. Or it may be possible to remove the circle/square of pupae and kill by freezing for 48 hours in a freezer and then returning the piece of frame.
5. Replace the frame into the hive.
6. Re-examine the frame 24 hours later.
7. If 95% or more of the pupae have been removed and the cells cleaned, the queen may be considered 'hygienic'.

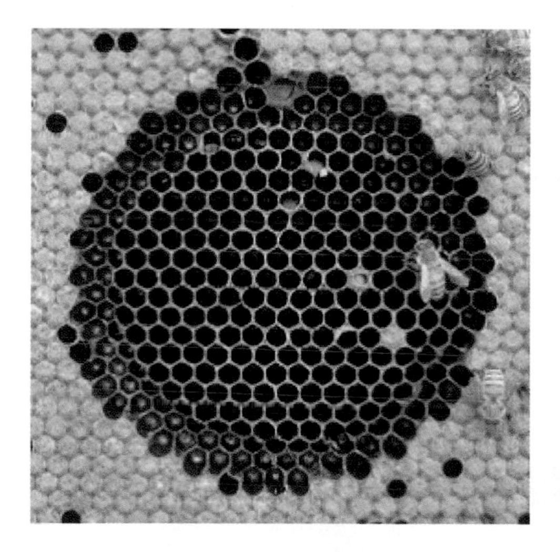

Figure 12.50 Hygienic bees have workers which remove dead pupae within 24 hours

Figure 12.51 Unhygienic bees have workers who only partly remove dead pupae within 24 hours

8. If the queen is considered hygienic she and her drones should be utilised as part of the future breeding programme.

Introduction of the queen

Queen bees or cells are normally introduced into a hive in spring, summer or autumn. You will achieve greater acceptance if the colony has access to a nectar and pollen flow at the time of introduction.

Find and kill the old queen the day before the new queen is to be introduced. Consider submitting her for analysis or do your own anatomical dissection to refresh your knowledge base. If you are uncomfortable offer her to the local bee group of which you are a member.

Purchased queen management

A queen cell should be transported to your home from the nucleus/supplier in a small foam esky filled with sawdust or wood shavings to minimise vibration and temperature fluctuations. But the dust must not be around the queen and her escorts.

It is essential to introduce the new queen as soon as possible to her new hive. Queens are usually accompanied in the cage by a few workers called escorts, who help to feed and groom the queen while she is caged. Queens may be safely mailed to their destination all over the world, although arrangements should be made to collect the package from the post office. Ensure that you have complied with all export or import regulations as required. Your local bee inspector or veterinarian will have details.

Queens that have been caged for several days should be given a little water. This can be done by placing two or three drops onto the gauze of the cage.

Queens should never be left in hot mailboxes or vehicles. Keep the cage free from any ants or other pests – including the pet cat or dog. Obviously keep them away from pesticides.

Mated queens through the post

Mated queens are sold in small wooden cages called mailing queen cages. One end of the queen cage has a queen candy plug, which acts as a time delay and enables the colony to gradually become accustomed to the odour of the new queen. The worker bees chew out the candy to eventually release the queen.

Place the queen cage between two frames of capped brood. Hold in place by gently squeezing the frames together so the top bars hold the queen cage. Wedge with the candy end tilted up slightly to prevent dead escorts from blocking the exit. The gauze may face up or down, but ensure honey doesn't leak over the cage, thereby drowning the queen.

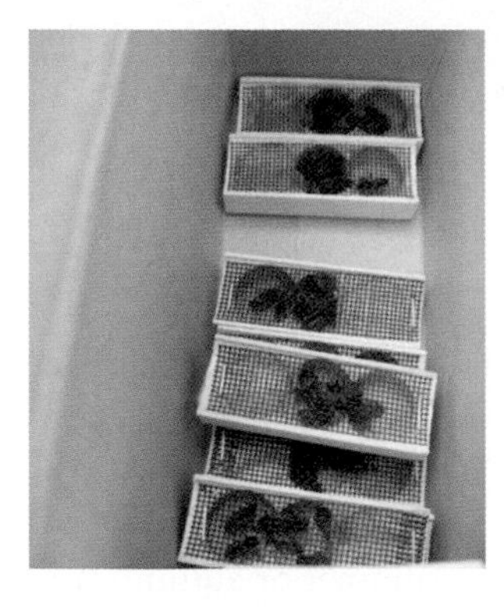

Figure 12.52 Queen cages in transit in an esky

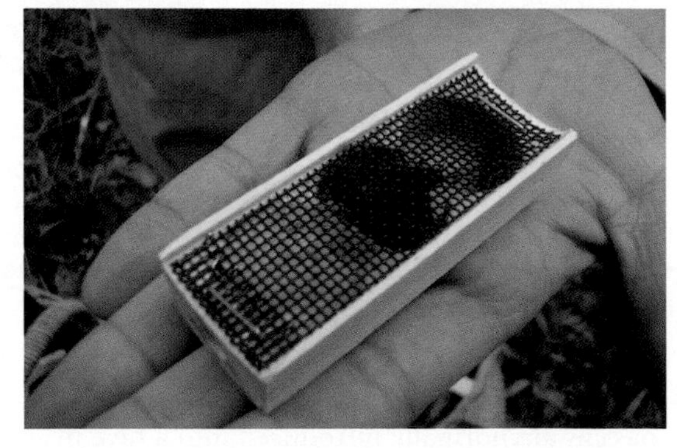

Figure 12.53 The mated queen cage

Figure 12.54 The queen cage in position in her new hive between two capped brood frames. Note the slight angle of the queen cage.

Do not disturb hives for at least three weeks after introducing a mated queen. Inspection, even then, should only be searching for signs of egg laying.

Do not try to find new queens at this stage, as the bees may reject and kill her if too much disturbance occurs.

Once a hive is successfully re-queened, you should notice a difference in temperament, production or disease resistance within 4–8 weeks.

Queen from a nucleus

If the beekeeper is new to beekeeping, purchasing a queen in a small nucleus may be a great option.

The bee is purchased from a breeder and is generally collected and transported back to her new hive by car.

1. Purchase a new hive with a brood box and super box and frames.
2. Make the new hive.
3. Seal the top board holes with simple metal gauze. This can be easily held in place by drawing pins.
4. Take the bottom board, brood box, two frames, the sealed top board and two poster pins.
5. Go to the breeder at the appropriate time.
6. The breeder will load your brood box with the frames from the nucleus. This will be about five frames of bees and obviously includes your new queen.
7. Place a brood frame and foundation on the open side of the brood frames and a super frame with foundation on the wall side of the brood frames.
8. Pin the frames in place using the poster pins.
9. Replace the top board.
10. Seal the entrance.
11. Secure the bottom board, brood box and top board so it is safe to travel.
12. Take the sealed brood box back to the new hive location.
13. Place the sealed brood box on the hive stand and replace the roof.
14. Leave for the evening.
15. The following morning gently remove the seal from the entrance and walk away.
16. Reduce the hive entrance. Leave enough entrance for only one bee to enter at a time.
17. Over the next week observe, from a distance, the landing board. Look for worker bees coming and going.

18 After two weeks open the hive. Open the top board and remove the tape sealing the brood box.
19 Quickly examine the new hive brood frames for the presence of eggs.
20 Remove the super and destroy and examine the trapped drone capped brood.
21 If the drone brood is not capped, leave and re-examine in seven days.
22 Fill the brood box with new frames.
23 Reseal the hive.
24 Open the hive entrance to full width.

Figure 12.55 The nucleus hives at the breeding apiary

Figure 12.56 Transfer the five frames with the queen into your brood box

Figure 12.57 Pin the frames into the brood box to stop them moving during transport

Figure 12.58 Place the closed top board onto the brood box. Seal the entrance to the hive

Figure 12.59 Tighten the bottom board, brood box and top board with a band. Tape the hive together and remove the wrapping

Figure 12.60 Transport back home in the car

Figure 12.61 Place on the hive stand and place the roof over the top board and leave

Figure 12.62 Next morning partially remove the seal in the hive entrance. Do nothing else. Leave the bees alone. Leave only enough room for only one bee to enter and leave

Fig.12.63 Your new hive in place

Figure 12.64 Over the week quietly observe activity on the landing board

Figure 12.65 After two weeks remove the roof and remove the tape around the brood box. Open the brood box and observe your new hive. Note the free-form comb. Examine the drone trap and destroy the drones and any trapped *Varroa* mites

Figure 12.66 Remove the pins holding the frames and move the whole block into the middle of the brood box. Quickly number the brood frames. Place new frames on either side and close the topbox and roof.
Leave undisturbed for another 2 weeks

Biosecurity for the introduction of a new queen and her attendants

1. Keep good records and a calendar of events.
2. Enquire about the health of the area from where the new queen originates.
3. Enquire if the source provides specific health information regarding their queen bees.
4. When bees arrive, keep them well separate from your hives and any bee equipment.
5. Examine the bees in detail – looking for any parasitic conditions and for any deformities especially in wing structure.
6. Sacrifice the workers and post mortem. Dissect the bees and look for tracheal mites (*Acarapis woodi*), and test for European foulbrood (*Melissococcus plutonius)* and American foulbrood (*Paenibacillus larvae*) by lateral flow. Macerate bees and examine for Nosema (*Nosema apis* and *Nosema ceranae*).
7. Allow the queen to walk on clean white paper and after a short while she will defecate. Examine her faeces for Nosema.
8. Set up the new hive as far as possible away from your current established apiaries.
9. After one month, examine the hive in detail. Review the health of the new hive. If no problems are noted, relocate the hive to your apiary.

The arrival of the new hive would be the time to eliminate as many *Varroa* mites as possible.

Place a super in the brood box to try and trap as many *Varroa* mites in the drone brood. After drone trapping use icing sugar to reduce surface mites and then watch the hive debris and destroy the fallen mites.

Queen manipulation

Make or obtain a queen muff.

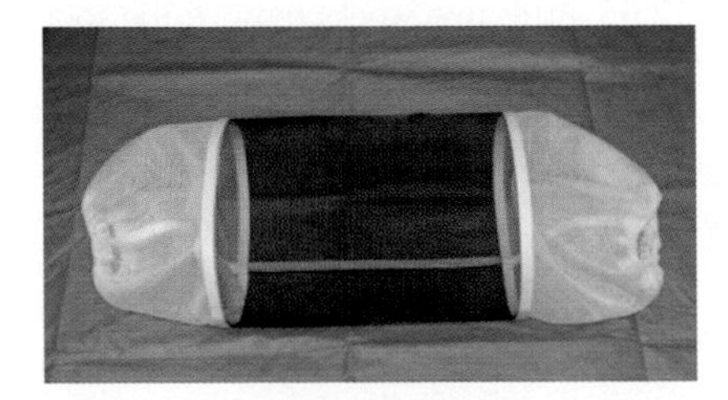

Figure 12.67 A queen muff.

This is a small cage where you can put all the items required to manipulate the queen.

Place your hands through the elastic armholes and you are ready to go. The queen is unlikely to sting you. Her sting will also not result in her death as her stinger is not barbed.

If the queen does escape from the queen cage, she cannot go far and you have a great chance of finding her and not harming her.

Queens are expensive and need to be treated like royalty!

The queen may have her wings clipped to stop her flying and reduces the chance of swarming.

Figure 12.68 Queen with her wings clipped

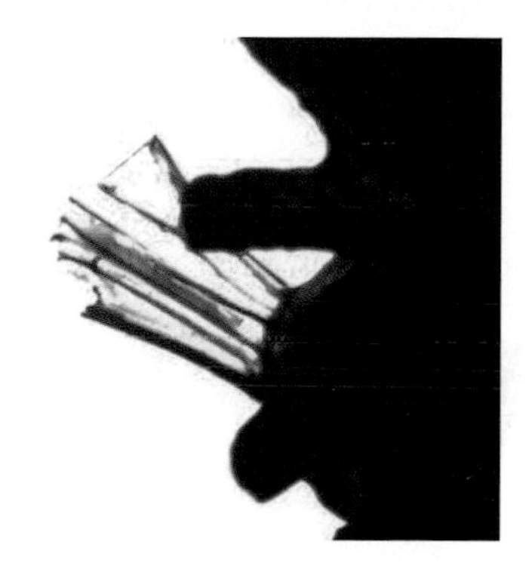

Figure 12.69 Detail of the clipped left wing down a dissecting microscope.

Queen identification

Manual queen marking

Experienced beekeepers mark a queen by holding onto her legs with one hand and quickly dabbing the paint with the other hand. It looks easy and takes only a moment. But it is extremely easy to damage the queen. It takes experience and confidence. Do not try it on your own. Work with the local beekeeper group and bee inspector to gain the necessary skills.

Catching and holding the queen

Use a queen catcher to trap your queen bee. The catcher is see-through so that you can watch the queen within the catcher. Put the catcher over the queen and her surrounding worker. The

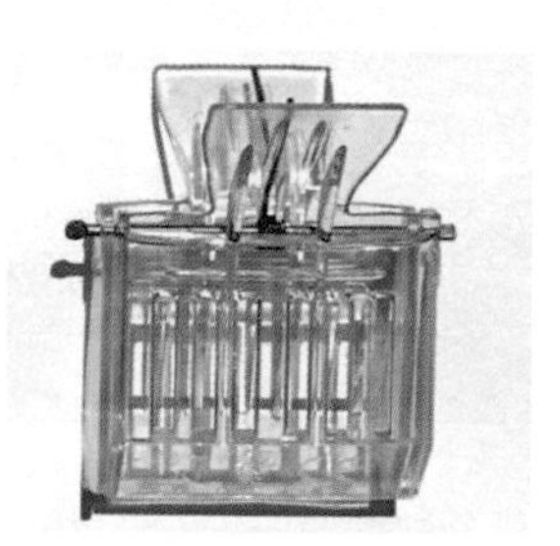

Figure 12.70 The queen catcher

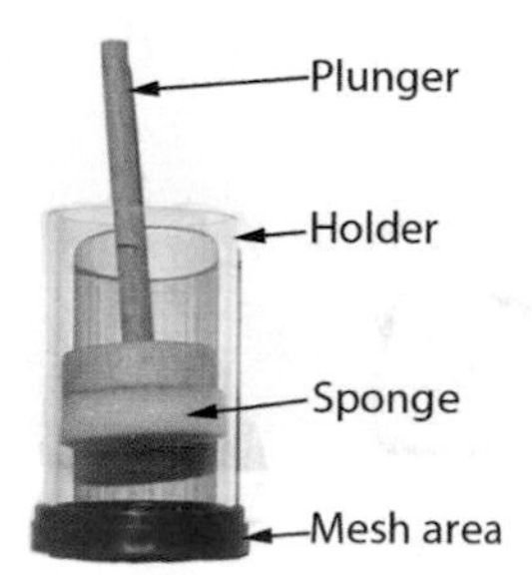

Figure 12.71 The queen holder

workers will walk through the sides of the catcher. The catcher is so designed that the queen's vital legs are not trapped or injured.

Marking the queen

1 The queen is caught in the queen catcher tool.
2 Drop her into the holder tube.
3 Insert the sponge-covered plunger.
4 Slowly push the queen to the mesh end, using the sponge-covered plunger part way.
5 Allow the queen to move until she is sitting on the sponge with the back towards the mesh cover.
6 Slowly push the plunger until the queen is captured between the mesh and the sponge. Be gentle.
7 You can mark the queen's thorax (back). Use the correct colour corresponding to the current year. Ensure your mark is only on her thorax and not on her head or eyes.

Marking	Year ending
White (gray)	1 or 6
Yellow	2 or 7
Red	3 or 8
Green	4 or 9
Blue	5 or 0

Queen excluder

A major advantage of the Langstroth type hive is the ability to separate the queen from the honey supers and keep her in the brood supers. This works on the principle that the honey bee queen is much larger than the workers or drones.

Figure 12.72 Queen excluder above the brood box note the bee space is uppermost

Figure 12.73 Detail of the queen excluder

Ensure that the queen excluder is the correct way round with the bee space uppermost. Mark the upper surface of the excluder to avoid mistakes.

Looking for the queen

Finding the queen can be more difficult than you would imagine. If there are eggs present the queen was certainly there three days ago. Sometimes that's the best you can achieve.

However a quick method to look for the queen is to remove selected frames from the hive. This uses the principles that

- The queen is likely to be in the middle of the brood area.
- The queen is likely to be near the current egg laying area.
- The queen is likely to be surrounded by a lot of worker bees.

Remove the first, fourth, fifth, eighth and ninth frames and examine them for the queen.

Figure 12.74 Looking for the queen using double frames

Is there a queen present?

If there are clear eggs, generally do not worry, the queen was there at least three days ago. In the autumn and winter or rainy season there will only be a few eggs.

If you have concerns and if you have two hives, place a numbered frame from an adjacent healthy hive containing eggs. If the workers do not raise queens then the original hive has a queen.

Integrated pest management

Artificial swarm technique

This technique needs advanced bee keeping skills and the ability to manipulate two colonies at the same time over a period of about 20 days.

The aim is to separate the flying bees from the majority of the *Varroa* mites. *Varroa* mites on the adults would then be susceptible to removal – using a pyrethroid, oxalic acid and/or icing sugar for example. The following technique cleans a colony from *Varroa* by providing an extended broodless period.

The protocol for artificial swarming

Time lapse	Colony A	Colony B
	Watch for development of queen cells.	Prepare new colony. Place a queen excluder above and below brood box – this will prevent the queen from absconding.
	When queen cells observed – capture original queen (A) and place her in a queen cage.	
Day 1	Move colony A 10 metres from original site	Place Colony B at the original site of colony A
	Remove queen from colony A.	Place colony A queen in colony B. The flying bees from colony A will enter colony B. Feed with spring 50% sugar syrup for one day to help production of brood. Take care to avoid robbing.
Day 2		Introduce icing sugar to remove any surface *Varroa* mites. Treat with a pyrethroid or oxalic acid.
Week 1	Remove all but one queen cell. Once the queen cell is sealed – place a cell protector cage over queen cell – to capture the queen upon emergence. Do not allow the queen to leave the hive and mate.	
Week 3	Transfer 2 frames with combs of unsealed brood from Colony B to Colony A. Clearly mark. These will act as bait for any *Varroa* mites.	

Time lapse	Colony A	Colony B
Week 4	When the 2 marked frame's combs are sealed – remove and destroy by incineration. Do not allow cells to become unsealed.	
	Introduce icing sugar to remove any surface *Varroa* mites. Treat with pyrethroids or oxalic acid.	
Week 6	Check for *Varroa* mites	Check for *Varroa* mites
Options	Remove virgin queen A and replace with a clean, mated and tested laying queen – thus producing two clean colonies.	
	Remove virgin queen from colony A. Combine cleaned colony A with colony B and original queen.	
	Remove the virgin queen from colony A. Re queen colony A and, when established, unite with de-queened colony B.	
Burn	All hive debris and sealed brood to destroy the *Varroa* mites.	

Artificial swarm technique over 30 days

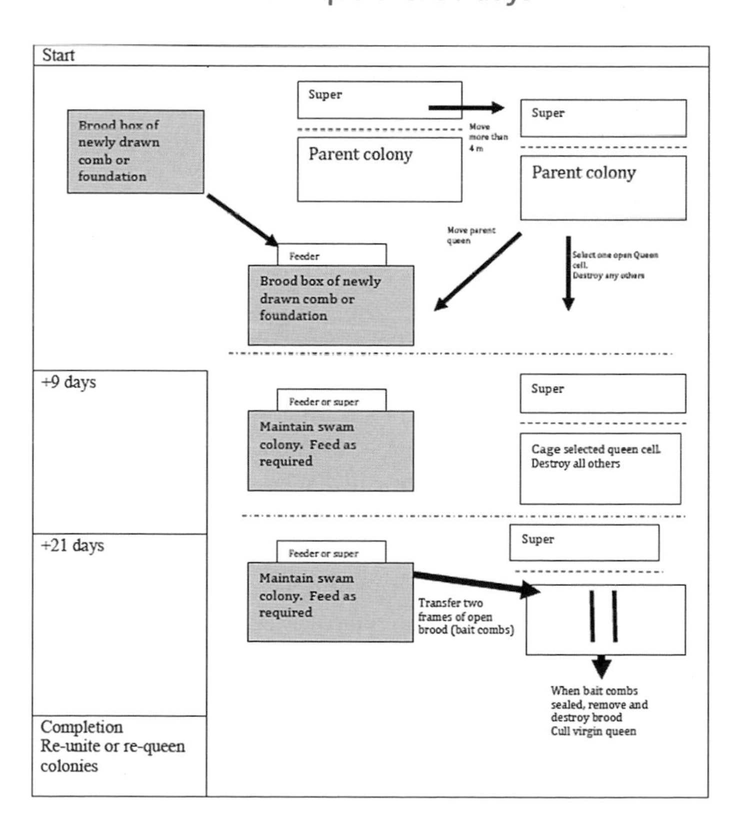

Combining two colonies

Concern

Identify why the weaker colony is weak. A clinically diseased weak colony will just pass on the infection to the 'stronger' colony.

Why combine two colonies?

You can make a stronger colony out of a weaker and a stronger colony. And two weak colonies put together can make a stronger one.

When to combine colonies?

Combine at the end of summer.

How to combine the two colonies?

Do not just shake the weaker colony into the stronger colony – the stronger colony will fight and kill the weaker bees.

The bees need to get to know each other. This takes about three days. An easy method is to use a newspaper separation.

1. Identify the stronger of the two colonies. This hive will provide the bottom brood box.
2. Smoke and open the weaker hive.
3. Examine each frame and select the frames containing the most capped brood, eggs or honey. Place into one brood box. Shake the worker bees into this brood box.
4. Do not move the weaker queen bee. If you have two queens one will be killed. This will be a distraction and may wound even the winning queen.
5. Post mortem the weaker queen.
6. Smoke and open the stronger hive.
7. Remove the roof and crown board and put a single sheet of newspaper over the top of the upper brood box.
8. Place the queen excluder above the newspaper. This keeps the stronger queen in the stronger brood box.
9. Provide odour contact between the two hives. Make several small holes with a nail through the newspaper. This allows hive odours to move between the two hives once they are combined.
10. Place the weaker brood box over the top of the stronger hive so that only the perforated sheet of newspaper separates the two colonies.
11. Add a feeder and fill it with autumn sugar syrup (61.7%). Place a super on top of the weaker brood box and place in some autumn sugar syrup or feeder.
12. Check the hive in a week. The newspaper will have been chewed away, and the two colonies will have joined.
13. Remove the queen excluder to the top of the second brood box. This gives the strong queen access to the whole brood area in the hive.

Figure 12.75 Review the strength and health of your hives in the autumn

Figure 12.76 Combine healthy weaker colonies with a healthy stronger colony. Identify the weaker colony's good brood frames and make up one good brood box

Figure 12.77 Place a newspaper over the stronger hive's brood box

Figure 12.78 Place queen excluder over stronger hive brood box and newspapaper

Figure 12.79 Add the good brood box to the stronger hive brood box, newspaper and queen excluder

Figure 12.80 One good colony which should survive the winter. Review in a week

Drone trapping

The principle is that the drone larvae take longer to pupate. The *Varroa* mite prefers drone brood.

In Langstroth hives

1 Two weeks into the beginning of summer, place a marked smaller super frame into the middle of the brood box. Or make a super frame which is half the size of a normal brood frame.

2 The bees will then build drone comb free-form underneath the 'super' frame. The super frame itself will be filled with worker brood.
3 Check seven days later.
4 Place another super in the brood box.
5 Check seven days later.
6 Remove the capped brood cells from the first super (day 6), cut the capped drone cells into the straight sided waste bucket.
7 The *Varroa* mites are trapped in the capped drone brood.
8 Place the super back into the super box. The workers will continue the development of the capped workers.
9 Destroy the removed free-form drone cells by incineration. Do not just throw into the nearby hedge. The mature female *Varroa* could escape from the damaged cells.
10 Check in seven days.
11 Remove the capped brood cells from the second super (day 6), cut the capped drone cells into the straight-sided waste bucket.
12 Place the super back into the super box. The workers will continue the development of the capped workers.
13 Destroy the removed free-form drone cells by incineration.

In Top Bar hives

1 Cut away the portions of the comb that are filled with capped drone brood.
2 The drones are often found at the cooler perimeter of the comb.
3 Remove the capped drone brood into the straight sided bucket.
4 Destroy the removed brood by incineration.

Figure 12.81 Drone cells being free-formed under a super frame placed in the brood box

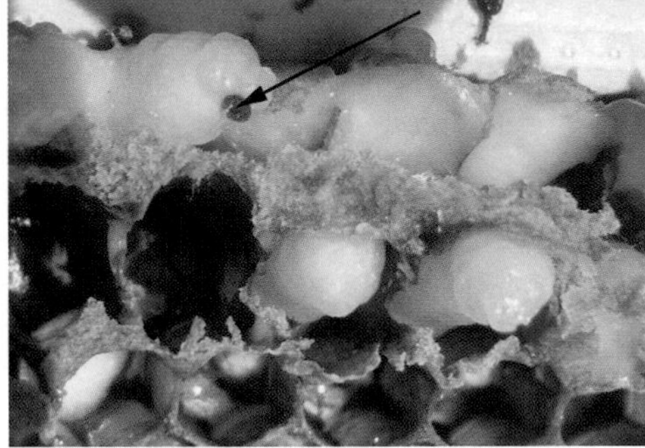

Figure 12.82 Demonstration of a *Varroa* mite (arrow) trapped in the removed capped drone cells

Enhancing hive health

Ensure bees go back to the same hive – reduce pathogen movement between hives.

Drifting is when bees fail to return to the correct colony and instead join an adjacent hive.

Reduce drifting by using painted hives, coloured entrances and using local landmarks for bees to locate their hive by, such as trees and bushes. Ensure hives are placed well enough apart (3m) and if possible have slightly different directions for the entrances.

Fitting a mouse guard

Purchase or make a mouse guard. In the autumn it is vital to keep mice out of the hive. Mice can wreak havoc in the hive eating brood, wax and honey.

The presence of wax outside the hive may be an indication of a mouse in the hive. Note the mouse may be killed and its mummified remains will be found in the hive during a routine examination. The bees may be unable to remove the dead mouse. This can then attract flies and other pathogens.

Figure 12.83 Different coloured shaped and directional hives allow bees to identify home easily reducing drifting

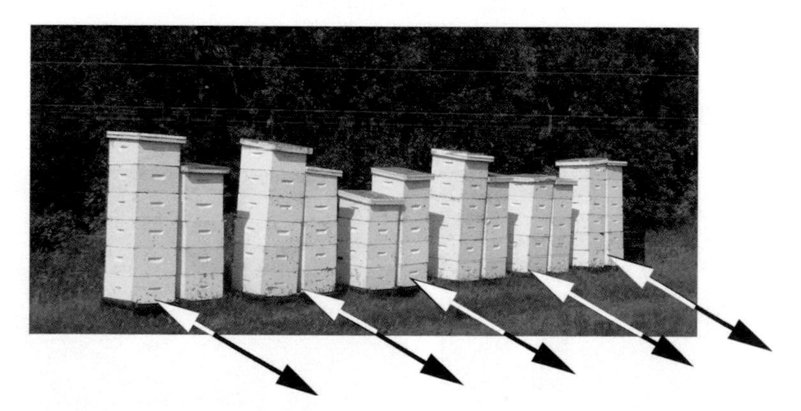

Figure 12.84 Neat and orderly hives, but drifting will occur between the hives, as they all look alike and have very similar flight paths

Figure 12.85 Commercial mouse guard. Held by drawing pins

Figure 12.86 Home-made mouse guard

Mice may also transmit pathogens. Septicaemia – Pseudomomas – is common in mice.

Shook swarm technique

This can be used to clean a hive from a number of brood issues.

1 Prepare a new sterilised brood box with undrawn foundation (hive 2).
2 Place this hive (hive 2) next to the original colony (hive 1).
3 Remove each of the frames of hive 1 and shake the bees into the new box (hive 2). It is essential to include the queen.
4 Remove the original hive (1).
5 Place the new hive (2) onto the original hive stand.
6 Treat the bees in hive 2 with icing sugar or pyrethroids.
7 Feed the bees in hive 2 with autumn feeding a 61.7% sugar solution, to help the bees draw out the combs.
8 Kill all remaining bees in hive 1.

Figure 12.87 Hive 1

Figure 12.88 Hive 1 full of bees

Original hive 1 with issues – Varroa for example. The picture on the left indicates the hive being initially examined the crown board and roof has been removed. Once the super is removed, the large number of bees in the brood (arrow) box is obvious

Figure 12.89 Hive 2

New hive (hive 2) with clean undrawn foundation

Figure 12.90 Bees shaken into hive 2

Shake bees from hive 1 into new hive (2)
Placing a super on top of the new brood box in hive 2 can be very helpful as this acts as funnel for the new bees. This reduces the loss of bees on the ground.

9 Remove and burn all the frames from hive 1.
10 Sterilise the brood boxes from hive 1.
11 Freeze components from hive 1 for at least seven days.

Note: If the queen is damaged, killed or fails to be transferred, the colony will be lost as the new colony has no brood present.

Medicine resistance test for *Varroa*

Basic review

The assumption is that your bees have a significant population of *Varroa* in the first place.

1 Place the cleaned *Varroa* tray in the hive to collect hive debris.
2 Place the medicine under test in a hive.
3 If after 24 hours there is a mite drop rate in the hundreds, the medicine's active ingredient will be having a sufficient effect to enable its use for that season.

Advanced review

1 Around 500 adult bees (one frame) are collected in a test container
2 Add the medicine under review.
3 After 3 hours the mite mortality is assessed

If the mite mortality level is high the mites are still sensitive to the medicine. If the mite mortality level is low, the mites are resistant to the medicine.

Collecting honey

Removing worker bees from supers

When the supers need to be removed or the supers need to be cleaned use a bee escape worker chamber. Place the bees escape board between the super boxes and the brood frame for 24 hours. The workers can enter the chamber but can only leave into the lower brood boxes.

Figure 12.91 Upper surface of bee escape chamber from the super

Figure 12.92 Lower surface of bee escape chamber into brood box

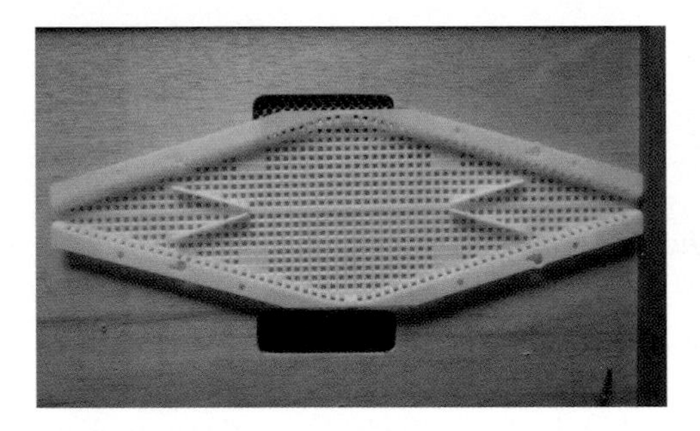

Figure 12.93 The workings on the bee escape chamber before fitting to the escape board (modified crown board)

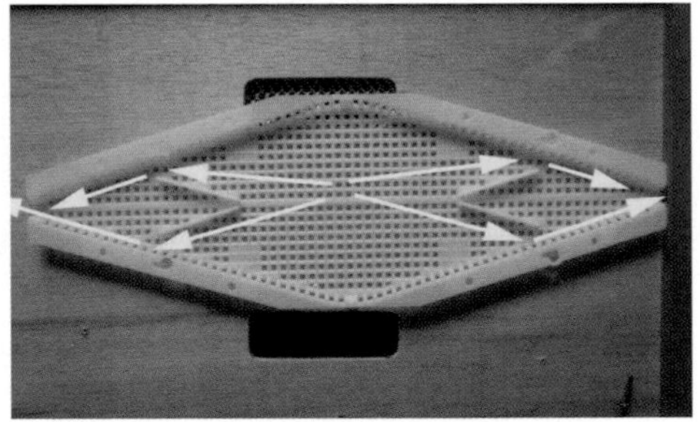

Figure 12.94 The path of the bees out of the bee escape. The V makes it difficult for bees to return

Miscellaneous

Killing bees

Individual/small groups of bees should be placed in the refrigerator for 20 minutes before euthanasia. Unfortunately from time to time it will be required to kill the hive.

1 Obtain 4.5 litres of water and add 250ml of washing-up liquid.
2 Allow the bees to enter the hive.

3 Seal the entrance.
4 Apply the strong washing-up liquid by a hand-held mister or pressure washer.
5 Destroy the frames.
6 Sterilise the brood box.
7 Freeze the brood box for at least seven days.

In emergency situations, it may be necessary to destroy the hive and all its surroundings either by burning or burying.

Figure 12.95 The careful incineration of old hives to destroy any pathogens

Figure 12.96 The burial of old hives to bury and destroy any pathogens

Wasp control

Wasps are major predators of bees and can terrorise hives. Wasps are also a major biosecurity concern especially when new species invade new areas.

Place a wasp trap near your hives so that wasps can be caught and removed from the locality of your hive.

Figure 12.97 Wasp trap in Northern Europe

Figure 12.98 Wasp trap in Asia

Further reading

Adey, M., Walker, P. and Walter, P.T. (1986) *Pest Control Safe for Bees. A manual and directory for the tropics and subtropics.* ISBN 0-86098-184-3

Aston, D. and Bucknall, S. (2010) *Keeping Healthy Honey Bees.* ISBN 978-1-904846-54-3

Bailey, L. (1981) *Honey Bee Pathology.* ISBN 0-12-073480-X

Basic Beekeeping Manual 1 (2009) Gregory Pam Manual sponsored by the waterloofoundation

Beekeeping Study Notes for the Bee Keeping Exams. ISBN 0-905652-33-9; ISBN 0-905652-34-7; ISBN 0-905652-21-5 and ISBN 0-905652-35-5

Boucias, D.G. and Pendland, J.C. (1998) *Principles of Insect Pathology.* ISBN 0-413-03591-X

Canadian Association of Professional Apriculturists (2013) *Honey bees' diseases and pests,* 3rd edn. ISBN 978-0-96933336-16, available online at www.capabees.org

Chapman, R.F. *The Insects: Structure and Function* (1998). ISBN 978-0-521-57048-0

Cullum-Kenyon, I.D. and Cullum-Kenyon, R. (2012) *The BBKA Guide to Beekeeping.* ISBN 978-1-4081-5458-8

Dade, H.A. (1994) *Anatomy and Dissection of the Honey Bee.* ISBN 978-0860982142

Davis, C. (2004) *The Honey Bee Inside Out.* ISBN 978-900147-10-4

Dietemann, V., Ellis, J.D. and Neumann, P. (2013) *Coloss Beebook Volumes I and II. Standard methods for Apis mellifera research.* ISBN 0-85098-274-2 2013

Edwards, M. and Jenner, M. (2005) *Field guide to the Bumblebees of Great Britain and Ireland.* ISBN 978-0-954971311

Erickson, E and Carlson, S.D. (1986) *A Scanning Electron Microscope Atlas of the Honey Bee.* ISBN 978-0813805467

Fisher, R.L. (2010) *Bee.* ISBN 978-61689-076-6

Flottum, D. (2005) *The Backyard Beekeeper.* ISBN 10-1-59253-118-0

Goodman, L. (2003) *Form and Function in the Honey Bee.* ISBN 0-86098-243-2

Goulson, D. (2010) *Bumblebees: Behaviour, Ecology and Conservation.* ISBN 978-0-19-955307-5

Graham, J. (1992) *The Hive and the Honey Bee.* ISBN0-915698-09-9

Grimaldi, D and Engle, M.S. (2005) *Evolution of the Insects.* ISBN 978-0-521-82149-0

Hepburn, H, Randall, R., and Radloff, E.E. (2011) *Honeybees of Asia.* ISBN 978-3-642-16421-7

Hepburn, H.R., Rirk, C.W.W. and Duangphakdee, O. (2014) *Honeybee Nest.* ISBN 978-3-642-54327-2

Herrod-Hempsall, W. (1943) *The Anatomy, Physiology and Natural History of the Honey Bee.* ASIN: B0007JNEJI

Honey Bee Health. USDA Website, available online at www.ars.usda.gov/ccd

Jalil, A.H. (2014) *Beescape for Meliponines.* ISBN 978-4828-2361-5
Kirk, W. A. (2006) *A Colour Guide to Pollen Loads of the Honey Bee.* ISBN 10-86098-248-3
Koeniger, K., Koeniger, G. and Tingek, S. (2010) *Honey bees of Borneo.* ISBN 978-983-812-128-0
Kugonza, D.R. (2009) *Beekeeping – Theory and Practice.* ISBN 978-9970-02-965-5
Mann, D.I. (1976) *Bees are wealth – a handy guide to bee-keeping in East Africa.* ISBN 9966-44-090-9
Marchand, D. and Marchard-Mayne, J. (2003) *Beekeeping. A practical guide for southern Africa.* ISBN 0-9584564-2-9
Maurer. B. (2012) *Practical Microscopy for Beekeepers.* ISBN 978-0-900147-13-5
McMullan, J. (2012) *Having Healthy Honey Bees. An integrated approach.* ISBN 978-0957135505
National Bee Keeping Training and Extension Manual (March 2012). Funded by the African development bank and government of Uganda
Oldroyd, B.P. and Wongsiri, S. (2006) *Asian Honeybees.* ISBN 0-674-02194-0
Peacock. P. (2008) *Keeping bees.* ISBN 978-0-7938-0669-0
Penn State University. (2011) *A field guide to Honey Bees and their Maladies*
Preston, C. (2006) *Bee.* ISBN 10-1-86189-256
Prys-Jones, O.E. and Corbet, S.A. (2011) *Bumblebees.* ISBN 978-1-907807-06-0
Root, A.I. (2007) *The ABC and XYZ of Bee Culture.* ISBN 978-0-936028-22-4 (from 1877)
Sammataro, D. and Avitabile, A. (1978) *The Beekeeper's Handbook.* ISBN 978-0-8014-8503-9
Sammataro, D. and Yoder, J.A. (2012) *Honey Bee Colony Health. Challenges and Sustainable Solutions.* ISBN 978-14398-794-05
Seeley, T.D. (2010) *Honeybee Democracy.* ISBN 978-0-691-14721-5
Snell, I. (2012) *Understanding Bee Anatomy: A Full Colour Guide.* ISBN 978-0-9574228-0-3
Snodgrass, R.E. (1953) *Anatomy of the Honey Bee.* ISBN 978-0-8014-9302-7
Tautz, J. (2008) *The Buzz About Bees.* ISBN 978-3-540-78727-3
Thomas, D. (2012) *The healthy bee hive.* ASIN: B008WARIZU
Varroosis, Tropilaelaps, Small Hive Beetle, Foulbrood. Series of booklets published by DEFRA–UK–FERA.
Vega, F.E. and Kaya, H.K. (2012) *Insect Pathology,* 2nd edn. ISBN 978-0-12-384984-7
Vidal-Naquet., N. (2015) *Honeybee Veterinary Medicine: Apis mellifera L.* ISBN 978-1-91945-504-3
Warring, C. and Warring, A. (2011) *Bee Manual.* ISBN 978-0-85733-057-4
Wilson-Rich, N. (2014) *The Bee. A Natural History.* ISBN: 978:1-78240-107-0
Wiscombe, D. and Blackiston, H. (2012) *Beekeeping for Dummies.* ISBN 978-1-119-97250-1
Woodward, D. (2010) *Queen Bee: Biology, Rearing and Breeding.* ISBN 978-0-473-11933-1

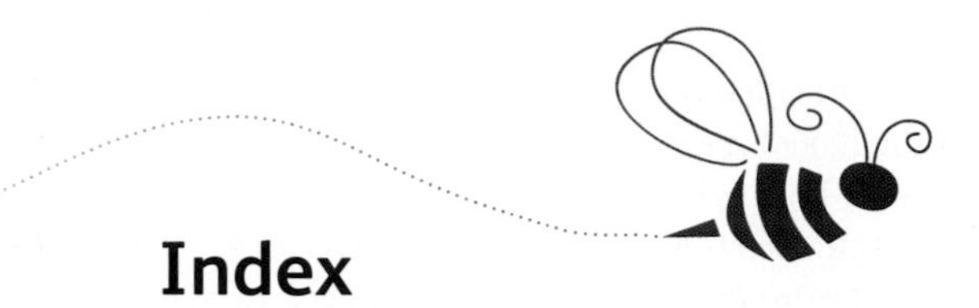

Index